ADHESION MOLECULES AND CHEMOKINES IN LYMPHOCYTE TRAFFICKING

Cell Adhesion and Communication

A series of books encompassing monographs on classes of adhesion molecules and monographs giving a broader functional synopsis on adhesion molecules on a particular system.

Edited by Christo Goridis

Volume 1
Cell Adhesion Molecules in Cancer and Inflammation
edited by A.A. Epenetos and M. Pignatelli

Volume 2
The Laminins
edited by P. Ekblom and R. Timpl

Volume 3
Tenascin and Counteradhesive Molecules of the Extracellular Matrix
edited by K.L. Crossin

Volume 4
Adhesion Molecules and Chemokines in Lymphocyte Trafficking
edited by A. Hamann

Other volumes in preparation

Ig Superfamily Adhesion Molecules
edited by P. Sonderegger

Epithelial Morphogenesis in Development and Disease
edited by W. Birchmeier and C. Birchmeier

Cell Adhesion Mediated by the Carcino Embryonic Antigen Family
edited by C. Stanners

This book is part of a series. The publisher will accept continuation orders which may be cancelled at any time and which provide for automatic billing and shipping of each title in the series upon publication. Please write for details.

ADHESION MOLECULES AND CHEMOKINES IN LYMPHOCYTE TRAFFICKING

Edited by

Alf Hamann
Department of Immunology
Medical Clinic
University Hospital Eppendorf
Hamburg
Germany

harwood academic publishers
Australia • Canada • China • France • Germany • India • Japan
Luxembourg • Malaysia • The Netherlands • Russia • Singapore
Switzerland • Thailand • United Kingdom

Published in The Netherlands by Harwood Academic Publishers.

 Printed in Singapore.

Amsteldijk 166
1st Floor
1079 LH Amsterdam
The Netherlands

British Library Cataloguing in Publication Data

A catalogue record for this book is available from the British Library.

CONTENTS

PREFACE TO THE SERIES

The development and normal functioning of all multicellular organisms is governed to a large part by the interactions cells undergo with neighbouring cells and with their acellular environment. Many of these interactions are mediated by cell-cell adhesion molecules and by extracellular matrix components and their cellular receptors, that is by molecules mediating direct cell-cell and cell-matrix contacts. These molecules are particularly important for determining whether a cell remains where it is or moves elsewhere and, if a cell moves, where it goes and when it stops migrating. These are of course key events during normal development, but they play equally crucial roles in adult physiology and pathology, such as the extravasation of white blood cells, inflammatory processes and wound healing, tumour invasiveness and metastasis. Moreover, recent advances in the field show that most, if not all cell adhesion molecules are capable of triggering intracellular events, in the same way as diffusible growth and differentiation factors and their cellular receptors do. It is thus hardly surprising that clinicians are devoting increasing attention to the molecular mechanisms underlying cell adhesion, and that cell adhesion molecules are now being considered as suitable targets for drug development.

This book series is aimed at scientists, both in academia and in industry, and at graduate students planning to move into the area, at the clinician, who wants to become familiar with a field with many clinical implications, at scientists already working in the field, who want to keep abreast with the recent developments outside their own speciality. Hence, each volume of the series provides a review of a particular segment of the field as well as critical assessment of recent discoveries and future developments. Each volume has a volume editor, who is an expert in the field and invites contributions to cover the different aspects of the topic. By keeping the number of contributors to each volume small, we hope to avoid overlaps and redundancies, common pitfalls of multi-author volumes. By looking at the previous volumes, I have the impression that we have been successful.

One part of the series concentrates on important structural families of molecules, another one addresses the role of cell adhesion in selected physiological and pathological phenomena. The previous volumes *The Laminins* and *Tenascin and Counteradhesive Molecules of the Extracellular Matrix* are examples of the first kind, the present volume is an excellent example of the second kind. It deals with probably the most important cell migration phenomenon in the adult organism: lymphocyte trafficking. As Sir James Gowan recalls in his authoritative foreword, the recirculation of lymphocytes and their homing to specific sites is at the heart of lymphocyte function and crucial for the continuous immune surveillance they exert in the body. Lymphocytes and the cells they encounter display a host of adhesion molecules which are essential for regulating lymphocyte migrations and whose activities are tightly regulated. The main part of the volume is devoted to presenting the latest advancements in this rapidly moving field. These adhesion molecules are obvious therapeutical targets, and the last two chapters deal with this issue, in the broader context of tumour surveillance and the control of inflammatory processes.

I am indebted to the staff at Harwood Academic Publishers, who made this enterprise possible, to the volume editors, each one an authority in his/her field, and last but not least to the many competent contributors.

Christo Goridis

FOREWORD

Lymphoid tissue is remarkable in that most of the cells which compose it are motile. If it were possible to inspect the interior of a lymph node microscopically *in vivo* it would appear as a writhing mass: cells would be entering from the blood, crawling through the node and leaving in the lymph, while internal migrations of differentiating cells would indicate the evolution of immune responses. Despite its dynamic interior, lymphoid tissue maintains a distinctive, gross morphology.

The motility of lymphocytes fits them for a vagrant life-style. They continually recirculate in large numbers from blood to lymph by way of the lymph nodes and gut-associated lymphoid tissue. There is also a large traffic through the spleen where lymphocytes enter and leave through blood vessels but, in the adult, circulating lymphocytes show little interest in the thymus. A smaller traffic takes place through all non-lymphoid tissues which possess lymphatics. This peripheral recirculation can be greatly increased in pathological conditions and, in the extreme case of a granuloma, the lymphoid mass can acquire its own intrinsic pattern of recirculation. These remarks apply to small lymphocytes, that is, those not yet in cell cycle. Even here there is some uncertainty. It is possible that not all small lymphocytes are in a recirculating mode at any one time. Certainly there are classes of lymphocytes which are arrested in the tissues after emigration from the blood, for example, certain effector cells including the larger, differentiating lymphocytes (blasts) which mediate local immunity.

There is currently much interest in the molecular mechanisms involved in the emigration of leukocytes from the blood. I remember in the 1950s watching cinemicrographs of cells passing through venules in inflamed tissue. First, polymorphonuclear leukocytes left the axial stream and rolled along the inner wall of the vessel before coming to rest and burrowing through the endothelium and its basement membrane. Later, blood monocytes emigrated into the tissue and became macrophages. Finally, lymphocytes joined the infiltrate. How is this sequence explained in terms of leukocyte receptors and their endothelial ligands and what controls the expression of the molecules which mediate these events? The entry of lymphocytes into lymph nodes is of special interest because the high endothelial venules through which they pass normally exclude all other leukocytes. Various candidates have been suggested for the lymphocyte homing receptor/vascular addressin pairs responsible for this transaction but, since some of these receptor molecules are shared with other leukocytes, it is likely that a particular repertoire of receptors is required for the specificity of the interaction. Some have questioned whether the *in vitro* binding assay for the identification of these adhesion molecules always mirrors events *in vivo*; others have pointed out that 'adhesion' is perhaps a strong word to use for the molecules mediating emigration because they must leave the lymphocytes free to exercise their motility and to move through the vessel wall. It should be noted that, notwithstanding the unique attraction of high endothelial venules for lymphocytes under normal conditions, if lymph nodes are inflamed then neutrophils will readily traverse them in large numbers. Presumably, the properties of the endothelium are radically changed by cytokines.

The early work on recirculation was undertaken to solve the problem of the high turnover of the blood lymphocytes. Small lymphocytes from the thoracic duct, which in the rat come mainly from the intestinal lymphatics, appeared to home to lymphoid tissue without preference for particular regional nodes; thus they migrated in large numbers into the cervical as well as the mesenteric nodes and Peyer's patches. Subsequent work has suggested that these bulk migrations may have obscured the existence of sub-populations with particular anatomical preferences, for example one with a predilection for homing to the gut and its lymphoid tissue. Candidates for the lymphocyte receptors which may identify this gut-seeking sub-population have been proposed. The homing into the *lamina propria* of the small intestine of blast cells derived from intestinal lymphoid tissue is a quite separate issue.

I referred at the outset to the vagrant life-style of lymphocytes but this does not imply that their wandering is aimless. On the contrary, it is central to their function. A lymphocyte only realises its potential when it meets the epitope for which it is specific. Once it is conceded that the immunological repertoire of any single lymphocyte is very restricted, an efficient method of selection requires a constant traffic through the regions where such epitopes may be located. It is not surprising that regional lymphoid tissue and the spleen, the major sites of antigen localisation, are also the major sites of lymphocyte recirculation. There is some experimental evidence for the operation of such selective mechanisms *in vivo* but this is a topic that requires further study.

The accumulation of lymphocytes in allografts and in many pathological conditions, for example, in the lesions of certain autoimmune diseases and in virus infected tissue, poses the question of the extent to which effector cells are specifically attracted from the blood by their particular targets. The answer is not known but it is possible that the vasculature permits a random, non-specific traffic of lymphocytes through these sites which allows the specific effectors to be arrested *en route.* It is unlikely that extravascular antigen alone could provide a specific homing signal for circulating lymphocytes unless it had also been captured by the endothelium itself.

The engagement of lymphocytes with antigen sets in train a series of migrations within lymphoid tissue about which there is little information. The activation of T cells by antigen-processing cells, the capture of antigen by B cells, the B/T collaborative events and finally, the differentiation of B lymphocytes into plasma cells are usually depicted in text books as if they all occurred within the same confined area. In fact, T and B cells occupy separate and distinct zones in lymphoid tissue while collections of plasma cells in the medullary cords of the nodes and in the red pulp of the spleen are in areas at a distance from them both. Some clarification of these events is coming from work on the generation of germinal centres. It is often forgotten that, in hyperimmune animals, a major site of antibody formation is in the bone marrow — a further migratory stream to add to the list.

These are some of the topics which are considered in this timely volume with its particular emphasis on the molecular mechanisms involved in lymphocyte traffic. But it will interest a wider audience than those concerned only with the wanderings of lymphocytes. The subject must be set in a wider context because families of adhesion molecules play a crucial part in the development and organisation of all the tissues of the body.

James L. Gowans

Sir James Gowans is a former director of the UK Medical Research Council (MRC) Cellular Immunology Unit and Secretary of the MRC, UK

CONTRIBUTORS

Paola Allavena
Department of Immunology and
Cell Biology, Mario Negri Institute
Via Eritrea, 62
20157 Milano
Italy

Robert F. Bargatze
Laboratory of Leukocyte-Endothelial
Cell Biology
Montana ImmunoTech Inc.
920 Technology Blvd., Suite C
Bozeman, MT 59715
USA

Alexandre Benmerah
Développement Normal et Pathologique
du Système Immunitaire
INSERM U-429
Hôpital Necker-Enfants Malades
149 rue de Sèvres
75743 Paris Cédex 15
France

Eugene C. Butcher
Laboratory Immunology and Vascular
Biology, Department of Pathology and
Stanford Digestive Disease Center
Stanford University School of Medicine
Stanford, CA 94305
and Center for Molecular Biology
in Medicine
Veterans Affairs Medical Center
Palo Alto, CA 94304
USA

Nadine Cerf-Bensussan
Développement Normal et Pathologique
du Système Immunitaire
INSERM U-429
Hôpital Necker-Enfants Malades
149 rue de Sèvres
75743 Paris Cédex 15
France

Dominique Dunon
Université Pierre et Marie Curie
CNRS URA
75252 Paris
France

Britta Engelhardt
Max-Planck-Institute for Physiological
and Clinical Research
W. G. Kerkhoff-Institute
Department of Molecular Cell Biology
Parkstr. 1, 61231 Bad Nauheim
Germany

Delphine Guy-Grand
Hôpital Necker-Enfants Malades
INSERM
149 rue de Sèvres
75743 Paris Cédex 15
France

Alf Hamann
Department of Immunology
Medical Clinic
University Hospital Eppendorf
Martinistr. 52
20246 Hamburg
Germany

Beat A. Imhof
Department of Pathology
Centre Médical Universitaite
1, rue Michel Servet
1211 Geneva 4
Switzerland

Peter E. Lipsky
Department of Internal Medicine
The University of Texas South-western
Medical School
5323 Harry Hines Boulevard
Dallas, TX 75235-8884
USA

Alberto Mantovani
Department of Immunology and
Cell Biology, Mario Negri Institute
Via Eritrea, 62
20157 Milano,
and Division of General Pathology
University of Brescia
25123 Brescia
Italy

William J. Murphy
Intramural Research and Support Programs
SAIC-Frederick
National Cancer Institute-Frederick
Cancer Research and Development Center
Frederick, MD 21702-1201
USA

Philippe Naquet
Center of Immunology
INSERM-CNRS
Parc Scientifique et Techologique
de Luminy
Case 906
13288 Marseille Cédex
France

Nancy Oppenheimer-Marks
Department of Internal Medicine
The University of Texas South-western
Medical School
5323 Harry Hines Boulevard
Dallas, TX 75235-8884
USA

Laurelee Osborn
Biogen, Inc
14 Cambridge Street
Cambridge, MA 02174
USA

Reinhard Pabst
Centre of Anatomy 4120
Medical School of Hannover
30623 Hannover
Germany

Louis J. Picker
Laboratories of Experimental Pathology
and Clinical Immunology
Department of Pathology
University of Texas South-western
Medical Center
Dallas, TX 75235-9072
USA

Silvano Sozzani
Department of Immunology and
Cell Biology, Mario Negri Institute
Via Eritrea, 62
20157 Milano
Italy

Dennis D. Taub
Clinical Services
SAIC-Frederick
National Cancer Institute-Frederick
Cancer Research and Development
Center
Frederick, MD 21702-1201
USA

Olli Vainio
Department of Medical Microbiology
Turku University
20520 Turku
Finland

Jürgen Westermann
Centre of Anatomy 4120
Medical School of Hannover
30623 Hannover
Germany

1. SPECIFIC TRAFFICKING: WHICH CELLS, WHICH FUNCTION?

ALF HAMANN

Dept. Immunology, Medical Clinic, Universityhospital Eppendorf, Martinistr. 52, 20246 Hamburg, Germany

Why does the immune system need mechanisms to regulate the migration of lymphocytes ? Would it not be sufficient to distribute the cells throughout the body, as with red blood cells ? As the complexity of subsets, interactions and local specializations within the immune system becomes increasingly elucidated, the importance of specific trafficking mechanisms becomes obvious. Trafficking helps to organize the immune system at several levels by targeting distinct populations into tissue at the appropriate time and site. In this chapter we will denominate these levels and address the functional context which can be associated with distinct migratory mechanisms.

I. ENDOTHELIAL DIVERSITY ENABLES THE SPATIAL AND TEMPORAL ORGANIZATION OF A DYNAMIC IMMUNE SYSTEM

The trapping and uptake of circulating cells by a given tissue requires specific recognition signals on the surface of the blood vessel. In general, it will be the endothelium which presents these signals to passing cells. Adhesion molecules (sometimes called "addressins" with respect to their function in this context) and signalling molecules expressed by endothelium are recognized by appropriate receptors on the circulating cells (some are termed "homing receptors"). These interactions determine at which place or at which time a given cell has the opportunity to leave the bloodstream and to enter the tissue. In some cases, such as lung or liver, the function of endothelium might be supplemented by macrophages lining the vasculature.

The diversity of endothelium is mirrored by a diversity of lymphocyte subsets, possessing distinct sets of receptors which allow them to target the different compartments. These reciprocal receptor systems are operative in several functional contexts:

- Recruitment of hematopoietic precursor cells into sites of maturation
- Guiding of lymphocytes into lymphoid organs in the process of recirculation
- Recruitment of effector cells into sites of inflammation
- Uptake of special subsets into functionally distinct compartments of the body

T cell Precursors are Selectively Recruited into the Thymus

In contrast to mammalian B-cells, which attain their final competence within the bone marrow, the differentiation of T cell precursors within this compartment stops before

Correpondence: Tel. ++49-40-4717-3613; Fax: ++49-40-4717-4243; e-mail: hamann@uke.uni-hamburg.de rearrangement of the T cell receptor and only proceeds to maturity after migration into the thymus. A significant body of data has accumulated which indicates that thymic precursor cells released by the bone marrow do not settle randomly but possess distinct adhesion molecules and respond to chemotactic stimuli, by which they gain access to the thymus. The specificity of these mechanisms, which are discussed in chapter 6, might guarantee that the rare precursor cells are preferentially selected for uptake into the thymus from the pool of circulating cells. Complementary mechanisms might guide thymus-independent hematopoietic T precursor cells into the gut or other sites as discussed in chapter 7.

Specific Mechanisms Mediate the Continual Recirculation of Naive Lymphocytes through Lymphoid Tissues

Naive lymphocytes recirculate continuously between blood and lymphoid organs. Their cycling is initiated by an extraction from the blood with a high efficiency within the spleen, lymph nodes and Peyer's patches (Ottaway, 1988; Washington *et al.*, 1988). Whereas the mechanisms mediating the uptake of lymphocytes into the spleen are unclear (see chapter 2), the constitutive expression of certain adhesion pathways on high endothelial venules (HEV) leads to adhesion and subsequent extravasation of lymphocytes in the lymph nodes and Peyer's patches. The respective endothelial molecules – those known to date – are ligands for L-selectin ("peripheral node addressin, PNAd"), the mucosal addressin MAdCAM-1, and ICAM-1 or -2.

The "peripheral node addressin" PNAd defines carbohydrate epitopes recognized by L-selectin. These epitopes decorate several distinct carrier proteins; they are synthesized in a tissue-restricted manner, almost exclusively on HEV-like endothelium in lymphoid tissue (including lymphoid follicles in granulomas; Streeter *et al.*, 1988; Berg *et al.*, 1991; Hanninen *et al.*, 1993). The most prominent expression of PNAd is found on HEV of peripheral and mesenteric lymph nodes. The carrier proteins in this tissue include CD34, which is ubiquitously expressed on almost all endothelia but functionally glycosylated only on HEV of lymphoid tissue (Baumhueter *et al.*, 1994), and GlyCAM-1 and others. The designation as "peripheral node addressin" is misleading; a significant level is also found on Peyer's patch HEV, where the carbohydrate groups decorate MAdCAM-1 (Berg *et al.*, 1993).

The expression of MAdCAM-1 is more restricted, as discussed below and in chapter 3 and 7. Its counterreceptor on lymphocytes is the integrin $\alpha_4\beta_7$. MAdCAM-1 and carbohydrate epitopes serving as ligands of L-selectin are largely reciprocally expressed on HEV of either mucosal sites (Peyer's patches) or subcutaneous lymph nodes (Berg *et al.*, 1989). Interestingly, this polarization is not present from birth on, but rather seems to be induced on HEV by unknown factors from the tissue enviroment. After some weeks, however, its specialization to a "mucosal" or "peripheral" HEV becomes imprinted and cannot be reversed by later transplantation of the lymhoid tissue into another environment (Mebius *et al.*, 1993). The mechanisms for this lock in the phenotype of endothelium (which, however, rapidly dedifferentiates upon *in vitro* culture of endothelial cells) is not known.

Further endothelial molecules involved in lymphocyte recirculation are ICAM-1 and ICAM-2, serving as ligand for LFA-1. This adhesion pathway serves the recirculation through peripheral and mucosal lymphoid tissues as well (Hamann *et al.*, 1988).

The vast majority of lymphocytes entering lymphoid tissue via HEV are of naive phenotype. For them, the differential expression of addressins is irrelevant since naive lymphocytes possess and use the receptors for both addressins, i.e. L-selectin and $\alpha_4\beta_7$ integrin, and recirculate randomly through all lymphoid organs. A polarized expression of addressins must play a role rather for the specific recruitment of subsets of memory or activated / effector cells, which can express more selective patterns of receptors as discussed in later parts of this chapter.

Inflammatory Stimuli Induce the Recruitment of Effector Populations into Tissue

In contrast to the constitutive recirculation of lymphocytes through secondary lymphoid organs, most other organs (except gut, lung and liver) do not encounter a significant influx of lymphoid cells in the healthy state. This changes drastically when local stimuli lead to an inflammatory reaction. Subsequently nonspecific leucocytes as well as lymphocyte populations become recruited into the tissue.

A key event in the initiation of an inflammatory infiltrate is the induction of adhesion molecules on endothelium. It has long been known that a variety of endothelial adhesion molecules are inducible by cytokines and other mediators, notably E-selectin, P-selectin, ICAM-1 and VCAM-1 (Pober and Cotran, 1990; Carlos and Harlan, 1994, and chapter 4, 9 and 11). This, together with the eventual upregulation of local chemokine production (chapter 5), is crucial to the development of a full inflammatory response.

It is conceivable that a primary encounter of antigen-specific T cells with their target induces the release of cytokines and chemokines, thereby inducing an adhesive phenotype of the vascular endothelium near the reaction site. This initiates a positive feedback loop leading to the accumulation of further effector cells in the tissue. Similarly, complement activation or IgE-mediated responses are able to release endothelium-activating mediators.

More recent data point to an additional mechanism initiating cellular infiltration: Tissue cells or resident leucocytes were found to react towards various unspecific "stress" signals imposed by infection, cell death or —damage with cytokine and chemokine release, thus replacing the need for an antigen-driven reaction as the initial event in an inflammatory response. Indeed, a irritative effect of skin sensitizing substances seems to play a crucial role in inducing classical T cell responses such as a cutaneous delayed type hypersensitivity (Grabbe *et al.*, 1996). In such reactions, the tissue cells play the role of receptors, telling the immune system that something has gone wrong and requires the generation of an active immune response; activation of antigen presenting cells as well as cell recruitment are the consequences and might be a prerequisite for the specific immune reaction (Matzinger, 1994).

Inflammatory reactions display a great variability with respect to localization, kinetics, the type of leucocytes recruited and the effector functions generated. It is conceivable that the diversity of damaging agents is answered by the triggering of specific signalling pathways. The subsequent induction of distinct adhesion molecules and chemokines paves the way for a selective recruitment of effector cell subsets, depending on the kind of infectious agent or other damaging signal, the tissue, and the time course of the process (see later sections of this chapter).

Tissue-specific Expression of Endothelial Adhesion Molecules Enables "Homing" of Distinct Populations of Lymphocytes

With the pioneering work of J. Gowans and colleagues it became not only evident that lymphocytes recirculate continuously but also that different populations might home to different sites of the body (Gowans and Knight, 1964); the resulting concept of organ-specific homing will be discussed in more detail below. A prerequisite for the immigration into distinct compartments is the tissue-restricted expression of recognition molecules on endothelium.

The best example for an endothelial adhesion molecule whose expression is restricted in a tissue-specific manner is, to date, the "mucosal addressin" MAdCAM-1, expressed on HEV in gut-associated lymphoid tissues (Peyer's patches, mesenteric lymph node) as well as on postcapillary venules in the lamina propria and some other mucosal sites (Nakache *et al.*, 1989). It plays a crucial role in both the above mentioned recirculation of naive lymphocytes through lymphoid sites and in the immigration of mucosal blasts and effector cells into non-lymphoid mucosal sites, as discussed in chapters 3 and 7. This central role of MAdCAM-1 for the recruitment of cells into the gut is also conserved in inflammatory bowel disease models (Allemand and Hamann, unpublished data).

The "peripheral node addressin" PNAd defines carbohydrate epitopes serving as ligand for L-selectin; in contrast to MAdCAM-1, its function seems largely restricted to the recirculation of naive cells through lymphoid tissues.

The interaction of E-selectin expressed in the inflamed skin with carbohydrate epitopes ("cutaneous lymphocyte antigen", CLA) expressed on distinct subsets of T-cells associated with skin localization has led to the suggestion that this adhesion pathway defines a skin-specific homing mechanism. The evidence for this is discussed in chapter 8. It is, as yet, unclear how the widespread expression of E-selectin in most inflammatory regions can be reconciled with a skin-specific function; perhaps further mechanisms restrict the tissue-specific properties of the subset of T cells expressing the CLA epitope.

The far-reaching expectation that other sites of the body might be associated with further, distinct mechanisms of recruitment still remains speculative. In various other sites such as inflamed synovia, pancreatic islets or brain (see chapter 9, 11) the identified adhesion mechanisms rather were variable sets of the adhesion molecules already known to be involved in trafficking to lymph nodes, mucosal sites or inflamed skin. In most of these locations, however, the basal trafficking of lymphocytes through the non-inflamed tissue is very low. In the diseased tissue, the expression of generic, inflammation-related receptors such as ICAM-1, VCAM-1 and endothelial selectins becomes prominent and can partially override the organ-specific differentiation of endothelium. Nevertheless, different patterns of endothelial adhesion molecule expression may be found, depending on tissue as well as disease-type and -stage.

Only few groups applied experimental strategies directed to the identification of new molecules; hence, further adhesion pathways operative in distinct compartments might have to be defined in future. This especially applies to the large organs lung and liver, which harbor a significant fraction of the total lymphocyte pool, as discussed in the next chapter. The molecular basis of lymphocyte interactions in lung and liver is not well known, apart from a role of LFA-1/ICAM-1 for lung localization (Hamann *et al.*, 1988; Hamann and Austrup, unpublished data) and some evidence for a role of VAP-1 in liver (Salmi and Jalkanen, 1992; McNab *et al.*, 1996). Notwithstanding, the trapping of cells in these organs is not random, but shows considerable selectivity towards distinct activation and differentiation stages of lymphocytes as discussed later below.

II. LYMPHOCYTES ACQUIRE SPECIFIC TRAFFICKING PROPERTIES UPON DIFFERENTIATION

Endothelial diversity is only of value for the system if an appropriate diversity in the specificity of migrating cells exists. What are the subsets displaying specific trafficking? Where and when do they acquire this specificity? In the following we will emphasize the importance of lymphocyte development and differentiation as key determinants in specific homing and discuss how lymphocyte trafficking changes as it goes through several developmental steps. These alterations in the migratory behavior along maturation and differentiation are depicted in Figure 1.

Naive, Mature Lymphocytes Recirculate Continuously Between Blood, Lymphoid Organs and Lymph

Although antigen presenting cells are distributed throughout the body, the population most effective for a primary activation is the interdigitating dendritic cells from lymphoid tissues. Accordingly, it is predominantly the secondary lymphoid organs which provide the right milieu and the abundancy of potential interaction partners which is required to initiate an antigen-specific reaction. The regular channeling of naive lymphocytes through the sophisticated tissue architecture of lymphoid organs enables the screeening of antigen presenting cells for the appropriate antigen as well as of thousands of other lymphocytes for a cognate T/B cell partner etc., as discussed in a recent review (Gretz *et al.*, 1996). This seems so far the most appealing hypothesis as to why lymphoid organs and recirculation might have developed in evolution, and why predominantly naive lymphocytes recirculate randomly through all lymphoid tissues.

Activation Drastically Changes the Trafficking of Lymphocytes

The most dramatic alterations in the migratory behavior of lymphocytes occur upon activation (Hamann and Rebstock, 1993). Shortly after application of a mitogenic stimulus (1 h or less), mechanisms enabling the entry into lymphoid tissues become downregulated (Hamann *et al.*, submitted). This is especially the case for L-selectin, which is shed from the surface of most lymphocytes upon activation by several stimuli (Bührer *et al.*, 1990; Jung and Dailey, 1990; Kishimoto *et al.*, 1990; Kishimoto *et al.*, 1995; Wroblewski and Hamann, 1997), but the integrin $\alpha_4\beta_7$ is also transiently downregulated, at least on $CD4^+$ T cells activated by anti CD3 within 24 h (Austrup and Hamann, unpublished data).

Also strongly reduced is the immigration of activated lymphocytes into the spleen, although the above receptor mechanisms are not relevant for entry into this organ (see next chapter). It is conceivable, although not proven, that this is caused by a redistribution into other compartments, especially due to the trapping of the activated cells within lung and liver.

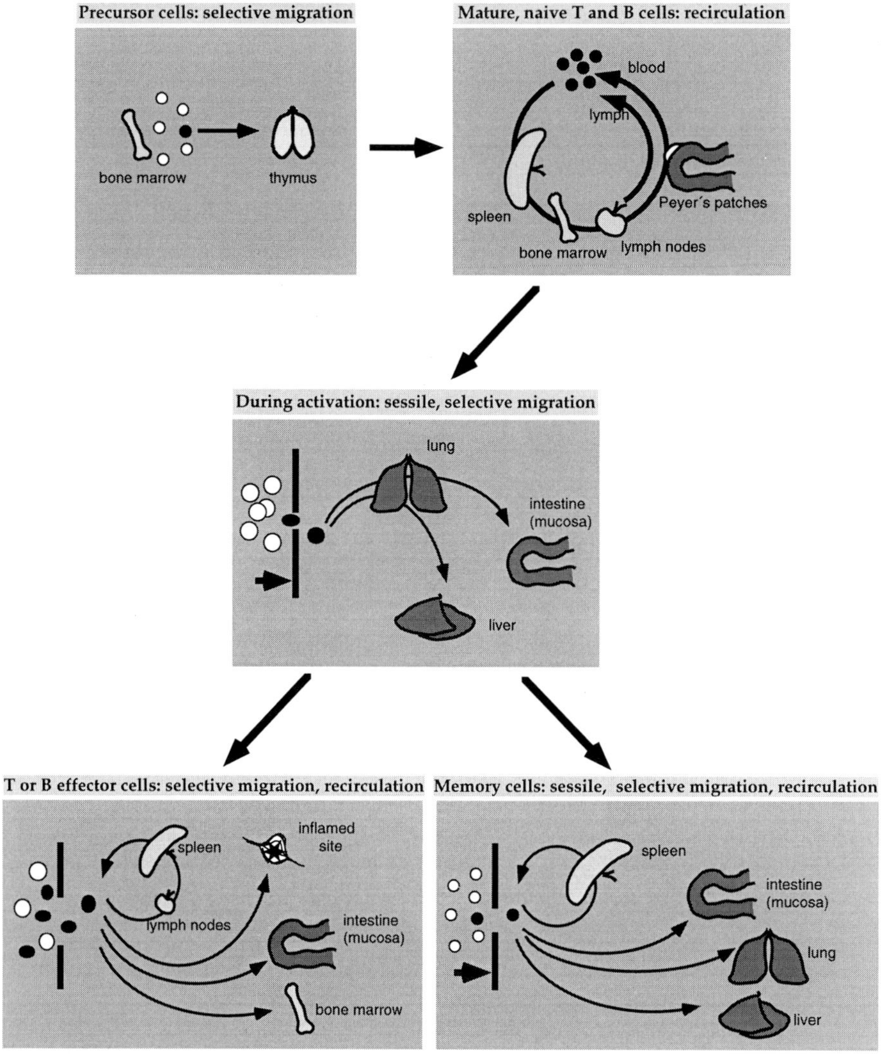

Figure 1 Alterations in the trafficking of lymphocytes along maturation, activation and differentiation:
a: After expansion and differentiation in the bone marrow, pre-T cells become migratory and express homing receptors for their major maturational environment, the thymus.
b: After proliferation and selection in the thymus or the bone marrow, mature T and B cells leave these tissues and join the recirculatory pool, which continuously cycles between the lymphoid tissues and blood. This lasts until an eventual encounter with a cognate antigen leads to activation.
c: Activation traps lymphocytes within the lymphoid tissues and lets them acquire a sessile phenotype; fractions of activated cells might access the blood and become redistributed into non-lymphoid tissues.
d and e: Differentiation into effector and memory stages generates subsets with the most contrasting properties: sessile fractions - circulating populations - subsets with the capacity to migrate specifically into distinct organs.
In black: migratory cells.

High Affinity of Activated Lymphocytes for Lung and Liver

Naive, resting cells are found for a short time after injection within the lung. However, most cells have left this organ within less than half an hour (Smith and Ford, 1983). In contrast, activated cells stay much longer in the lung, some up to 24 hours (Smith *et al.*, 1980). In addition, subsequent accumulation in the liver is characteristic for many activated lymphocytes. Parallel to the downregulation of receptors involved in recirculation, mechanisms causing a trapping of these cells in non-lymphoid organs such as lung or liver are upregulated. Activation of constitutively expressed receptors, as well as mechanisms requiring prior protein synsthesis, mediate the increased trapping of activated cells in liver and lung; at least in the latter organ, LFA-1 plays a significant, albeit not dominant role (Hamann *et al.*, 1988; Hamann *et al.*, submitted). The affinity of LFA-1, as of other integrins, has been shown to be upregulated independent of its expression upon T cell activation (Dustin and Springer, 1989; Hynes, 1992); the integrins are therefore good candidates for the early alterations in trafficking, although a role of β_1 or α_4-integrins has not been demonstrated so far. It is conceivable, however, that further adhesion mechanisms such as VAP-1 (McNab *et al.*, 1996), lectins (Crispe and Mehal, 1996) or others might mediate the interaction of activated cells with the vascular bed of lung and liver, as also discussed in the next chapter of the book.

The Majority of Activated Cells Might be Sessile

An increased localization in lung and liver after i.v. injection is a hallmark of almost every type of activated lymphocyte, short term as well as long term activated, *ex vivo* isolated or adapted to cell culture (Smith *et al.*, 1980; Parrott and Wilkinson, 1981; Dailey *et al.*, 1982; Hamann and Rebstock, 1993; Hamann *et al.*, submitted). But, *in vivo*, do activated cells actually have a chance to reach lung and liver? Trapping of circulating activated cells within the large vascular beds of lung and liver might just be an expression of an increased adhesiveness for ubiquitous ligands and it could be argued that these cells are, rather, fixed in place during proliferation by the same mechanisms, constituting a sessile phenotype.

Unfortunately, quantitative data on the turnover and recirculation rate of activated cells are scarce. *In vitro* experiments provide evidence that activated lymphocytes develop an increased adhesion to resident cells, such as the dendritic cells of lymphoid tissue (Hamann *et al.*, submitted). This interaction is not unlikely to prevent their exit from the tissue. It is also conceivable that this is, in addition, the mechanism which explains the long known phenomenon of antigen-dependent recruitment or trapping of lymphocytes within stimulated lymph nodes and spleen (Sprent, 1977; Ager and Drayson, 1988).

Potential Roles of Lung and Liver Localization

On the other hand, a siginificant number of activated lymphocytes can be detected in lymph and blood, either recirculating or recently produced by local proliferation and release. Their fate is not known. It is tempting to suggest that the filtering out of activated cells by lung and liver might fulfil distinct functions in the regulation of immune responses.

We found that effects of activation on T cell trafficking to lung and liver were not congruent: early after activation an increased accumulation within the liver was observed, which was not affected by protein synthesis inhibitors; 6–24 hours after activation their localization within the lung was strongly increased, but only in the absence of inhibitors (Hamann *et al.*, submitted). This indicates that disparate adhesion pathways are involved, serving a differential targeting of specific subsets/differentiation stages into lung or liver. The transient accumulation of activated lymphocytes within the lung will focus effector cells in this area, representing a large entry port for potential pathogens. In contrast, the trapping of activated T cells in the liver seems rather to lead to a downregulation of effector functions. It is known from, for example, transplantation studies that the liver exerts an immunosuppressive effect (Kobayashi *et al.*, 1994). Recently it has been shown that immunosuppressive cytokines such as Il-10 or TGFβ are produced in the liver (Lohse *et al.*, 1996). Moreover, the liver eliminates T cells undergoing antigen-triggered apoptosis (Huang *et al.*, 1994). The downregulation or killing of activated cells accumulating in the liver would therefore represent a regulatory mechanism able to control (excessive, "wrong" ?) T cell activation within the organism (Crispe and Mehal, 1996). It should be mentioned, however, that a variety of studies has shown that not every cell arriving in the liver is eliminated; a significant number return to the circulation via lymph or blood (Smith *et al.*, 1980; Parrott and Wilkinson, 1981; Young *et al.*, 1994; see also the following chapter).

Migrating Subsets of Activated or Effector Lymphocytes?

The actual fraction of cells emigrating from lymphoid or non-lymphoid tissues during activation is not known. It therefore is unclear whether circulating blasts (e.g. in human blood, 5–10% of T cells express activation markers) are a major fraction of all activated cells or whether they represent a distinct, small subset or differentiation stage which becomes exported. What speaks for the latter argument is the differences in trafficking behavior between the total blasts from lymphoid tissues and the blasts isolated from the thoracic duct; the latter display a markedly lower affinity for lung and liver and a high capacity to migrate directly into the gut wall (Guy-Grand *et al.*, 1974; Smith *et al.*, 1980).

Lymph-derived blasts are an exception to the dominant association of lung/liver seeking properties with the activated stage of lymphocytes. Moreover, lymph-derived blasts were found to display the most striking patterns of organ-specific trafficking (Hall *et al.*, 1977; Smith *et al.*, 1980). It suggests a peculiar stage in the differentiation in which the cells are able to circulate. In the case of mucosal IgA$^+$ B-cells, it has been found that the capacity to home to the lamina propria requires several days to develop after priming within the Peyer's patches and is only acquired upon differentiation of surface IgA$^+$ blasts into a stage expressing intracellular IgA (Guy-Grand *et al.*, 1974). This indicates that blasts have to adopt a specific phenotype which enables them to leave the tisşue as well as to reach the organ of destination. When lymphocytes activated *in vitro* are allowed to return in a resting state, L-selectin becomes re-expressed and the cells regain at least partially their capacity to recirculate (Jung and Dailey, 1988; Austrup *et al.*, 1995; Hamann *et al.*, submitted), contrasting in this respect to cells of memory phenotype. It might be assumed that this stage correlates to a circulating phase of effector cells.

It is as yet completely unknown which events determine the switch from the dominant phenotype of blasts being sessile and lung/liver seeking, to the stage of a migrating blast or effector cell. It has to be stressed that the latter is likely to be only a transient phase; memory cells again can adopt the sessile phenotype with lung/liver seeking properties, as discussed below.

In conclusion, the data available argue for a strong dependency of homing capacities from the differentiation stage, even among activated cells.

Memory Lymphocytes: Distinct Migratory Properties?

Work by C. Mackay showed that naive and memory lymphocytes differ in migration pathways. Whereas lymphocytes entering a lymph node via HEV display almost exclusively the phenotype of naive lymphocytes in sheep, the lymph draining the skin was found to contain large numbers of cells expressing the CD45 R0 marker associated in man and sheep with memory or activated cells (Mackay *et al.*, 1990). This led to the conclusion that only naive lymphocytes recirculate via HEV, whereas memory cells extravasate within the non-lymphoid peripheral tissues and travel into the draining lymph, from which they enter lymph nodes via the afferent lymph vessels (reviewed in: Mackay, 1993; Westermann and Pabst, 1996).

Recent work in our lab confirmed in the mouse that cells of memory phenotype have a low capacity to enter lymphoid tissues, especially the lymph nodes and Peyer's patches. Interestingly, they share this property, linked with a high affinity for lung and liver, with activated lymphocytes (Tietz and Hamann, in press). In mouse $CD4^+$ T cells, memory is found almost exclusively among a L-selectin negative subset (Bradley *et al.*, 1992); as L-selectin is absolutely essential for entry into peripheral lymph nodes (Arbones *et al.*, 1994; Bradley *et al.*, 1994; Lepault *et al.*, 1994) and of major importance for mesenteric nodes and even Peyer's patches (Hamann *et al.*, 1991), alone for this reason their recirculation via HEV of lymph nodes must be defective. L-selectin expressing memory cells are found, however, in the blood (Steeber *et al.*, 1996), among $CD8^+$ and B-cells (Kraal *et al.*, 1988; Mobley *et al.*, 1994), and among $CD4^+$ and $CD8^+$ memory cells in human PBL (Tedder *et al.*, 1990). Even more confusing, studies in rat clearly showed a capacity of CD45RC negative, supposed memory cells to recirculate via HEV through lymph nodes (Westermann *et al.*, 1994; Westermann and Pabst, 1996). Thus, it seems that not all memory populations are defective in recirculatory capacities and that differences in the species must be considered (Westermann and Pabst, 1996).

The abundancy of memory cells in the lymph of peripheral tissues and among the lymphocyte populations found in various non-lymphoid tissues led to the suggestion that memory lymphocytes recirculate actively, albeit by the route peripheral tissue – afferent lymph – to lymph nodes. We could not confirm this in the mouse by using purified memory cell populations from lymphoid tissues: these cells entered predominantly lung, liver and other non-lymphoid tissues within 1–24 hours after i.v. injection, resembling in that respect activated lymphocytes; the number of cells returning to lymph nodes via skin and afferent lymph was extremely small, indicating that this is a very inefficient pathway, at least in the absence of large inflammatory sites (Tietz and Hamann, in press). The kinetics of lymphocyte depletion from lymphoid organs upon longterm thoracic duct drainage indicated that the majority of memory cells rather might be a sessile population indeed (L. Bradley, pers. comm.).

An alternative explanation for the abundancy of memory cells in the afferent lymph has not yet been addressed in detail: it is not excluded that a significant fraction is continuously produced within and exported from local tissues; in peripheral tissues such as the skin or gut a significant degree of proliferation may take place (Fritz *et al.*, 1990).

Among the memory cells found in the lymph, a significant fraction expresses increased levels of certain molecules involved in organ-specific trafficking, such as either $\alpha_4\beta_7$ integrin or the E-selectin ligand CLA (Andrew *et al.*, 1996; Picker *et al.*, 1994; Rott *et al.*, 1996). This suggested that among the memory cells a remarkable degree of organ-specifity must exist. However, among memory cells residing in lymphoid tissues only a fraction expresses high levels of distinct receptors (Andrew *et al.*, 1996; Tietz and Hamann, in press), indicating that few cells actually are ready to travel into specific organs. Accordingly, the differences in tissue localization after injections of total memory cells of either mesenteric or subcutaneous lymph nodes is not very impressive (Tietz and Hamann, in press), in contrast to cells obtained from lymph.

Memory Cells and Their Trafficking — Still Controversial

The partially contradictory findings make thoughts about the biological significance of a distinct trafficking behavior of memory cells highly speculative. Moreover, not all authors believe that long-living, resting memory cells actually exist; alternatively, memory might be contained in slowly cycling cells (Gray and Matzinger, 1991; Michie *et al.*, 1992); with respect to the similarity in trafficking properties between activated lymphocytes and subsets considered as memory cells, this hypothesis is quite appealing.

The following conclusion is therefore rather subjective; another and more comprehensive view is to be found in a recent review (Westermann and Pabst, 1996).

- Not all memory cells are equal with respect to trafficking properties.
- The majority of memory cells might sit sessile in lymphoid and non-lymphoid tissues or only slowly recirculates. This could serve the same purpose as supposed for organ-specific trafficking, namely to concentrate antigen-specific cells in regions bearing the largest risk of repeated encounter with potential pathogens.
- Some subsets of memory cells (or differentiation stages), especially of $CD8^+$ T cells or B-cells can recirculate via HEV; differences between species in this respect must be considered.
- Organ-specific trafficking is found only on small subsets (or differentiation stages) of memory cells.
- A higher preference of memory cells for non-lymphoid tissues such as the lung or for inflamed sites might help to target them into potential entry ports of microorganisms. On the other hand, the trapping of these cells, especially in the liver, might be a way to control an excess of generated memory cells by induction of anergy or apoptosis.

Organ-specific Trafficking: Dependent on Origin?

As mentioned in the first part of this chapter, early studies on lymphocyte trafficking already provided evidence for subsets of cells with selective "homing" properties. Both the mesenterial IgA-blasts returning from blood into the gut mucosa and the skin derived blasts avoiding mucosal sites gave rise to the view that lymphocytes home back to their origin (see chapter 7).

The evolving concept of organ-specific homing (Gowans and Knight, 1964; Griscelli *et al.*, 1969; Guy-Grand *et al.*, 1974; Hall and Smith, 1970) was based predominantly on the impressive capacity of gut-derived T- blasts or IgA-producing B-blasts to return into the gut. Findings with skin-derived blasts which re-appeared preferentially within skin lymphatics or other non-mucosal sites (Rose *et al.*, 1976; Smith *et al.*, 1980; Chin and Hay, 1980) strengthened the view that lymphocytes home back into sites related to their origin. The term "organ-specific homing" is largely taken to be synonymous to the reciprocal trafficking of mucosa-specific and skin- or subcutaneous lymph node-specific subsets; the key receptors on endothelium involved in these processes and the concept itself are discussed in the chapters 3, 7, 8 and several excellent reviews (Butcher, 1986; Mackay, 1993; Picker, 1994; Butcher and Picker, 1996). It has long been clear that naive lymphocytes do not show organ preferences and rather recirculate randomly between peripheral and mesenteric lymph nodes, Peyer's patches and spleen, independent of their origin (Guy-Grand *et al.*, 1974; Freitas *et al.*, 1980; Stevens *et al.*, 1982). The concept of organ-specific homing behavior was therefore refined and now assumes that lymphocytes acquire an organ-specific homing behavior upon activation and eventual differentiation into effector or memory cells, the site of initial antigen encounter determining the later homing specificity.

However, neither the total blasts nor the bulk of memory cells isolated from mucosal or skin-associated lymphoid tissue display a strong degree of tissue-specificity, in contrast to cells traveling within the lymph (Smith *et al.*, 1980; Tietz and Hamann, in press). Moreover, a careful analysis of the contribution of immigration vs. survival to the organ-specific accumulation of lymphocytes indicates that part of the published evidence for organ-specific homing may indeed rather be caused by tissue-dependent differential proliferation and survival (Peters, Wonigeit, Pabst and Westermann, submitted; see also the next chapter). Among cells obtained from Peyer's patches or mesenteric lymph nodes, only a fraction of the memory cells shows a selective expression of the $\alpha_4\beta_7$-integrin, mediating their entry into mucosal sites (Andrew *et al.*, 1996; Tietz and Hamann, unpublished data). These findings indicate that a specific "homing phenotype" is not associated with the majority of cells undergoing activation and differentiation within a given microenvironment, but displayed only by subsets of cells, presumably representing distinct differentiation stages.

Differential Trafficking of Lymphocyte Subsets

These findings could give rise to the consideration that tissue-specific properties might rather be associated with distinct functional subsets than with the origin of the cells. Various investigations have presented evidence that already among small, resting lymphocytes differences in entry rate and transit time through lymphoid or non-lymphoid organs can be associated with the major subsets, namely T vs. B cells and CD4$^+$ vs. CD8$^+$ cells. CD4$^+$ T cells were found to have the highest capacity to enter peripheral lymph nodes (Mackay *et al.*, 1988; Washington *et al.*, 1988; Kimpton *et al.*, 1989; Abernethy *et al.*, 1990), B cells were reported to show a preference for Peyer's patches (Stevens *et al.*, 1982) but a low capacity to circulate through the gut (Mackay *et al.*, 1992), CD8$^+$ cells showed a higher recirculation through the gut

(Mackay *et al.*, 1992) compared with CD4$^+$ cells and a higher accumulation in lung and liver (Tietz, Austrup and Hamann, unpublished data). However, these differences are mainly of quantitative nature and some of them could not be confirmed by other investigators, in a different species (Westermann *et al.*, 1994). The lack of a separate analysis of naive versus memory cells among the populations in most of these studies, different techniques and different cell sources may account for the conflicting findings.

Better evidence for subset-specific migration patterns comes from cells in the effector phase. Indeed, many of the findings on mucosa-specific homing relate to IgA-expressing blasts. Although probably not all IgA$^+$ blasts are mucosa-specific (McDermott and Bienenstock, 1979), some authors believe that an association of mucosa-specific homing with the IgA$^+$ phenotype rather than with the origin of the cells is likely (Meeusen *et al.*, 1996).

A distinct tissue-specific localization is also found for the plasma cell stage of B cells. Although there are very few studies which directly address their trafficking properties, available data show a major plasma cell-specific localization in bone marrow and, surprisingly, in muscle (Binns *et al.*, 1992; Pabst and Pötschick, 1983). The preferential localization of plasma cell tumors within the bone marrow supports the former finding.

Selective Trafficking of Th1 Effector Cells into the Inflamed Skin

In T cells, studies on the trafficking of effector stage cells have hardly been performed until recently. We studied CD4$^+$ T cells induced *in vitro* to differentiate into either Th1 or Th2 cells by activation under the influence of cytokines. The experiments demonstrated a clearcut difference between Th1 and Th2 cell trafficking with respect to their entry into sites of an acute, Th1–dominated inflammation: the capacity of CD4$^+$ T cells to enter inflamed skin is restricted to the Th1 subset. Th2 cells, in contrast, are unable to enter the inflamed skin (Austrup *et al.*, 1997). These functional differences are paralleled by a differential expression of adhesion receptors: Th1 cells express functional ligands for E- and P-selectin and use both to enter the cutaneous site (Borges *et al.*, 1997); Th2 cells lack these ligands. This clearly shows that functional subsets of T cells with the same origin can be endowed with distinct trafficking properties.

Selective Recruitment of Effector Cell Subsets May Shape the Character of Local Immune Responses

The above findings are of considerable importance for the understanding of immune reactions *in vivo*. Effector cells of Th1 type activate macrophages and provide efficient cellular defense reactions but also can give rise to self-destructive processes; reactions dominated by Th2 cells induce humoral responses and, rather, downregulate cellular inflammatory processes (Mosmann and Coffmann, 1989; Mosmann and Sad, 1996). Up to now the kind of cytokines present within the tissue was seen to be the major determinant in the balance between a Th1 and Th2 -like response. The present data show that selective trafficking properties of the subsets may equally contribute to the development of a specific local reactivity. In this way, the selective expression of adhesion molecules in a given tissue may shape the character of a local immune reaction by recruiting specific subsets of effector cells (Figure 2).

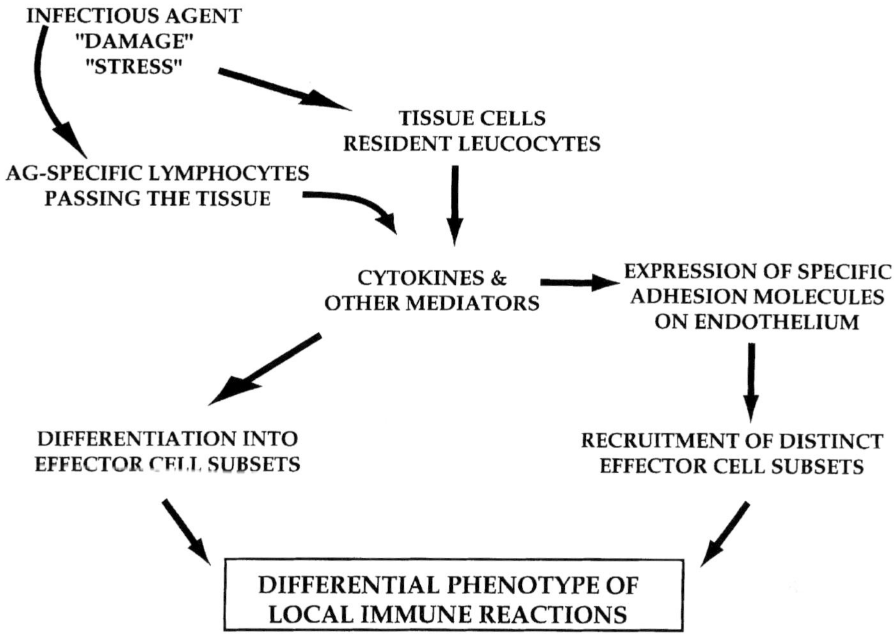

Figure 2 Selective recruitment targets distinct types of effector cells into inflammatory sites. This, together with the regulatory effect of cytokines and mediators on the differentiation and activation of effector cell populations helps to shape the character of an immune response (e.g. the bias towards Th1/Th2 dominated reactions).

Does the Expression of Ligands for Endothelial Selectins Determine Skin-specific or Inflammation-specific Trafficking?

The most selective recruitment of Th1 over Th2 cells was found in inflamed skin. However, we found also a preferential capacity of *in vitro* generated Th1 cells to enter other inflammatory regions such as the synovium (Austrup *et al.*, 1997; Allemand and Hamann, unpublished data). As in skin, their entry into the synovium is dependent on E-and P-selectin (Austrup, Bräuer and Hamann, in preparation). The trafficking of Th1 cells which express the appropriate ligands for endothelial selectins therefore seems to display a specificity for Th1-dominated, acute inflammatory regions rather than for specific tissues.

This is in some contrast to the assumed role of carbohydrate epitopes serving as E-selectin ligands in the human. These are detected by mAb Heca 452 and called CLA (cutaneous lymphocyte antigen), which is found on skin-associated lymphocytes but rarely on T cells in inflammatory sites of other tissues, as discussed in chapter 8. It binds to E-selectin and might therefore be equivalent to the ligands detected by soluble E/P-selectin constructs as used in our studies in the mouse. So far, the importance of kind (Th1/Th2-dominated) and stage of the inflammation for the differential recruitment of T cell subsets has not been studied carefully; we would suggest that the expression of E- (and P-) selectin ligands predicts a preference for acute, Th1-dominated inflammatory sites rather than a strict specificity for the organ skin. In fact, delayed-type hypersensitivity reactions in skin are Th1-, whereas allergic mucosal responses such as asthma are Th2-dominated (Mosmann and Sad, 1996).

Simultaneous Induction of Polarized Effector Functions and Migratory Phenotypes

As discussed in chapter 8, the milieu in which T cells undergo activation and differentiation is likely to be crucial for the acquisition of specific patterns of adhesion molecule expression, e.g. CLA. The same factors might determine the development of polarized effector functions such as cytokine profiles. Indeed, Il-12 plays an essential role in the generation of the Th1 effector subset (Germann *et al.*, 1993) as well as in the induction of CLA in humans (Leung *et al.*, 1995) or of ligands for endothelial selectins in mice (Austrup *et al.*, 1997; Borges *et al.*, 1997). Th1 cells and Il-12 were found to be crucial for cutaneous delayed-type hypersensitivity reactions (Cher and Mosmann, 1987; Riemann *et al.*, 1996); in contrast, a bias towards Il-4 production, the predominant Th2–inducing cytokine, was found in T-cells from mucosa-associated lymphoid tissues (Daynes *et al.*, 1990). If a dominant role of Il-12 in the cutaneous microenvironment is assumed, a linkage between Th1 differentiation and preferential skin-homing will necessarily result. Organ-specific trafficking may than be re-interpreted as a consequence of a tissue-specific induction of distinct functional subsets as discussed recently (Meeusen *et al.*, 1996). Indeed, most of the data on the homing of Th1 cells into acute inflammatory regions, especially of the skin, as well as data on the expression of CLA on skin-derived lymphocytes seem to be compatible with this view.

Functional T cell Subsets Associated with Mucosal Homing are Still Elusive

The expectation that Il-4 induced Th2 cells might display a preferential migration into mucosal sites (Meeusen *et al.*, 1996) has not been confirmed by our studies. *In vitro* generated Th2 cells were not found to enter the gut or Peyer's patches consistently better than Th1 cells (Austrup *et al.*, 1997).

Mucosa-specific homing is also not induced by another cytokine, TGFβ, which drives the differentiation of B- as well as T-cells towards a mucosal phenotype; TGFβ contributes to Ig switch towards IgA (Kim and Kagnoff, 1990) and induces the marker α_E-integrin typical for "mucosal" (more specifically: intraepithelial) T cells (Kilshaw and Murant, 1991). It does not affect, however, the homing behavior of T cells activated in its presence (Austrup *et al.*, 1995). Thus, the kind of functional T cell subsets associated with mucosal homing is still elusive, as are factors inducing it.

CONCLUSION

Although a significant part of the molecular mechanisms regulating lymphocyte recirculation and entry into inflamed regions are quite well understood by now, the context in which organ-specific trafficking is acquired and used by subsets or differentiation stages of lymphocytes remains a controversial issue. It is to be expected that an increasing understanding of the flexible reactions of lymphocytes to different activating signals will shed more light on the generation and variability of functional and phenotypic subsets of lymphocytes. It is not unlikely that, as a result, organ-specific homing will be recognized as a property of distinct differentiation stages and functional subsets, whose tissue preferences are adapted to various specific functions rather than guaranteeing the mere return into a tissue of origin.

REFERENCES

Abernethy, N.J., Hay, J.B., Kimpton, W.G., Washington, E.A. and Cahill, R.N. (1990). Non-random recirculation of small, CD4$^+$ and CD8$^+$ T lymphocytes in sheep: evidence for lymphocyte subset-specific lymphocyte- endothelial cell recognition. *Int. Immunol.*, **2**, 231–8.

Ager, A., and Drayson, M.T. (1988). Lymphocyte migration in the rat. In: A.J. Husband (ed.), *Migration and homing of lymphoid cells*, CRC Press, Boca Raton, pp. 19–49.

Andrew, D.P., Rott, L.S., Kilshaw, P.J. and Butcher, E.C. (1996). Distribution of alpha 4 beta 7 and alpha E beta 7 integrins on thymocytes, intestinal epithelial lymphocytes and peripheral lymphocytes. *Eur. J. Immunol.*, **26**, 897–905.

Arbones, M.L., Ord, D.C., Ley, K., Ratech, H., Maynard Curry, C., Otten, G., Capon, D.J. and Tedder, T.F. (1994). Lymphocyte homing and leukocyte rolling and migration are impaired in L-selectin-deficient mice. *Immunity*, **1**, 247–60.

Austrup, F., Rebstock, S., Kilshaw, P.J. and Hamann, A. (1995). TGF beta1–induced expression of the mucosa-related integrin alpha E on lymphocytes is not associated with mucosa-specific homing. *Eur. J. Immunol.*, **25**, 1487–1491.

Austrup, F., Vestweber, D., Borges, E., Löhning, M., Bräuer, R., Herz, U., Renz, H., Hallmann, R., Scheffold, A., Radbruch, A. and Hamann, A. (1997). P-and E-selectin mediate recruitment of T helper 1 but not T helper 2 cells into inflamed tissues. *Nature*, **385**, 81–83.

Baumhueter, S., Dybdal, N., Kyle, C. and Lasky, L.A. (1994). Global vascular expression of murine CD34, a sialomucin-like endothelial ligand for L-selectin. *Blood*, **84**, 2554–65.

Berg, E.L., Goldstein, L.A., Jutila, M.A., Nakache, M., Picker, L.J., Streeter, P.R.W.N., Zhou, D. and Butcher, E.C. (1989). Homing receptors and vascular addressins: cell adhesion molecules that direct lymphocyte traffic. *Immunol. Rev.*, **108**, 5–18.

Berg, E.L., McEvoy, L.M., Berlin, C., Bargatze, R.F. and Butcher, E.C. (1993). L-selectin-mediated lymphocyte rolling on MAdCAM-1. *Nature*, **366**, 695–698.

Berg, E.L., Robinson, M.K., Warnock, R.A. and Butcher, E.C. (1991). The human peripheral lymph node vascular addressin is a ligand for LECAM-1, the peripheral lymph node homing receptor. *J. Cell Biol.*, **114**, 343–9.

Binns, R.M., Licence, S.T. and Pabst, R. (1992). Homing of blood, splenic and lung emigrant lymphoblasts: comparison with the behaviour of lymphocytes from these sources. *Int. Immunol.*, 4, 1011–1019.

Borges, E., Tietz, W., Steegmaier, M., Moll, T., Hallmann, R., Hamann, A. and Vestweber, D. (1997). PSGL-1 on T helper 1 but not on T helper 2 cells binds to P-selectin and supports migration into inflamed skin. *J. Exp. Med.*, **185**, 573–578.

Bradley, L.M., Atkins, G.G. and Swain, S.L. (1992). Long-term CD4$^+$ memory T cells from the spleen lack MEL-14, the lymph node homing receptor. *J. Immunol.*, **148**, 324–31.

Bradley, L.M., Watson, S.R. and Swain, S.L. (1994). Entry of naive CD4 T cells into peripheral lymph nodes requires L-selectin. *J. Exp. Med.*, **180**, 2401–6.

Bührer, C., Berlin, C., Thiele, H.-G. and Hamann, A. (1990). Lymphocyte activation and expression of the human leucocyte-endothelial cell adhesion molecule 1 (Leu-8/TQ1 antigen). *Immunology,* 71, 442–448.

Butcher, E.C. (1986). The regulation of lymphocyte traffic. *Curr. Top. Microbiol. Immunol.,* **128** , 85–122.

Butcher, E.C. and Picker, L.J. (1996). Lymphocyte homing and homeostasis. *Science,* **272,** 60–66.

Carlos, T.M. and Harlan, J.M. (1994). Leukocyte-endothelial adhesion molecules. *Blood,* **84,** 2068–101.

Cher, D.J. and Mosmann, T.R. (1987). Two types of murine helper T cell clones II. Delayed type hypersensitivity is mediated by Th1 clones. *J. Immunol.,* **138,** 3688–94.

Chin, W. and Hay, J.B. (1980). A comparision of lymphocyte migration through intestinal lymph nodes, subcutaneous lymph nodes and chronic inflammatory sites of sheep. *Gastroenterology,* 79, 1231–1242.

Crispe, I.N. and Mehal, W.Z. (1996). Strange brew: T cells in the liver. *Immunol. Today,* 17, 522–525.

Dailey, M.O., Fathmann, C.G., Butcher, E.C., Pillemer, E. and Weissman, I.L. (1982). Abnormal migration of T cell clones. *J. Immunol.,* **128,** 2134–2136.

Daynes, R.A., Araneo, B.A., Dowell, T.A., Huang, K. and Dudley, D. (1990). Regulation of murine lymphokine production in vivo. III. The lymphoid tissue microenvironment exerts regulatory influences over T helper cell function. *J. Exp. Med.,* **171,** 979–96.

Dustin, M.L. and Springer, T.A. (1989). T-cell receptor cross-linking transiently stimulates adhesiveness through LFA-1. *Nature,* **341,** 619–24.

Freitas, A.A., Rose, M. and Rocha, B. (1980). Random recirculation of small T lymphocytes from thoracic duct in the mouse. *Cell. Immunol.,* **56,** 29–39.

Fritz, F.J., Pabst, R. and Binns, R.M. (1990). Lymphocyte subsets and their proliferation in a model for a delayed-type hypersensitivity reaction in the skin. *Immunology,* 71, 508–16.

Germann, T., Gately, M.K., Schoenhaut, D.S., Lohoff, M., Mattner, F., Fischer, S., Jin, S.C., Schmitt, E. and Rude, E. (1993). Interleukin-12/T cell stimulating factor, a cytokine with multiple effects on T helper type 1 (Th1) but not on Th2 cells. *Eur. J. Immunol.,* **23,** 1762–70.

Gowans, J.L. and Knight, E.J. (1964). The route of recirculation of lymphocytes in the rat. *Proceed. Roy. Soc. London, B,* **159,** 257–282.

Grabbe, S., Steinert, M., Mahnke, K., Schwarz, A., Luger, T.A. and Schwarz, T. (1996). Dissection of antigenic and irritative effects of epicutaneously applied haptens in mice. *J. Clin. Invest.,* **98,** 1158–1164.

Gray, D. and Matzinger, P. (1991). T cell memory is short-lived in the absence of antigen. *J. Exp. Med.,,* **174,** 969–974.

Gretz, J.E., Kaldijan, E.P., Anderson, A.O. and Shaw, S. (1996). Sophisticated strategies for information encounter in the lymph node. The reticular network as a conduit of soluble information an a highway for cell traffic. *J. Immunol.,* **157,** 495–499.

Griscelli, C., Vassalli, P. and McCluskey, R.T. (1969). The distribution of large dividing lymph node cells in syngeneic recipient rats after intravenous injection. *J. Exp. Med.,* **130,** 1427.

Guy-Grand, D., Griscelli, C. and Vassalli, P. (1974). The gut associated lymphoid system: nature and properties of the large dividing cells. *Eur. J. Immunol.,* **4,** 435.

Hall, J.G., Hopkins, J. and Orlans, E. (1977). Studies on the lymphocytes of sheep. III. Destination of lymph-borne immunoblasts in relation to their tissue of origin. *Eur. J. Immunol.,* 7, 30–37.

Hall, J.G. and Smith, M.E. (1970). Homing of lymph-borne immunoblasts to the gut. *Nature,* **226,** 262–4.

Hamann, A., Austrup, F., Jablonski-Westrich, D., Jonas, P. and Tietz, W. (submitted). Alterations in the trafficking of T cells upon activation.

Hamann, A., Jablonski-Westrich, D., Duijvestijn, A., Butcher, E.C., Baisch, H., Harder, R. and Thiele, H.-G. (1988). Evidence for an accessory role of LFA-1 in lymphocyte-high endothelium interaction during homing. *J. Immunol.*, **140**, 693–9.

Hamann, A., Jablonski-Westrich, D., Jonas, P. and Thiele, H.-G. (1991). Homing receptors re-examined: mouse LECAM-1 (MEL-14 antigen) is involved in lymphocyte migration into gut-associated lymphoid tissue. *Eur. J. Immunol.*, **21**, 2925–2929.

Hamann, A. and Rebstock, S. (1993). Migration of activated lymphocytes. In D. Dunon, C.R. Mackay and B. Imhof (eds.), *Adhesion in leucocyte homing and differentiation*, Springer Verlag, Berlin-Heidelberg-New York, pp. 109–124.

Hanninen, A., Taylor, C., Streeter, P.R., Stark, L.S., Sarte, J.M., Shizuru, J.A., Simell, O. and Michie, S.A. (1993). Vascular addressins are induced on islet vessels during insulitis in nonobese diabetic mice and are involved in lymphoid cell binding to islet endothelium. *J. Clin. Invest.*, **92**, 2509–15.

Huang, L., Soldevilla, G., Leeker, M., Flavell, R. and Crispe, I.N. (1994). The liver eliminates T cells undergoing antigen-triggered apoptosis *in vivo. Immunity*, **1**, 741–749.

Hynes, R. (1992). Integrins: versatility, modulation and signaling in cell adhesion. *Cell*, **69**, 11–25.

Jung, T.M. and Dailey, M.O. (1990). Rapid modulation of homing receptors (gp90MEL-14) induced by activators of protein kinase C. Receptor shedding due to accelerated proteolytic cleavage at the cell surface. *J. Immunol.*, **144**, 3130–6.

Jung, T.M. and Dailey, M.O. (1988). Reversibility of loss of homing receptor expression following activation. *Adv. Exp. Med. Biol.*, **237**, 519–24.

Kilshaw, P.J. and Murant, S.J. (1991). Expression and regulation of beta 7(beta p) integrins on mouse lymphocytes: relevance to the mucosal immune system. *Eur. J. Immunol.*, **21**, 2591–7.

Kim, P.H. and Kagnoff, M.F. (1990). Transforming growth factor beta 1 increases IgA isotype switching at the clonal level. *J. Immunol.*, **145**, 3773–8.

Kimpton, W.G., Washington, E.A. and Cahill, R.N. (1989). Recirculation of lymphocyte subsets ($CD5^+$, $CD4^+$, $CD8^+$, $T19^+$ and B cells) through fetal lymph nodes. *Immunology*, **68**, 575–9.

Kishimoto, T.K., Jutila, M.A. and Butcher, E.C. (1990). Identification of a human peripheral lymph node homing receptor: a rapidly down-regulated adhesion molecule. *Proc. Natl. Acad. Sci. USA*, **87**, 2244–8.

Kishimoto, T.K., Kahn, J., Migaki, G., Mainolfi, E., Shirley, F., Ingraham, R. and Rothlein, R. (1995). Regulation of L-selectin expression by membrane proximal proteolysis. *Agents Actions Suppl*, 47, 121–34.

Kobayashi, E., Kamanda, N., Enosawa, S., Toyama, N., Walker, N. I. and Miyata, M. (1994). Prevention by liver transplantation of the graft-versus-host reaction and allograft rejection in a rat model of small bowel transplantation. *Transplantation*, 57, 177–81.

Kraal, G., Weissman, I.L. and Butcher, E.C. (1988). Memory B cells express a phenotype consistent with migratory competence after secondary but not short-term primary immunization. *Cell. Immunol.*, **115**, 78–87.

Lepault, F., Gagnerault, M.C., Faveeuw, C. and Boitard, C. (1994). Recirculation, phenotype and functions of lymphocytes in mice treated with monoclonal antibody MEL-14. *Eur. J. Immunol.*, 24, 3106–12.

Leung, D.Y., Gately, M., Trumble, A., Ferguson Darnell, B., Schlievert, P.M. and Picker, L.J. (1995). Bacterial superantigens induce T cell expression of the skin-selective homing receptor, the cutaneous lymphocyte-associated antigen, via stimulation of interleukin 12 production. *J. Exp. Med.*, **181**, 747–53.

Lohse, A.W., Knolle, P.A., Bilo, K., Uhrig, A., Waldmann, C., Ibe, M., Schmitt, E., Gerken, G. and Meyer Zum Buschenfelde, K.H. (1996). Antigen-presenting function and B7 expression of murine sinusoidal endothelial cells and Kupffer cells. *Gastroenterology*, **110**, 1175–81.

Mackay, C.R. (1993). Homing of naive, memory and effector lymphocytes. *Curr. Opin. Immunol.*, **5**, 423–7.

Mackay, C.R., Kimpton, W.G., Brandon, M.R. and Cahill, R.N. (1988). Lymphocyte subsets show marked differences in their distribution between blood and the afferent and efferent lymph of peripheral lymph nodes. *J. Exp. Med.*, **167**, 1755–65.

Mackay, C.R., Marston, W.L. and Dudler, L. (1990). Naive and memory T cells show distinct pathways of lymphocyte recirculation. *J. Exp. Med.*, **171**, 801–17.

Mackay, C.R., Marston, W.L., Dudler, L., Spertini, O., Tedder, T.F. and Hein, W.R. (1992). Tissue-specific migration pathways by phenotypically distinct subpopulations of memory T cells. *Eur. J. Immunol.*, **22**, 887–95.

Matzinger, P. (1994). Tolerance, danger and the extended family. *Ann. Rev. Immunol.*, **12**, 991–1045.

McDermott, M.R. and Bienenstock, J. (1979). Evidence for a common mucosal immunologic system. *J. Immunol.*, **122**, 1892–8.

McNab, G., Reeves, J.L., Salmi, M., Hubscher, S., Jalkanen, S. and Adams, D.H. (1996). Vascular adhesion protein 1 mediates binding of T cells to human hepatic endothelium. *Gastroentrology*, **110**, 522–8.

Mebius, R.E., Breve, J., Kraal, G. and Streeter, P.R. (1993). Developmental regulation of vascular addressin expression: a possible role for site-associated environments. *Int. Immunol.*, **5**, 443–9.

Meeusen, E.N.T., Premier, R.R. and Brandon, M.R. (1996). Tissue-specific migration of lymphocytes: a key role for Th1 and Th2 cells? *Immunol. Today*, **17**, 421–424.

Michie, C.A., McLean, A., Alcock, C. and Beverly, P.C.L. (1992). Lifespan of human lymphocyte subsets defined by CD45 isoforms. *Nature*, **360**, 264–5.

Mobley, J.L., Rigby, S.M. and Dailey, M.O. (1994). Regulation of adhesion molecule expression by CD8 T cells, *in vivo*. II. Expression of L-selectin (CD62L) by memory cytolytic T cells responding to minor histocompatibility antigens. *J. Immunol.*, **153**, 5443–52.

Mosmann, T.R. and Coffmann, R.L. (1989). TH1 and TH2 cells: Different patterns of lymphokine sekretion lead to different functional properties. *Ann. Rev. Immunol.*, **7**, 145–173.

Mosmann, T.R. and Sad, S. (1996). The expanding universe of T-cell subsets: Th1, Th2 and more. *Immunol. Today*, **17**, 138–146.

Nakache, M., Lakey-Berg, E., Streeter, P.R. and Butcher, E.C. (1989). The mucosal vascular addressin is a tissue-specific endothelial adhesion molecule for circulating lymph nodes. *Nature*, **337**, 179–181.

Ottaway, C.A. (1988). Dynamic aspects of lymphoid cell migration. In A.J. Husband (ed.), *Migration and homing of lymphoid cells*, CRC Press, Boca Raton, pp. 167–194.

Pabst, R. and Pötschick, K. (1983). Proliferation and emigration of newly formed lymphocytes from pig spleens during an immune response. *Immunology*, **50**, 281–288.

Parrott, D.M.V. and Wilkinson, P.C. (1981). Lymphocyte locomotion and migration. *Prog. Allergy*, **28**, 193–284.

Picker, L.J. (1994). Control of lymphocyte homing. *Curr. Opin. Immunol.*, **6**, 394–406.

Picker, L.J., Martin, R.J., Trumble, A., Newman, L.S., Collins, P.A., Bergstresser, P.R. and Leung, D.Y. (1994). Differential expression of lymphocyte homing receptors by human memory/effector T cells in pulmonary versus cutaneous immune effector sites. *Eur. J. Immunol.*, **24**, 1269–77.

Pober, J.S. and Cotran, R.S. (1990). Cytokines and endothelial cell biology. *Physiol. Rev.*, **70**, 427–51.

Riemann, H., Schwarz, A., Grabbe, S., Aragane, Y., Luger, T. A., Wysocka, M., Kubin, M., Trinchieri, G. and Schwarz, T. (1996). Neutralization of IL-12 in vivo prevents induction of contact hypersensitivity and induces hapten-specific tolerance. *J. Immunol.*, **156**, 1799–803.

Rose, M.L., Parrott, D.M.V. and Bruce, R.G. (1976). Migration of lymphoblasts to the small intestine: II. Divergent migration of mesenteric and peripheral immunoblasts to sites of inflammation in the mouse. *Cell. Immunol.*, **27**, 36–46.

Rott, L.S., Briskin, M.J. Andrew, D.P., Berg, E.L. and Butcher, E.C. (1996). A fundamental subdivision of circulating lymphocytes defined by adhesion to mucosal addressin cell adhesion molecule-1. Comparison with vascular cell adhesion molecule-1 and correlation with beta 7 integrins and memory differentiation. *J. Immunol.*, **156**, 3727–36.

Salmi, M. and Jalkanen, S. (1992). A 90–kilodalton endothelial cell molecule mediating lymphocyte binding in humans. *Science*, **257**, 1407–9.

Smith, M.E. and Ford, W.L. (1983). The recirculating lymphocyte pool of the rat: a systematic description of the migratory behaviour of recirculating lymphocytes. *Immunology*, **49**, 83–94.

Smith, M.E., Martin, A.F. and Ford, W.L. (1980). Migration of lymphoblasts in the rat. *Monogr. Allergy*, **16**, 203–232.

Sprent, J. (1977). Migration and lifespan of lymphocytes. In F. Loor and G. E. Roelants, eds., *B and T cells in immune response*, John Wiley & Sons, Basel, pp. 59–82.

Steeber, D.A., Green, N.E., Sato, S. and Tedder, T.F. (1996). Lyphocyte migration in L-selectin-deficient mice. Altered subset migration and aging of the immune system. *J. Immunol.*, **157**, 1096–106.

Stevens, S., Weissman, I.L. and Butcher, E.C. (1982). Differences in the migration of B and T lymphocytes: organ-selective localization in vivo and the role of lymphocyte-emndothelial cell recognition. *J. Immunol.*, **128**, 844–851.

Streeter, P.R., Rouse, B.T. and Butcher, E.C. (1988). Immunohistologic and functional characterization of a vascular addressin involved in lymphocyte homing into peripheral lymph nodes. *J. Cell Biol.*, **107**, 1853–62.

Tedder, T.F., Matsuyama, T., Rothstein, D., Schlossman, S.F. and Morimoto, C. (1990). Human antigen-specific memory T cells express the homing receptor (LAM-1) necessary for lymphocyte recirculation. *Eur. J. Immunol.*, **20**, 1351–5.

Tietz, W. and Hamann, A. (1997). The migratory behavior of murine $CD4^+$ cells of memory phenotype. *Eur. J. Immunol.*, in press.

Washington, E.A., Kimpton, W.G. and Cahill, R.N. (1988). $CD4^+$ lymphocytes are extracted from blood by peripheral lymph nodes at different rates from other T cell subsets and B cells. *Eur. J. Immunol.*, **18**, 2093–6.

Westermann, J., Nagahori, Y., Walter, S., Heerwagen, C., Miyasaka, M. and Pabst, R. (1994). B and T lymphocyte subsets enter peripheral lymph nodes and Peyer's patches without preference in vivo: no correlation occurs between their localization in different types of high endothelial venules and the expression of CD44, VLA-4, LFA-1, ICAM-1, CD2 or L-selectin. *Eur. J. Immunol.*, **24**, 2312–6.

Westermann, J. and Pabst, R. (1996). How organ-specific is the migration of "naive" and "memory" T cells? *Immunol. Today*, **17**, 278–282.

Westermann, J., Persin, S., Matyas, J., van-der-Meide, P. and Pabst, R. (1994). Migration of so-called naive and memory T lymphocytes from blood to lymph in the rat. The influence of IFN-gamma on the circulation pattern. *J. Immunol.*, **152**, 1744–50.

Wroblewski, M. and Hamann, A. (1997). CD45–mediated signals can trigger shedding of lymphocyte L-selectin. *Int. Immunol.*, **9**, 555–562.

Young, A.J., Hare, G.M. and Hay, J.B. (1994). Blood-to-lymph migration of small lymphocytes through the liver of the sheep. *Hepatology*, **19**, 758–63.

2. LYMPHOCYTE TRAFFIC TO LYMPHOID AND NON-LYMPHOID ORGANS IN DIFFERENT SPECIES IS REGULATED BY SEVERAL MECHANISMS

REINHARD PABST and JÜRGEN WESTERMANN

Centre of Anatomy, Medical School of Hannover, 30623 Hannover, Germany

INTRODUCTION

About 110 years ago Walter Flemming postulated that lymphocytes in the large lymphatics consist not only of newly produced lymphocytes but also of lymphocytes on their way back to the blood which they had left via capillaries and venules in lymph nodes. This idea of a continuous circuit of lymphocytes leaving and returning to the blood was proven in the sixties by Gowans in rats and from then on called lymphocyte recirculation. With respect to the blood cells, this phenomenon is unique to lymphoid cells, as it has never been shown that granulocytes and monocytes can return to the blood after leaving by migrating into a tissue. Thus, trafficking lymphocytes continuously integrate the immune system using the blood and lymph vessels as traffic routes. The basic experiments on lymphocyte traffic up to 1980 were reviewed excellently by Ford (1980). Sometimes the term lymphocyte "homing" is used, describing a selective emigration from the blood into an organ. However, this term does not mean that the lymphocytes come from this "home" or will stay in their "home". After a lymphocyte has entered the parenchyma of a lymph node, for example, it does not mean that this lymphocyte will mature to an effector cell, e.g. a plasma cell, in this specific node, and will not leave the organ again.Thus, it would be better to replace the term "homing" by migration or traffic. Similarly, experiments documenting the entry of lymphocytes into an organ should not refer to lymphocyte "recirculation", as long as the return of these cells to the blood either directly via venules or via lymphatics has not been demonstrated.

In the last 10 years an increasing number of adhesion molecules on the surface of lymphocytes and endothelial cells has been characterized and their role in lymphocyte migration described (for review see Picker and Butcher, 1992; Ager, 1994; Springer, 1994; Girard and Springer, 1995; Imhof and Dunon, 1995). At this point in time several questions are pertinent: Is another review on lymphocyte traffic really necessary? Why are contradictory data so often published? Why are widely accepted hypotheses sometimes challenged by different groups and what are the areas of future research? There are several reasons for the current discrepancies: differences between species are often neglected, and data obtained in one species are mistakenly taken as representative for all other species. A number of technical aspects of experiments on lymphocyte migration strongly influence the results, e.g. the type of cells used to study traffic, the labelling techniques and evaluations are critical. All too often only one or a few time points were used to follow the route of migrating lymphocytes. This is usually insufficient to describe the kinetics of lymphocyte migration. *In vitro* experiments are very helpful in exploiting molecular mechanisms of the interaction between lymphocytes and endothelial cells, but the physiological relevance of such observations has to be documented *in vivo.* Relative changes in lymphocyte numbers found in one organ after injecting labelled cells can be caused by alterations in the total number, therefore as far as possible relative and absolute numbers should be determined.

The main aims of this chapter are to describe the quantitative aspects of the migratory routes of lymphocytes; to point out the differences between organs with specialized venules as entry sites for migrating lymphocytes and those with conventional venules; to mention species differences and the effects of the cell source and labelling techniques and to point to areas of contradictory findings and to organs often neglected in reviews on lymphocyte traffic. For details on the role of adhesion molecules, migration to the gut wall and skin and the migratory route of activated lymphoid cells the reader should refer to the other chapters of this book.

QUANTITATIVE ASPECTS OF LYMPHOCYTE LOCALIZATION AND MIGRATION

In a healthy young man about 450×10^9 lymphocytes (mainly small mature cells) are distributed all over the lymphoid and non-lymphoid organs of the body (for review see Westermann and Pabst, 1990). Only about 2% are in the blood at any given time. In disease, when the numbers of lymphocytes in the blood are decreased by 50% (equivalent to approx. 5×10^9 lymphocytes), this may be due to a maldistribution between the blood and lymphoid organs, and will hardly be measurable in other organs. As long as there are no kinetic data available for several lymphoid organs the decreases/increases of lymphocytes in one organ cannot be extrapolated to indicate the situation in the whole lymphoid system. A good clinical example is the decrease of $CD4^+$ lymphocytes in the blood of HIV-infected patients, which does not necessarily indicate a reduction of $CD4^+$ lymphocytes in lymph nodes. On average lymphocytes stay in the blood for only about 30 min before they enter lymphoid and non-lymphoid organs; this results in the circulating blood pool being exchanged 48 times a day. It has long been accepted that great numbers of granulocytes are found near to the vessel wall, the so-called marginal pool, and that they can be mobilized by different stressors causing rapid leukocytosis. There is growing evidence of a comparable marginal pool of lymphocytes, e.g. in the lung capillary bed (Pabst and Tschernig, 1995). In a recent study differences in the expression of adhesion molecules on lymphocytes from the circulating vascular pool and the marginal pool were documented in the rat (Klonz *et al.*, 1996). The vast majority of lymphocytes leave the blood in organs via normal venules and sinusoids, as in the lung, liver, spleen and bone marrow (Figure 1). Most lymphocytes return to the blood after a short transit time via venules and only few lymphocytes migrate via afferent lymphatics to the regional lymph nodes and finally return to the blood via the thoracic duct.

Less than 10% of all lymphocytes in the blood migrate into organs such as the lymph nodes, tonsils, and Peyer's patches. In these organs the lymphocytes use specialized venules which have a cobblestone-like endothelium and are therefore called high endothelial venules (HEV). In most species the lymphocytes return to the blood via efferent lymphatics, e.g. the thoracic duct. The transit time of lymphocytes from the

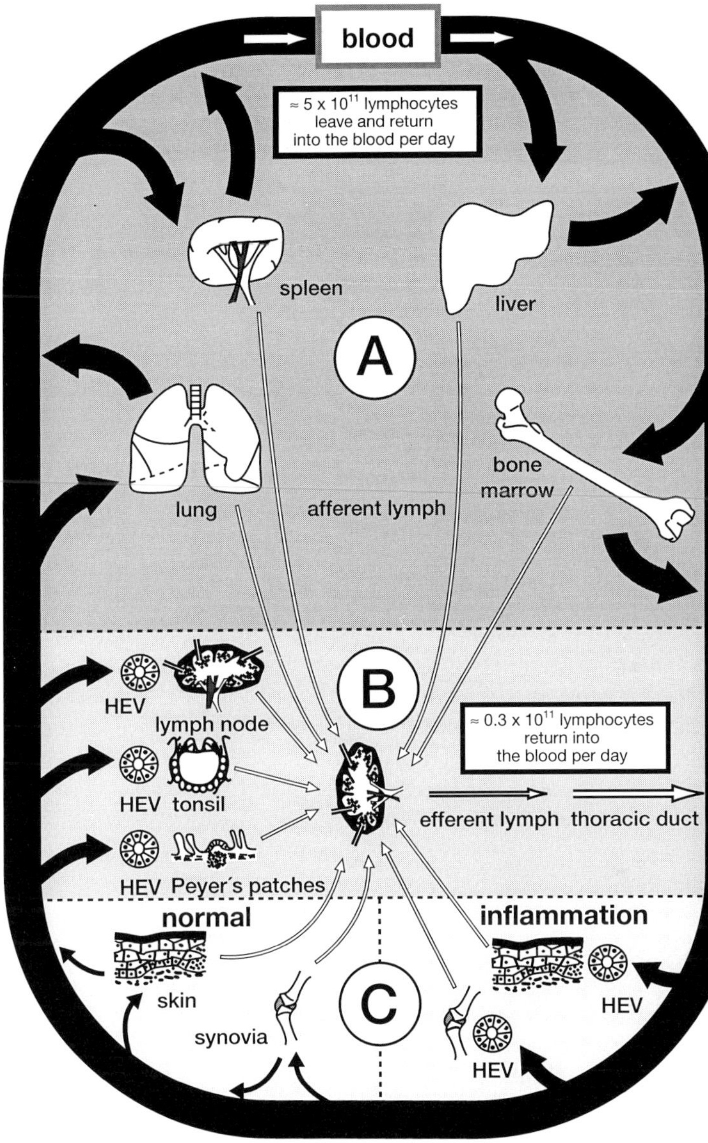

Figure 1 Schematic illustration of the migratory routes of lymphocytes between lymphoid and non-lymphoid organs. The size of the arrows indicates the extent of lymphocytes entering the different organs. The white arrows stand for afferent and efferent lymphatics. In the upper section **A**, organs without high endothelial venules (HEV) as entry sites into the organ are shown. Large numbers of lymphocytes enter and also leave the organs via the blood. In section **B** the organs with HEV, such as lymph nodes, tonsils and Peyer's patches, are shown. The lymphocytes enter via the blood but in most species leave via the efferent lymph and reach the blood via the thoracic duct. In the lower section **C**, organs with very low lymphocyte entry (under normal circumstances) are shown, e.g. skin and synovia of joints. During acute or chronic inflammation many more lymphocytes migrate to these tissues (indicated by the larger arrows). In total about 5×10^{11} lymphocytes leave the blood per day in a healthy young man. Only a small proportion of these return via the thoracic duct to the blood (0.3×10^{11} per day).

blood to the thoracic duct via this route is about 5 times longer than that through organs without HEV. This route is the classical pathway of lymphocyte recirculation and has attracted much interest over the last 20 years. It has to be stressed, however, that on a quantitative basis only a minority of lymphocytes leave the blood via this route (Figure 1). Under normal circumstances very few lymphocytes migrate into organs such as the skin, synovia, muscle and brain (Figure 1). They enter these tissues via normal venules and return to the blood either via normal venules or via lymphatics. During acute or chronic inflammation, however, the venules can develop a high endothelium similar to HEV in lymph nodes, and this results in a large increase in lymphocyte influx. Cytokines produced locally seem to play a critical role in the regulation of lymphocyte entry into inflammatory sites (for details see Chapter 11).

ADVANTAGES AND LIMITATIONS OF DIFFERENT TECHNIQUES USED IN LYMPHOCYTE MIGRATION STUDIES

To study migrating lymphocytes, the route of the cells has to be followed *in vivo*, therefore, the cells have to be labelled. The first critical aspect is the source of lymphocytes used for the experiments. In small rodents, e.g. the mouse, cell suspensions are often prepared from lymphoid organs such as the spleen or lymph nodes. Not only the cell preparation procedure will affect the lymphocytes, but it also has to be considered that a mixture of sessile and migrating cells will be injected. Therefore, physiologically migrating lymphocytes, such as those recovered from the blood or lymph, have several advantages. To label the cells, a series of radioactive and fluorescent labels have been used for all cells or precursors for proliferating cells (for review see Pabst *et al.*, 1993). Fluorescent dyes have proven valuable as they can be combined with surface markers to identify migrating cells not only in suspensions but also in tissue sections. These characterizations are made after the cells migrate *in vivo*, thus potentially noxious *in vitro* separation procedures to enrich lymphocyte subsets are not necessary. Some fluorescent dyes are, however, initially too bright for the fluorescent cell sorter and at later time points sometimes too dim to be identified after several hours or days. In species in which congenic strains exist, a further advantage is obvious: lymphocytes can be transferred without any *in vitro* labelling procedure and their route through lymphoid organs can be followed by identifying the injected cells with monoclonal antibodies. Thus, using adequate techniques lymphocyte subsets can be localized on their traffic routes through the compartments of lymphoid organs, e.g. the spleen (Willführ *et al.*, 1990) and lymph nodes (Blaschke *et al.*, 1995). *In vitro* tests, e.g. the frozen section assay, have been successfully used as an alternative to study in detail the interaction of lymphocytes with endothelial cells. Meanwhile, high endothelial venular cells from rodent lymph nodes can be cultured *in vitro* for such studies. For experiments with human lymphocytes, cultured human umbilical cord venous endothelial cells (HUVEC) are often used to study the mechanisms involved in lymphocyte binding. However, it has to be kept in mind that this endothelium is hardly representative of the endothelium in other organs, e.g. that of HEV. The locomotor phenomenon of lymphocyte subsets can also be studied in three dimensional collagen lattices (Friedl *et al.*, 1994). A major technical breakthrough in studying lymphocyte migration was the application of *in vivo* videomicroscopy, allowing migrating lymphocytes to be analysed as well as the effects of blocking specific adhesion molecules under physiological shear forces (e.g. Miura *et al.*, 1995; Finger *et al.*, 1996).

In conclusion, there is not one single optimal technique for lymphocyte migration experiments, but a combination of different methods is often most advantageous. In interpreting data and comparing conflicting results between different groups the potential effects of the techniques have to be considered.

ENTRY OF LYMPHOCYTES INTO LYMPHOID ORGANS VIA HEV: SEVERAL ADHESION MOLECULES ARE INVOLVED

It is now generally accepted that a series of interactions between lymphocytes and endothelial cells takes place in HEV, described as a cascade-type, partially overlapping sequence of interactions (Picker and Butcher, 1992; Springer, 1994; Imhof and Dunon, 1995). As other chapters in this book have focused on this aspect, no details will be given here. There are at least three phases: lymphocytes come into contact with the endothelial surface and start to roll, also called "tethering". The following tight adhesion is the result of the activation of further adhesion molecules by factors from the endothelium and the underlying cells (e.g. chemokines and cytokines). Finally, the lymphocytes transmigrate between endothelial cells, also called diapedesis (Figure 2). There is an overlap of effects of the adhesion molecules such as selectins, integrins and the Ig super family. Recent data on the preferential localization of adhesion molecules on the surface of lymphocytes are of particular interest; e.g. L-selectin on the tip of the microvilli, LFA-1 at the base of the microvilli (Berlin *et al.*, 1995; Hasslen *et al.*, 1995). Thus, knowledge about the spatial interaction between the lymphocyte and the endothelial cell surface is growing. Also of great interest is the time needed for the different migration steps into a lymph node: adhesion and activation take seconds, strong adhesion a few minutes and transmigration several minutes. These different lengths of time seem to be important in the regulation of lymphocyte traffic.

In recent years evidence has emerged that there are not only specific pairs of adhesion molecules for migration into Peyer's patches (PP) and peripheral lymph nodes (pLN) for example. L-selectin was long taken as the pLN homing receptor. The results from experiments with L-selectin knockout mice have helped to modify this concept: in these animals lymphocyte migration to pLN was drastically reduced but that to PP was also dramatically affected (Arbones *et al.*, 1994; Steeber *et al.*, 1996). Thus, L-selectin is not only important for the entry of lymphocytes into pLN, but also into other lymphoid organs.

TRANSMIGRATION OF LYMPHOCYTES THROUGH THE LYMPHOID PARENCHYMA

After a lymphocyte has left the blood and transmigrated between endothelial cells it has to pass the basal lamina and will come into contact with the extracellular matrix, e.g. lamellin, fibronectins, collagen but also interstitial cells such as macrophages, dendritic cells, fibroblasts and other lymphoid cells. These cells can produce chemokines and cytokines, and activate adhesion molecules in a paracrine fashion. As transit through a lymphoid organ lasts several hours, there is much more time for such interactions or the activation of specific lymphocyte subsets. Recently, many previously published data were compiled and a new concept was proposed as to how factors

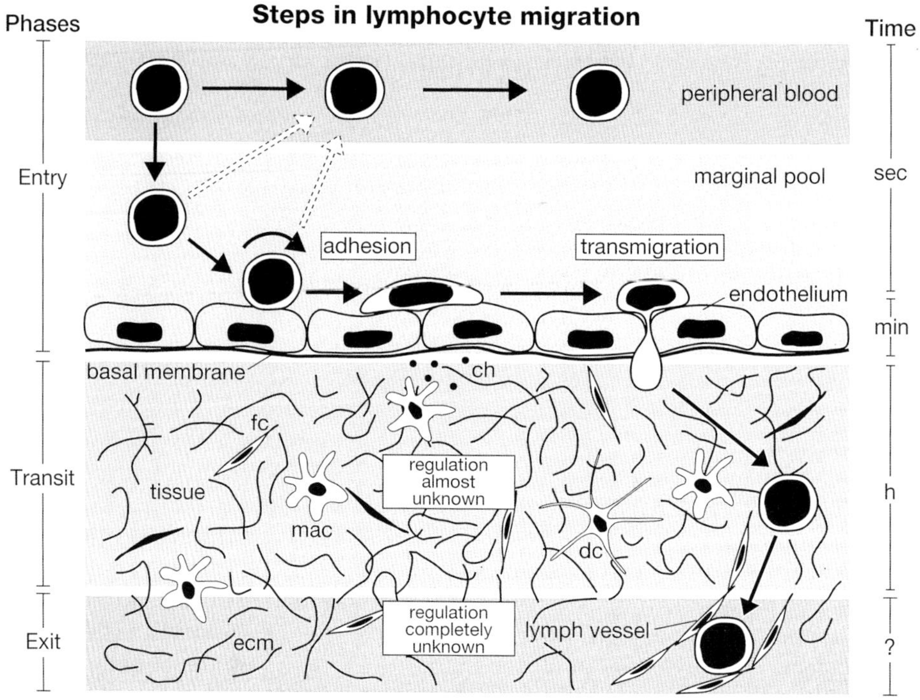

Figure 2 The different phases of lymphocyte migration are shown. The time needed for entry into the tissue, transit through the organ and exit from the organ are indicated on the right. There is a continuous exchange between lymphocytes of the circulating and the marginal pool. During the transit phase lymphocytes come into contact with non-lymphoid cells (fc = fibrocyte, mac = macrophage, dc = dendritic cell), and in the extracellular matrix (ecm) with many different fibres. Chemokines (ch) not only influence the adhesion to endothelial cells but also the transmigration of lymphocytes.

reaching a lymph node via afferent lymphatics can quickly regulate lymphocyte entry via HEV and probably also within the parenchyma (Gretz *et al.*, 1996). A conduit system of extracellular matrix fibres seems to connect the afferent lymphatics with the HEV. The most important aspect of this concept is that more interest is focused on the interaction and regulation mechanism within the parenchyma and not only on the adhesion phenomena between HEV and lymphocytes (Figure 2).

HOW IS THE EXIT OF LYMPHOCYTES FROM A LYMPHOID ORGAN REGULATED?

Little is known about which factors are relevant for the exit of a lymphocyte from a lymph node, i.e. entering the efferent lymphatic. A comparative aspect is interesting in this respect. Pigs have so-called inverted lymph nodes: the medulla is more peripheral and the cortex in the centre. What is more unusual, the efferent lymph of pigs is almost devoid of lymphocytes, while lymphocytes enter lymph nodes via HEV at a similar rate as in other species. The HEV in pig lymph nodes are also sites of lymphocyte exit, thus the lymphocytes return to the blood via the same structure they had used to enter the node. So far it is not clear whether certain HEV are the entrance sites and others the exit sites, or whether the HEV in pig lymph nodes work like a revolving door (Binns and Pabst, 1994). There are indications that the exit is carefully regulated. For example, lymphocytes taken from sheep efficiently used the HEV in pig lymph nodes to exit into the blood directly, which is a route not taken by sheep lymphocytes in sheep lymph nodes. In addition, pig lymphocytes migrated from lymph nodes of fetal lambs into efferent lymphatics, a route not taken by pig lymphocytes but which is typical for lymphocytes in sheep (summarized in Binns and Pabst, 1994, 1996). Taken together, these data show that the mechanisms regulating lymphocyte migration are well conserved at least between these two species which diverged from a common ancestor about 50 million years ago. Comparisons between species are very helpful in drawing general conclusions on the mechanisms at work in lymphocyte migration.

THE SPLEEN IS QUANTITATIVELY THE PREDOMINANT ORGAN IN LYMPHOCYTE RECIRCULATION

Many more lymphocytes migrate through the spleen per day than through all lymph nodes together (Pabst, 1988a). Several techniques have been used to demonstrate this phenomenon in rats, pigs and also in humans (for review see Pabst, 1988b; Pabst and Binns, 1989). The transit time of lymphocytes through the spleen is several fold shorter than through lymph nodes, and the vast majority of lymphocytes leave the spleen via the splenic vein and not via lymphatics (Figure 1). The route of lymphocyte subsets through the different splenic compartments has been described for the mouse and rat spleen. The marginal zone is the primary site where lymphocytes leave the blood. The lymphocyte subsets all seem to use the same site of entry into the spleen, but T and B lymphocytes then home to specific compartments (for review see Willführ *et al.*, 1990).

For many years mechanisms regulating lymphocyte entry into the spleen were largely unknown (see Pabst and Binns, 1989). Several procedures failed to show any effect on lymphocyte entry into the spleen, e.g. the pretreatment of lymphocytes with trypsin to modify the cell surface or with antibodies against many different adhesion molecules. The study of Hamann *et al.* (1988), for instance, with anti-LFA-1 treatment of lymphocytes *in vitro* had no effect on lymphocytes entering the spleen in a short term *in vivo* lymphocyte homing test. Weston and Parish (1991) documented that mannose-containing structures are involved in the splenic entry of lymphocytes. Kraal *et al.* (1994) showed that the removal of sialic acid moieties from lymphocytes results in fewer cells gaining entry into the white pulp although the total entry was similar. In a recent series of experiments, Lyons and Parish (1995) focused on the use of fucoidan, a polysulfated polysaccharide that interacts with the binding of lymphocytes to HEV, and pertussis toxoid, which does not interact with the HEV but with the emigration, the final step of extravasation. An interesting aspect of these experiments was the combination of looking at splenic entry in general with the more precise localization of the injected lymphocytes in the different compartments of the spleen at several time points. The conclusion for the mouse spleen was that here different steps of lymphocyte migration also exist: the initial entry into the marginal zone might be mediated by fucoidan influencing the interaction between lymphocytes and marginal zone macrophages. The lymphocyte motility essential for further migration from the marginal zone to the white pulp is restricted by pertussis toxin. However, nothing is known about which molecules are involved in directing B and T lymphocyte subsets through the different compartments of the spleen or regulating the exit into the veins and to a lesser extent into the splenic lymphatics.

The increase in lymphocytes in the blood after physical exercises and/or acute psychological stress had been interpreted as a mobilization of lymphocytes from the spleen by homonal effects such as catecholamines. In a recent study including splenectomized patients no evidence for an effect of catecholamines was found, thus factors regulating the emigration of lymphocytes from the spleen are still unknown (Schedlowski *et al.*, 1996). Recently Kraal *et al.* (1995) published data on the expression of an adhesion molecule previously thought to be only important in lymphocyte traffic to the gut-associated lymphoid system. The adhesion molecule MAdCAM-1 is expressed on endothelial cells of Peyer's patches and venules of the gut lamina propria. A receptor for this adhesion molecule is the lymphocyte integrin $\alpha_4\beta_7$. Using cryostat sections and EM studies, MAdCAM-1 expression was demonstrated on the sinus-lining cells of the marginal zone of the mouse spleen, with a preference for the site of the cells facing the sinus, away from the splenic T cell zone, the periarteriolar lymphatic sheath. Had the additional functional studies not been performed one could have speculated about an important function of this adhesion molecule in lymphocyte migration to the spleen. When an antibody against MAdCAM-1 was injected and the migration of lymphocytes to the spleen tested *in vivo*, no effect was shown, while migration to Peyer's patches was blocked by more than 70%. In a further series of experiments lymphocytes were incubated with anti-$\alpha_4\beta_7$ and still no effect on lymphocyte entry into the spleen was seen. Two important aspects of these experiments are firstly, it is still largely unknown which adhesion molecule pairs on lymphocytes and marginal zone sinusoidal cells regulate lymphocyte entry into the spleen, and secondly, one should not extrapolate a functional role of an adhesion molecule in lymphocyte traffic from the sole evidence of the expression of the molecule on a section.

HETEROGENEITY IN LYMPHOCYTE MIGRATION TO DIFFERENT LYMPHOID ORGANS SUCH AS PEYER'S PATCHES AND LYMPH NODES

There are not only species differences in the extent of lymphocytes migrating to different organs, but there is also a great variability between lymph nodes and between Peyer's patches. Migration to the mandibular and portal lymph nodes in pigs is significantly higher and to the prescapular lymph nodes markedly lower than to average lymph nodes (see Binns and Pabst, 1988). This phenomenon does not depend on the antigen load from the draining area, as a similar pattern exists in the antigen-free pig fetus well before birth. After autotransplantation of parts of these different lymph nodes to other regions of the body the basic pattern persists (for review see Pabst and Binns, 1989). The obvious heterogeneity in lymphocyte entry into lymph nodes illustrates that similarities in a lymphoid structure do not necessarily mean comparable function in respect to lymphocyte migration.

Similarly, differences in lymphocyte entry exist between the discrete Peyer's patches in the pig and sheep jejunum and the continuous Peyer's patch in the terminal ileum (reviewed in Pabst and Binns, 1989; Griebel and Hein, 1996). No differences in adhesion molecules on the venular endothelium or on lymphocyte subsets have been found so far to explain this heterogeneity. Thus, extrapolation of results from one type of Peyer's patch to the other or to other species should be avoided.

THE THYMIC MEDULLA IS INCLUDED IN THE TRAFFIC ROUTE OF LYMPHOCYTES

The production and export of T lymphocytes from the thymus is such a central aspect of immunology that the entry of mature lymphocytes from secondary lymphoid organs into the thymus is often overlooked. It has been summarized before (Pabst and Binns, 1989) that a few labelled lymphocytes were found in the thymus in all studies labelling individual lymphoid organs (e.g. the spleen, mesenteric lymph nodes or the gut) in pigs and sheep using the innocuous selective perfusion technique. These lymphocytes were located only in the medulla. Near to the corticomedullary junction there are venules with a somewhat higher endothelium than in normal postcapillary venules; these seem to be the entry site for lymphocytes from the blood. In young pigs' blood, lymphocytes were labelled with ^{51}Cr and considerable counts were determined for the comparatively large thymus. Entry into the thymus can be inhibited by incubating the lymphocytes with trypsin, a technique which reduces traffic to organs with HEV.

We have recently summarized data from several experiments in the rat, all showing an influx of different lymphocyte subsets (e.g. small resting lymphocytes from the lung, spleen and also proliferating cells) into the thymic medulla (Westermann *et al.*, 1996). Scollay and Coward (1996) infused large numbers of splenic and lymph node lymphocytes into rats. The lymphocytes recovered from the thymus were $CD4^+$, and $CD8^+$ single positive cells, but also some $NK1.1^+$ T cells and B cells. Once again the thymic medulla was the location. Thus, there is no doubt that lymphocytes from the blood and peripheral lymphoid organs enter the thymic medulla. The function of these cells is unclear, likewise the regulatory mechanisms, e.g. pairs of adhesion molecules which are expressed on these venules and the corresponding lymphocytes. The only available data show that the α_4 integrin–VCAM-1 interaction plays a role in the entrance of activated $CD4^+$ T lymphocytes into the thymic medulla (Bell *et al.*, 1995).

MIGRATION OF LYMPHOCYTES TO THE LIVER NOT ONLY SERVES TO REMOVE EFFETE CELLS BUT IS ALSO A PHYSIOLOGICAL PHENOMENON

The localization of labelled lymphocytes in the liver after intravenous injection has often only been taken as an indication of the damage done to the cells during their separation and labelling. Without any doubt, lymphocytes damaged *in vitro* are found to a large extent in the liver. However, over the years increasing evidence has been presented that lymphocyte entry into the liver is a physiological phenomenon (Smith and Ford, 1983 and summarized in Pabst and Binns, 1989). The liver with its considerable proportion of body mass is an excellent example to demonstrate that the number of lymphocytes entering an organ has to be calculated not only per section or unit weight but also for the whole organ. There might be only a few cells per section, but because of the huge volume of the liver, in comparison to lymph nodes, a large total number results. In a series of experiments in pigs using ^{51}Cr labelled lymphocytes, about 15–20% of all injected cells were recovered from the liver, with little variation between 1 hour and 4 days after injection (summarized in Binns and Pabst, 1988). The great majority of lymphocytes do not enter the liver parenchyma, but must be in the hepatic marginal pool, which can be mobilized by *in vitro* perfusion of the liver (for review see Pabst and Binns, 1989). The migration of lymphocytes to the liver is comparable with that to other organs such as the lung and spleen which likewise do not contain high endothelial venules. This is documented by the lack of effects on lymphocyte localization of protease treatment of lymphocytes before their injection (Binns and Licence, 1989).

Recently, adhesion molecules have been shown on hepatic endothelium: ICAM-1 and ICAM-2 were expressed on sinusoidal endothelium, while the portal endothelium was negative (for details see McNab *et al.*, 1996). The novel vascular adhesion protein-1 (VAP-1), which mediates adhesion of lymphocytes to HEV in lymph nodes, is expressed on non-inflamed human liver sinusoids (Salmi *et al.*, 1993). In a recent study, an *in vitro* adhesion assay of lymphocytes to human hepatic tissue was tested: anti-VAP-1 antibody blocked the binding while anti-ICAM-1 and anti-CD18 combined with anti-CD29 had no effect (McNab *et al.*, 1996). This is an initial indication that adhesion molecule interactions play a role in lymphocyte entry into the liver. Previous studies in which adhesion molecules for migration to LN were blocked showed no effect on lymphocyte migration to the liver, e.g. blocking of LFA-1 (Hamann *et al.*, 1988). Recently, Adams (1996) summarized the lymphocyte-endothelial interactions in hepatic inflammation and also mentioned some future therapeutic applications, including liver transplantation (Steinhoff *et al.*, 1993).

LYMPHOCYTE MIGRATION TO THE LUNG IS A PHYSIOLOGICAL PHENOMENON AND NOT AN ARTIFACT

A frequent finding in lymphocyte migration studies was an early high number of lymphocytes in the lung. This was often taken as an artifact, because the lung capillary bed is the first capillary filter after i.v. injection of labelled lymphocytes. Meanwhile, these lymphocytes have been localized in different compartments of the lung (Pabst and Tschernig, 1995). The potential role of adhesion molecules in immune reactions in diseases of the lung has recently been summarized (Wegner *et al.*, 1992; Pilewski and Albelda, 1993; Hamacher and Schaberg, 1994). Adhesion molecules such as ICAM-1 on lung endothelial cells can be upregulated by cytokines or during inflammation (Burns *et al.*, 1994; Hill *et al.*, 1994). Antibodies against adhesion molecules were applied with the effect of preventing the accumulation of granulocytes and eosinophils (Montefort *et al.*, 1992; Wegner *et al.*, 1992; Hamacher and Schaberg, 1994). Although there is a great deal of data on the role the adhesion molecules play in lymphocyte migration to lymphoid organs, hardly anything is known about the role of these adhesion molecules in the lung vascular bed. One difference is that in lymphoid organs lymphocytes leave the blood compartment in venules, but in the lung the capillaries are the site of diapedesis. When the effect of antibodies against the integrin $\alpha_4\beta_7$ and anti-L-selectin was tested in the mouse, major differences were observed in the entry capacity to Peyer's patches and peripheral lymph nodes, respectively, but the number of lymphocytes in the lung was not affected (Hamann *et al.*, 1994). Thus, other yet undefined adhesion molecules must be working in the lung vasculature. In future studies the different compartments of the lung should be examined separately, as an adhesion molecule might effect the adhesion to endothelial cells in the bronchial epithelium (Pretolani *et al.*, 1994) but not to the lung capillary bed. Research should in future focus on therapeutic aspects, e.g. blocking the entry of lymphocytes to the lung in allergic diseases. There have been some indications that allergen challenge in asthma patients tended to decrease the localization specificity expressed by typical homing receptors on lymphocytes in the bronchoalveolar space (Picker *et al.*, 1994).

THE BONE MARROW IS ALSO INCLUDED IN THE ROUTE OF LYMPHOCYTE TRAFFIC

In many studies on lymphocyte traffic the bone marrow was neglected. Either data have been given for one bone only, e.g. the femur, providing no results for the total bone marrow distributed in many bones; or the bone marrow has been completely omitted. This neglect of the bone marrow as a site of lymphocyte recirculation is surprising, because in the detailed experiments of Rannie and Donald (1977) and Gowans and Knight (1964) a high proportion of intravenously injected radioactively labelled lymphocytes was recovered from the bone marrow and a rather short transit time was documented for most cells. In young pigs up to 20% of the injected ^{51}Cr labelled blood lymphocytes migrated to the total bone marrow, which constitutes approx. 4% of the body weight at that age (Binns and Pabst, 1988). In the rat, the maximum recovery was at 1 hour after injection of thoracic duct lymphocytes, with a gradual release over the next 23 hours (Westermann *et al.*, 1989). It had been argued that these data are due to the *in vitro* preparation and labelling procedure or to contaminating blood. When lymphocytes were labelled in their normal microenvironment by selective, temporal, extracorporeal perfusion high numbers of lymphocytes also migrated to the bone marrow, as summarized previously (Pabst and Binns, 1989). Evidence of the physiological role of the bone marrow in lymphocyte traffic was also obtained when the bone marrow of pigs (Pabst *et al.*, 1983) and sheep (Pabst *et al.*, 1986) was labelled *in situ* with FITC by selective perfusion. Considerable numbers of small lymphocytes were found one day later in all lymphoid organs.

Very little is known whether and (if so) which adhesion molecules regulate the entry into, transit through and exit from the bone marrow. The only data known to us on the expression of adhesion molecules on the sinus-lining cells of the bone marrow is that of VCAM-1 in the mouse (Jacobsen *et al.*, 1996). In other organs VCAM-1 is only expressed on endothelial cells after induction by cytokines, e.g. at sites of inflammation, and not constitutively, as on the sinus-lining cells of the mouse bone marrow. The lack of data can be explained by the problems encountered in preparing adequate cryostat or even paraffin sections and thereby preserving the sinusoids for immunohistology. Many monoclonal antibodies do not work on plastic-embedded sections.

In future experiments blocking adhesion molecules on lymphocytes or on the surface of endothelial cells, the bone marrow should be included to further characterize its local regulatory mechanisms.

DO B LYMPHOCYTES PREFERENTIALLY MIGRATE TO PEYER'S PATCHES AND T LYMPHOCYTES TO PERIPHERAL LYMPH NODES?

Several years ago lymphocytes prepared from cell suspensions of lymphoid organs showed such a preference in an *in vitro* lymphocyte adhesion assay (Stevens *et al.*, 1982). There are now three lines of evidence contradicting these data. 1. Physiologically migrating lymphocytes from the thoracic duct and not cell suspensions from lymph nodes were used in the same *in vitro* assay (Walter *et al.*, 1995). 2. Their *in vivo* migration was studied after labelling the cells with FITC and characterizing the subsets by immunohistochemistry (Walter *et al.*, 1995). 3. Lymphocyte subsets were characterized in the wall of venules with a high endothelium (Westermann *et al.*, 1994). In all these experiments B and T lymphocytes as well as $CD4^{+}$ and $CD8^{+}$ lymphocytes were found to have a similar frequency. In addition, lymphocyte subsets were characterized on their route into and through rat Peyer's patches using intravital videomicroscopy (Miura *et al.*, 1995). There was no apparent difference in the population of lymphocytes in transit between B and T or $CD4^{+}$ and $CD8^{+}$ lymphocytes. However, the tempo and the routes through the Peyer's patches showed marked differences between lymphocyte subsets, and only few B cells in contrast to many T cells were identified in the efferent lymphatics. Another interesting observation was that $CD4^{+}$ T lymphocytes adhered to defined segments of HEV while B lymphocytes tended to attach to the endothelium of a wider range of vessels and appeared as often in smaller venules and capillaries.

In conclusion, more physiological techniques demonstrate that there is no preference of B lymphocytes for PP or T lymphocytes for peripheral lymph node HEV. Data obtained by *in vitro* assays alone and using cell suspensions of heterogenous origin should not be used for extrapolation to the *in vivo* situation.

ACTIVATED LYMPHOID CELLS DIFFER IN THEIR MIGRATION PATTERN

Following presentation of the respective antigen in coordination with co-stimulatory factors naive lymphocytes become activated and finally lymphoblasts. The traffic of proliferating large lymphoid cells shows a different pattern compared to that of small lymphocytes (for review see Hamann and Rebstock, 1993). Proliferating lymphoid cells labelled by the DNA precurser I^{125}UdR showed major differences in their organ distribution when the cell sources were the marginal pool of the lung, or of the spleen in comparison to lymphoblasts from the peripheral blood. It was surprising to find blasts not only in lymphoid and mucosal organs but also in the bone marrow, lung, liver, skin and muscles (Binns *et al.*, 1992). The route of IgA^{+} lymphoblasts from the thoracic duct has frequently been studied. These plasma cell precursors migrate preferentially to Peyer's patches and the lamina propria of the gut, as well as to other mucosal sites such as the lamina propria of the respiratory tract, the mammary gland etc. (Brandtzaeg *et al.*, 1989). In all experiments preferential migration is seen rather than an exclusive location. It has to be stressed that not only B lymphoblasts but also stimulated T lymphocytes migrate from the gut to other mucosal organs such as the lung (for review see Dunkley *et al.*, 1995).

CONCLUSIONS

Lymphocytes in the blood, although only about 2% of all lymphocytes in the body, integrate the immune organs into one system with their continuous traffic through lymphoid and non-lymphoid organs. The short blood transit time of lymphocytes has to be considered when interpreting data on lymphocyte subsets and adhesion molecules on their surfaces. Many more lymphocytes migrate through the spleen, lung and liver per day than through organs with HEV, e.g. lymph nodes, Peyer's patches and tonsils. The entry of lymphocytes into organs via HEV depends on a cascade-type sequence of interactions of adhesion molecules on lymphocytes and endothelial cells. The route of lymphocyte subsets through lymphoid organs such as the spleen and lymph nodes has been documented, but the role of adhesion molecules in this transit phase and in the exit into lymphatics has not been defined so far. The physiological role of lymphocyte migration to the lung, liver and bone marrow as well as the thymic medulla has been documented. There are major species differences in lymphocyte traffic and therefore extrapolation from one species to another has to be performed with extreme caution. The expression of an adhesion molecule on a lymphocyte surface does not mean that this molecule plays a role in lymphocyte migration. Kinetic data have to be obtained in relative and absolute numbers to document the effects of blocking adhesion molecules or stimulating their expression. The dynamics of the lymphoid system with all its variations in individual organs, lymphocyte subsets and the role of adhesion molecules continues to be a fascinating topic of research. There are still many questions, which when answered might finally lead to new therapeutic approaches.

ACKNOWLEDGEMENTS

The authors' own studies were supported by the Deutsche Forschungsgemeinschaft (SFB 244 A7, SFB 280 C1 and Pa 240/7–1). Correction of the English text by Sheila Fryk is gratefully acknowledged.

REFERENCES

Adams, D.H. (1996). Lymphocyte-endothelial cell interactions in hepatic inflammation. *Hepato-Gastroenterology*, **43**, 32–43.

Ager, A. (1994). Lymphocyte recirculation and homing: roles of adhesion molecules and chemoattractants. *Trends in Cell Biology*, **4**, 326–333.

Arbones, M.L., Ord, D.C., Ley, K., Radich, H., Maynard-Curry, C., Capon, D.J., Tedder, T.F. (1994). Lymphocyte homing and leukocyte rolling and migration are impaired in L-selectin (CD62L)-deficient mice. *Immunity*, **1**, 247–260.

Bell, E. B., Sparshott, S.M. and Ager, A. (1995). Migration pathways of CD4 T cell subsets *in vivo*: the $CD45RC^-$ subset enters the thymus via α_4 integrin–VCAM-1 interaction. *International Immunology*, 7, 1861–1871.

Berlin, C., Bargatze, R.F., Campbell, J.J., von Andrian, U.H., Szabo, M.C., Hasslen, S.R., Nelson, R.D., Berg, E.L., Erlandsen, S.L. and Butcher, E.C. (1995). α_4 integrins mediate lymphocyte attachment and rolling under physiological flow. *Cell*, **80**, 413–422.

Binns, R.M. and Licence, S.T. (1989). Xenogeneic lymphocyte migration in the foetal lamb and unsuckled neonatal piglet. Evidence for the evolutionary conservation and heterogeneity of migration determinants. *Advances in Experimental Medical Biology*, **237**, 533–539.

Binns, R.M., Licence, S.T. and Pabst, R. (1992). Homing of blood, splenic, and lung emigrant lymphoblasts: comparison with the behaviour of lymphocytes from these sources. *International Immunology*, 4, 1011–1019.

Binns, R.M. and Pabst, R. (1988). Lymphoid cell migration and homing in the young pig: alternative immune mechanisms in action. In A.J. Husband, (ed.), *Migration and Homing of Lymphoid Cells*, Vol. II, CRC Press, Boca Raton, pp. 137–174.

Binns, R.M. and Pabst, R. (1994). Lymphoid tissue structure and lymphocyte trafficking in the pig. *Veterinary Immunology and Immunopathology*, **43**, 79–87.

Binns, R.M. and Pabst, R. (1996). The functional structure of the pig's immune system, resting and activated. In M.E. Tumblesom and L.B. Schook (eds.), *Advances in Swine in Biomedical Research*, Vol. 1, Plenum, New York, pp. 253–265.

Blaschke, V., Micheel, B., Pabst, R. and Westermann, J. (1995). Lymphocyte traffic through lymph nodes and Peyer's patches of the rat: B- and T-cell-specific migration patterns within the tissue, and their dependence on splenic tissue. *Cell and Tissue Research*, **282**, 377–386.

Brandtzaeg, P., Halstensen, T.S., Kett, K., Krajci, P., Kvale, D., Rognum, T.O., Scott, H. and Sollid, L.M. (1989). Immunobiology and immunopathology of human gut mucosa: humoral immunity and intraepithelial lymphocytes. *Gastroenterology*, **97**, 1562–1584.

Burns, A.R., Takei, F. and Doerschuk, C.M. (1994). Quantitation of ICAM-1 expression in mouse lung during pneumonia. *Journal of Immunology*, **153**, 3189–3198.

Dunkley, M., Pabst, R. and Cripps, A. (1995). An important role for intestinally derived T cells in respiratory defence. *Immunology Today*, **16**, 231–236.

Finger, E.B., Puri, K.D., Alon, R., Lawrence, M.B., von Andrian, U.H. and Springer, T.A. (1996). Adhesion through L-selectin requires a threshold hydrodynamic shear. *Nature*, **379**, 266–269.

Ford, W.L. (1980). The lymphocyte — its transformation from a frustrating enigma to a model of cellular function. In M.M. Wintrobe, (ed *.), Blood, Pure and Eloquent: a Story of the Discovery of People and Ideas*, McGraw Hill, New York, pp. 457–508.

Friedl, P., Noble, P.B., Shields, E.D. and Zänker, K.S. (1994). Locomotor phenotypes of unstimulated CD45RA^{high} and CD45RO^{high} CD4^{+} and CD8^{+} lymphocytes in three-dimensional collagen lattices. *Immunology*, **82**, 617–624.

Girard, J.P. and Springer, T.A. (1995). High endothelial venules (HEV): specialized endothelium for lymphocyte migration. *Immunology Today*, **16**, 449–457.

Gowans, J.L. and Knight, E.J. (1964). The route of recirculation of lymphocytes in the rat. *Proceedings of the Royal Society of London (Biology)*, **159**, 257–282.

Gretz, J.E., Kaldjian, E.P., Anderson, A.O. and Shaw, S. (1996). Sophisticated strategies for information encounter in the lymph node. The reticular network as a conduit of soluble information and a highway for cell traffic. *Journal of Immunology*, **157**, 495–499.

Griebel, P.J. and Hein, W.R. (1996). Expanding the role of Peyer's patches in B cell ontogeny. *Immunology Today*, **17**, 30–39.

Hamacher, J. and Schaberg, T. (1994). Adhesion molecules in lung diseases. *Lung*, **172**, 189–213.

Hamann, A., Andrew, D.P., Jablonski-Westrich, D., Holzmann, B. and Butcher, E.C. (1994). Role of α_4-integrins in lymphocyte homing to mucosal tissues *in vivo*. *Jounal of Immunology*, **152**, 3282–3293.

Hamann, A., Jablonski-Westrich, D., Duijvestijn, A., Butcher, E.C., Baisch, H., Harder, R. and Thiele, H.G. (1988). Evidence for an accessory role of LFA-1 in lymphocyte-high endothelium interaction during homing. *Journal of Immunology*, **140**, 693–699.

Hamann, A. and Rebstock, S. (1993). Migration of activated lymphocytes. *Current Topics in Microbiology and Immunology*, **184**, 109–124.

Hasslen, S.R., von Andrian, U.H., Butcher, E.C., Nelson, R.D. and Erlandsen, S.L. (1995). Spatial distribution of L-selectin (CD62L). on human lymphocytes and transfected murine LI-2 cells. *Histochemistry Journal*, **27**, 547–554.

Hill, P.A., Lan, H.Y., Nikolic-Paterson, D.J. and Atkins, R.C. (1994). Pulmonary expression of ICAM-1 and LFA-1 in experimental Goodpasture's syndrome. *American Journal of Pathology*, **145**, 220–227.

Imhof, B.A. and Dunon, D. (1995). Leukocyte migration and adhesion. *Advances in Immunology*, **58**, 345–416.

Jacobsen, K., Kravitz, J., Kincade, P.W. and Osmond, D.G. (1996). Adhesion receptors on bone marrow stromal cells: *in vivo* expression of vascular cell adhesion molecule-1 by reticular cells and sinusoidal endothelium in normal and γ-irradiated mice. *Blood*, **87**, 73–82.

Klonz, A., Wonigeit, K., Pabst, R. and Westermann, J. (1996). The marginal blood pool of the rat contains not only granulocytes but also lymphocytes, NK-cells and monocytes: a second intravascular compartment, its cellular composition, adhesion molecule expression and interaction with the peripheral blood pool. *Scandinavian Journal of Immunology*, **44**, 461–469.

Kraal, G., Hoeben, K., Brevé, J. and van den Berg, T.K. (1994). The role of sialic acid in the localization of lymphocytes in the spleen. *Immunobiology*, **190**, 138–149,

Kraal, G., Schornagel, K., Streeter, P.R., Holzmann, B. and Butcher, E.C. (1995). Expression of the mucosal vascular addressin, MAdCAM-1, on sinus-lining cells in the spleen. *American Journal of Pathology*, **147**, 763–771.

Lyons, A.B. and Parish, C.R. (1995). Are murine marginal-zone macrophages the splenic white pulp analog of high endothelial venules? *European Journal of Immunology*, **25**, 3165–3172.

McNab, G., Reeves, J.L., Salmi, M., Hubscher, S., Jalkanen, S. and Adams, D.H. (1996). Vascular adhesion protein 1 mediates binding of T cells to human hepatic endothelium. *Gastroenterology*, **110**, 522–528.

Miura, S., Tsuzuki, Y., Fukumura, D., Serizawa, H., Suematsu, M., Kurose, I., Imaeda, H., Kimura, H., Nagata, H., Tsuchiya, M. and Ishii, H. (1995). Intravital demonstration of sequential migration process of lymphocyte subpopulations in rat Peyer's patches. *Gastroenterology*, **109**, 1113–1123.

Montefort, S., Roche, W.R., Howarth, P.H., Djukanovic, R., Gratziou, C., Carroll, M., Smith, L., Britten, K.M., Haskard, D., Lee, T.H. and Holgate, S.T. (1992). Intercellular adhesion molecule-1 (ICAM-1). and endothelial leucocyte adhesion molecule-1 (ELAM-1). expression in the bronchial mucosa of normal and asthmatic subjects. *European Respiratory Journal*, **5**, 815–823.

Pabst, R. (1988a). The spleen in lymphocyte migration. *Immunology Today*, **9**, 43–45.

Pabst, R. (1988b). The role of the spleen in lymphocyte migration. In A.J. Husband, (ed.), *Migration and Homing of Lymphoid Cells*, Vol. I, CRC Press, Boca Raton, pp. 63–84.

Pabst, R. and Binns, R.M. (1989). Heterogeneity of lymphocyte homing physiology: several mechanisms operate in the control of migration to lymphoid and non-lymphoid organs *in vivo*. *Immunological Reviews*, **108**, 83–109.

Pabst, R., Binns, R.M., Rothkötter, H.J. and Westermann, J. (1993). Quantitative analysis of lymphocyte fluxes *in vivo*. *Current Topics in Microbiology and Immunology*, **184**, 151–159.

Pabst, R., Kaatz, M. and Westermann, J. (1983). *In situ* labelling of bone marrow lymphocytes with fluorescein isothiocyanate for lymphocyte migration studies in pigs. *Scandinavian Journal of Haematology*, **31**, 267–274.

Pabst, R., Miyasaka, M. and Dudler, L. (1986). Numbers and phenotype of lymphocytes emigrating from sheep bone marrow after *in situ* labelling with fluorescein isothiocyanate. *Immunology*, **59**, 217–222.

Pabst, R. and Tschernig, T. (1995). Lymphocytes in the lung: an often neglected cell. Numbers, characterization and compartmentalization. *Anatomy and Embryology*, **192**, 293–299.

Picker, L.J. and Butcher, E.C. (1992). Physiological and molecular mechanisms of lymphocyte homing. *Annual Review of Immunology*, **10**, 561–591.

Picker, L.J., Martin, R.J., Trumble, A., Newman, L.S., Collins, P.A., Bergstresser, P.R. and Leung, D.Y.M. (1994). Differential expression of lymphocyte homing receptors by human memory/effector T cells in pulmonary versus cutaneous immune effector sites. *European Journal of Immunology*, **24**, 1269–1277.

Pilewski, J.M. and Albelda, S.M. (1993). Adhesion molecules in the lung. An overview. *American Review of Respiratory Disease*, **148**, (Suppl.). S31–S37.

Pretolani, M., Ruffié, C., Lapa e Silva, J.R., Joseph, D., Lobb, R.R. and Vargaftig, B.B. (1994). Antibody to very late activation antigen 4 prevents antigen-induced bronchial hyperreactivity and cellular infiltration in the guinea pig airways. *Journal of Experimental Medicine*, **180**, 795–805.

Rannie, G.H. and Donald, K.J. (1977). The migration of thoracic duct lymphocytes to non lymphoid tissues. A comparison of the distribution of radioactivity at intervals following i.v. transfusion of cells labelled with ^{3}H, ^{14}C, ^{75}Se, ^{99m}Tc, ^{125}I and ^{51}Cr in the rat. *Cell and Tissue Kinetics*, **10**, 523–541.

Salmi, M., Kalimo, K. and Jalkanen, S. (1993). Induction and function of vascular adhesion protein-1 at sites of inflammation. *Journal of Experimental Medicine*, **178**, 2255–2260.

Schedlowski, M., Hosch, W., Oberbeck, R., Benschop, R.J., Jacobs, R., Raab, H.R. and Schmidt, R.E. (1996). Catecholamines modulate human NK cell circulation and function via spleen-independent β_2-adrenergic mechanisms. *Journal of Immunology*, **156**, 93–99.

Scollay, R. and Coward, A. (1996). Return of the T cells: studies on the circulation of mature T cells back into the thymus. In Abstract Volume *12th International Conference on Lymphoid Tissues and Germinal Centers in Immune Reactions in Graz*, p. 55.

Smith, M.E. and Ford, W.L. (1983). The recirculating lymphocyte pool of the rat: a systematic description of the migratory behaviour of recirculating lymphocytes. *Immunology*, **49**, 83–94.

Springer, T.A. (1994). Traffic signals for lymphocyte recirculation and leukocyte emigration: the multistep paradigm. *Cell*, **76**, 301–314.

Steeber, D.A., Green, N.E., Sato, S. and Tedder, T.F. (1996). Lymphocyte migration in L-selectin-deficient mice. Altered subset migration and ageing of the immune system. *Journal of Immunology*, **157**, 1096–1106.

Steinhoff, G., Behrend, M., Schrader, B. and Pichlmayr, R. (1993). Intercellular immune adhesion molecules in human liver transplants: overview on expression patterns of leukocyte receptor and ligand molecules. *Hepatology*, **18**, 440–453.

Stevens, S.K., Weissman, I.L. and Butcher, E.C. (1982). Differences in the migration of B and T lymphocytes: organ-selective localization *in vivo* and the role of lymphocyte endothelial cell recognition. *Journal of Immunology*, **128**, 844–851.

Walter, S., Micheel, B., Pabst, R. and Westermann, J. (1995). Interaction of B and T lymphocyte subsets with high endothelial venules in the rat: binding *in vitro* does not reflect homing *in vivo*. *European Journal of Immunology*, **25**, 1199–1205.

Wegner, C.D., Gundel, R.H., Rothlein, R. and Letts, L.G. (1992). Expression and probable roles of cell adhesion molecules in lung inflammation. *Chest*, **101**, 34S-39S.

Westermann, J., Nagahori, Y., Walter, S., Heerwagen, C., Miyasaka, M. and Pabst, R. (1994). B and T lymphocyte subsets enter peripheral lymph nodes and Peyer's patches without preference *in vivo*: no correlation occurs between their localization in different types of high endothelial venules and the expression of CD44, VLA-4, LFA-1, ICAM-1, CD2 or L-selectin. *European Journal of Immunology*, **24**, 2312–2316.

Westermann, J. and Pabst, R. (1990). Lymphocytes in the blood. A diagnostic window on the lymphoid system? *Immunology Today*, **11**, 406–410.

Westermann, J., Smith, T., Peters, U., Tschernig, T., Pabst, R., Steinhoff, G., Sparshott, S.M. and Bell, E.B. (1996). Both activated and nonactivated leukocytes from the periphery continuously enter the thymic medulla of adult rats: phenotypes, sources and magnitude of traffic. *European Journal of Immunology*, **26**, 1866–1874.

Westermann, J., Willführ, K.U., Rothkötter, H.J., Fritz, F.J. and Pabst, R. (1989). Migration pattern of lymphocyte subsets in the normal rat and the influence of splenic tissue. *Scandinavian Journal of Immunology*, **29**, 193–201.

Weston, S.A. and Parish, C.R. (1991). Modification of lymphocyte migration by mannans and phosphomannans. Different carbohydrate structures control entry of lymphocytes into spleen and lymph nodes. *Journal of Immunology*, **146**, 4180–4186.

Willführ, K.U., Westermann, J., Pabst, R. (1990). Absolute numbers of lymphocyte subsets migrating through the compartments of the normal and transplanted spleen. *European Journal of Immunology*, **20**, 903–911.

3. THE MULTISTEP PARADIGM AND THE CONTROL OF LYMPHOCYTE TRAFFICKING: HOMING TO MUCOSAL TISSUES AS A MODEL

ROBERT F. BARGATZE[1,*] and EUGENE C. BUTCHER[2]

[1]*Laboratory of Leukocyte-Endothelial Cell Biology, Montana ImmunoTech Inc., 920 Technology Blvd., Suite C, Bozeman, MT 59715, USA*

[2]*Laboratory of Immunology and Vascular Biology, Department of Pathology, and the Stanford Digestive Disease Center, Stanford University School of Medicine, Stanford California 94305; and the Center for Molecular Biology in Medicine, Veterans Affairs Medical Center, Palo Alto, California 94304, USA*

INTRODUCTION

Leukocytes traffic through the bloodstream in their passage from their tissue of origin to extravascular sites of inflammation or immune responses. The recruitment of leukocytes from the blood is an exquisitely specific process, controlled in large part by mechanisms of selective leukocyte-endothelial cell recognition and diapedesis. Classic examples of specific leukocyte recruitment include selective eosinophil localization to sites of allergic or parasitic insult, neutrophil recruitment during acute inflammation and monocyte recruitment during subacute and chronic inflammation. Lymphocytes recirculate as naive cells from one lymphoid organ to another, a process of shuffling and reshuffling which is thought to allow the full repertoire of clonal lymphocyte specificities to be available to respond to antigen entering any site in the body (and presented in the secondary lymphoid organs, the lymph nodes, mucosal Peyer's patches, and spleen). Upon differentiation into effector precursors (immunoblasts) or mature memory cells, antigen-reactive lymphocytes return to the circulation displaying highly specialized homing/trafficking properties: in the words of Steve Shaw, memory lymphocytes are an "army of specialists", displaying unique patterns of homing-associated adhesion and trafficking molecules that target their migration differentially to the various organs, tissues, and microenvironments in the body (see Figure 1). *In vitro* and *in vivo* physiologic studies demonstrate specialized lymphocyte trafficking to intestinal mucosal surfaces, to the skin, to lymph nodes and Peyer's patches, and probably to the joints and lungs. Lymphocyte trafficking to other extralymphoid sites of inflammation is less well characterized, and may involve either more generalized (less specialized) inflammation-selective recruitment, or additional specialized homing mechanisms which remain to be characterized. Antigen specificity

*Correspondence: Robert F. Bargatze, Montana ImmunoTech Inc., 920 Technology Blvd., Suite C, Bozeman, MT 59715, USA. Tel: (406) 585-2733; Fax: (406) 585-2766; E-mail: rpbfb@imt.net.

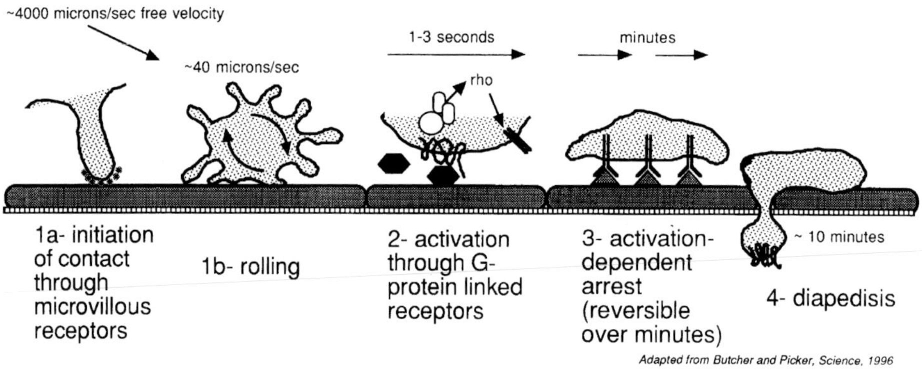

Figure 1 The multi-step model of lymphocyte-EC recognition and recruitment from the blood. Circulating lymphocytes enter postcapillary venules in sites of inflammation or in lymphoid organs under high flow. Functional contact with the endothelium ("tethering", step 1a) is initiated through constitutively functional receptors presented on lymphocyte microvilli (see Table 1). Transient sticking or slow rolling (step 1b) permits sampling of the vessel for activating factors. Activation through $G\alpha_i$ protein-linked (probably chemoattractant family) receptors (step 2) stimulates rho-dependent (Laudanna, 1996) triggering of lymphocyte integrins. Activation-dependent lymphocyte arrest (step 3) is mediated by integrin binding to vascular Ig family ligands. Arrest is stable under the shear stress of flowing blood, but reverses spontaneously within several minutes unless followed by transendothelial migration (diapedesis – step 4) in response to haptotactic or chemoattractant gradients. The sequential requirement for at least four independently regulated receptor-ligand interactions allows combinatorial determination of the specificity of lymphocyte homing (Butcher, 1991). The probability of successful recruitment is thus the product of probabilities of passing each decision point, implying that the specificity of the overall process can greatly exceed that of its component steps and that recruitment can be regulated or blocked at any one of the four steps. (adapted from Butcher and Picker, 1996)

plays no direct role in this process (but can influence the subsequent retention, viability, and local proliferation of antigen-specific lymphocytes in a site of antigen deposition; reviewed in Butcher (1986).

In vivo studies have revealed that leukocyte-endothelial (EC) cell interactions and diapedesis involve several sequential if somewhat overlapping steps, each representing a key decision point in determining the probability of recruitment of the leukocyte into the surrounding tissues. In the case of neutrophils, these steps have been characterized as 1) a selectin-dependent primary adhesion, separable into a) contact initiation or "tethering" and b) "rolling", 2) activation of the rolling neutrophil which triggers $\beta 2$ integrins, and 3) subsequent $\beta 2$ integrin-mediated firm adhesion and arrest. Recent studies reveal that lymphocyte interactions with HEV can also involve very similar sequential rolling, activation, and activation-dependent sticking events. These events leading to leukocyte arrest and shear-stable attachment constitute the process of leukocyte-EC "recognition". 4) The recruitment of leukocytes from the blood into sites of inflammation or into lymphoid tissues must involve an additional step (diapedesis) which requires the presence of specific chemoattractant gradients and appropriate substrates for leukocyte migration.

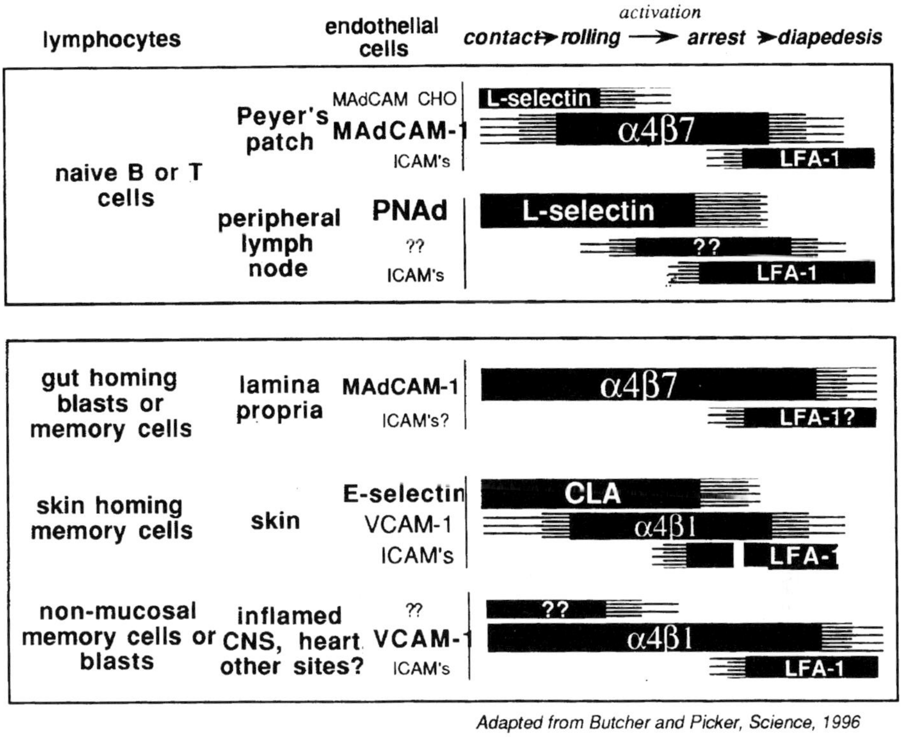

Figure 2 Known or hypothesized adhesion/decision cascades conferring specificity of lymphocyte homing in different sites. Naive lymphocytes home to Peyer's patches and lymph nodes; distinct subsets of memory and circulating effector lymphocytes can traffic selectively to different organs or tissues. Targeted homing is a function of patterns and levels of expression of homing receptors, and of complementary vascular "addressins" or adhesion ligands in different tissues, and is determined by their sequential participation in the multi-step extravasation process. The phenotype of postcapillary venules in different tissues is summarized, and the relative roles of each homing receptor-vascular counterreceptor pathway in contact, rolling, sticking and diapedesis is indicated. Overlap of bars indicates overlapping functions in particular settings, and the vertical width of bars indicates the relative expression/function levels of each component in the cascade, which may of course be variable. Activating signals are unknown, and may not be required for arrest of immunoblasts expressing preactivated integrins. As indicated, the homing receptor phenotype of naive lymphocytes allows them access to lymph nodes and PP, but not to inflammatory extralymphoid tissues. In contrast, distinctive subsets of memory cells express phenotypes that allow their targeted access to skin, lamina propria, and potentially other sites such as joints. It should be mentioned that tissue-selective cascades can be altered both developmentally, and as a function of severe local inflammation which can lead to more promiscuous lymphocyte recruitment. (adapted from Butcher and Picker, 1996)

We and others have proposed that these sequential steps represent an active and even interactive decision process that controls leukocyte recruitment from the blood (Butcher, 1991; Shimizu, 1992; Carlos and Harlan; Springer, 1994). The decision process is illustrated schematically in Figure 2. The implications of this model for the physiologic control of leukocyte-EC interactions and diapedesis have been explored previously (Butcher, 1991). In general, the model provides for the generation of specificity and diversity through combinatorial association/utilization of the different receptor ligand pairs available at each of the 4 or more steps involved. More specifically, the model implies that . . . 1) Recruitment can be regulated or blocked at any one of the 4 steps depicted. 2) The specificity of the overall process can greatly exceed that of its component steps, because the probability of successful diapedesis is the product of probabilities of passing each decision point. Thus if a skin-homing memory lymphocyte has twice the probability of adhering or being activated or chemotaxing at each of the steps compared to a neutrophil, it will be 16-fold more likely to extravasate successfully. 3) Highly useful adhesion receptors, such as the L-selectin which appears well-specialized for initiation of contact under shear, can be utilized in multiple independently regulated leukocyte-EC interactions. 4) New highly specific leukocyte trafficking events can arise during evolution without the need for novel adhesion receptors or chemoattractant factors.

Consistent with these concepts, normal lymph node lymphocyte populations exhibit an extensive range of adhesive phenotypes during transit through PP HEV *in vivo.* Observation and analysis of lymphocyte HEV interactions *in situ,* employing video microscopy, facilitates a detailed evaluation of individual lymphocyte interactions with the endothelium. Varied behaviors early after infusion are displayed, for example, by preactivated lymphocytes (from the *in vivo* lymphocyte prep) which exhibit immediate sticking upon HEV entry, whereas most cells (predominantly naive lymphocytes) roll and arrest or roll without stopping, and some cells are completely non-interactive.

These behaviors are generated through the molecular diversity and state of their adhesion and signaling receptor receptors. It has been generally accepted that naive lymphocytes prior to encountering antigen can traffic through both peripheral and mucosal secondary lymphoid organs and upon encountering antigen differentiate into memory cells establishing defined tissue selective trafficking circuits (peripheral and mucosal) (Butcher, 1986; Mackay, 1993). These circuits are determined through active decision making mechanisms (by both the lymphocyte and the endothelial cell) that direct homing to specific tissues and thereby define the character of inflammatory and local immune responses. The regulation of trafficking via adhesion receptors, signaling molecules, and their ligands during vascular wall rolling, sticking, and diapedesis has the potential for actively directing distinct memory cells that are specific for tissues and antigens as in the skin or mucosa (Imhof and Dunon, 1995; Butcher and Picker, 1996). The result is likely a more efficient immune surveillance where memory cells experience a higher probability of encountering their cognate antigens in an environment optimized for support. All of these refinements in specificity add to a precise definition of the immune response in a given tissue. Tissue-selective trafficking of memory cells may also help control their competition for supportive microenvironments, and thus their homeostasis (Butcher and Picker, 1996).

MOLECULAR MECHANISMS

In vivo and *in vitro* studies have begun to define specialized features of receptors utilized at each of the steps involved.

Step 1

The selectins and their carbohydrate ligands appear specialized for primary transient interactions—tethering and rolling (Vestweber, 1993; McEver, 1995). These interactions occur under flow, and are initiated by contacts between leukocyte microvilli and the vascular wall (von Andrian, 1995). The leukocyte or L-selectin is highly effective at initiating contact with the endothelium under flow, perhaps because it is concentrated on the tips of microvillous processes where it can participate both in presentation of carbohydrates to the vascular selectins, and in recognition of vascular carbohydrates (as in homing via the peripheral node addressin) (von Andrian, 1991). Skin-homing lymphocytes express the cutaneous lymphocyte antigen (CLA), comprising a carbohydrate ligand for E-selectin; the carbohydrate epitope is widely displayed on the cell surface including on microvilli, and it can mediate lymphocyte rolling (Berg, 1991; Picker, 1992). Microvillous presentation appear critical to initiation of functional cellular adhesive interactions under flow (von Andrian, 1995). The P-selectin glycoprotein ligand-1 (PSGL-1) is also presented on microvilli, and may permit rolling of additional lymphocyte subsets on P-selectin (McEver, 1995).

The $\alpha4$ integrins (again microvilli expressed) and their Ig family counterreceptors on endothelium can also initiate lymphocyte contact under flow (Berlin, 1995). $\alpha4\beta1$:VCAM-1 is thought to be principally involved in lymphocyte trafficking to non-mucosal sites of tissue inflammation. $\alpha4\beta7$ but not $\alpha4\beta1$ binds the mucosal vascular addressin MAdCAM-1, and this interaction is critical to targetting subsets of lymphocytes to mucosal sites, both the mucosal lymphoid organs Peyer's patches and the intestinal sub-epithelial lamina propria (Berlin, 1995; Bargatze, 1995). $\alpha4\beta1$ vs. $\alpha4\beta7$ are differentially expressed by subpopulations of memory T and B cells that are thought to carry non-mucosal and mucosal memory, respectively (Shaw, 1993; Erle, 1994). Like the L-selectin and PSGL-1, the $\alpha4$ integrins are displayed on the tips of microvilli (Berlin, 1995). (In contrast, the $\beta2$ integrins are excluded from microvilli (Erlandsen, 1993), and play little or no role in the initiation of adhesion under flow (Springer, 1994; Bargatze, 1995). Physiologic involvement of other adhesion receptors in primary attachment and transient rolling remains to be reported, although CD44 on long-term *in vitro*-activated lymphoblasts can support rolling via hyaluronate on cultured endothelium in *in vitro* systems (DeGrendele, J. Exp. Med., in press). VAP1 also mediates 1° adhesion during shear and may suppport lymphocyte attachment from the blood flow. VAP1, a sialomucin-like protein expressed on tonsil HEV, supports PNAd-L-selectin independent lymphocyte attachment via non-selectin lymphocyte ligands (Salmi and Jalkanen, 1992; Salmi and Jalkanen, J. Exp. Med., in press).

Step 2

The activation step in leukocyte-endothelial cell interaction is perhaps least well characterized in *in situ* studies, but it is clear that activation can occur as rapidly as 1–3 sec. after primary adhesion, and that $G\alpha_i$ protein-linked receptors play a critical role in those instances examined. For example, in the case of lymphocyte interactions with Peyer's patch-HEV, activation involves pertussis toxin (PTX)-sensitive $G\alpha_i$-protein signaling pathways, so that treatment of lymphocytes with PTX yields an exclusive rolling phenotype, abrogating subsequent sticking and stable binding of lymphocytes to the HEV (Bargatze, 1993). The chemokines and other chemoattractants, acting through 7-transmembrane $G\alpha_i$ linked receptors, are attractive candidates for molecules that might provide specificity at this activation step in leukocyte-EC recognition (and/or in subsequent diapedesis and chemotaxis within tissues – see below) (Butcher, 1991; Tanaka, 1993). Indeed, in the case of neutrophils, classical chemoattractants like platelet-activating factor, complement component C5A, and formyl peptides as well as the chemokine IL-8 were able to fulfill the activating role in step 2 (Oppenheim, 1991; Kelvin, 1993; Schall, 1991, 1994, Carlos and Harland, 1994). Similarly, the chemokine eotaxin appears able to trigger rapid arrest and diapedesis of eosinophils (Griffiths-Johnson, 1993). The physiologically relevant activating factors involved in lymphocyte-EC interactions, however, remain to be identified, and it is worth mentioning that none of the chemokines available to date supports robust rapid (1–5 sec.) activation of lymphocyte adhesion to vascular Ig family members. This is not a limitation of the ability of this class of receptor to signal in lymphocytes, but instead is thought to reflect their very low levels of expression on circulating lymphocytes (J. Campbell, in press) (see also below).

Step 3

Integrins are thought to dominate activation-dependent adhesion and arrest, the third step in the decision process leading to diapedesis. In the case of neutrophils, $\beta 2$ integrins mediate firm attachment and arrest following activation of rolling neutrophils *in vivo* (Arfors, 1989; von Andrian, 1991) and *in vitro* (Lawrence and Springer, 1991). In the case of lymphocyte homing interactions, the $\beta 2$ integrin LFA1, as well as $\alpha 4$ integrins ($\alpha 4\beta 1$:VCAM-1 and $\alpha 4\beta 7$:MAdCAM-1) may play critical roles in activation-dependent adhesion in certain sites. Subsets of B cells and memory T cells also express MAC1 ($\alpha m\beta 2$). The differential expression of these integrins by defined memory and immunoblast populations provides an important mechanism, in conjunction with the specific activating signals, for the regulation of lymphocyte subset-specific traffic to particular sites of inflammation in the body. It should be emphasized, however, that the involvement of other adhesion receptors in primary and secondary adhesion remains likely.

Step 4

After lymphocytes bind to HEV, they diapedese within ~10–15 min., and rapidly migrate into specific B or T cell microenvironments within lymphoid tissues. For most lymphocytes, the mechanisms responsible for this avid chemotaxis or haptotaxis remain unknown. *In vitro* models suggest a role for $\beta 2$ integrins (Hamann, 1988) and also suggest that members of the chemokine family can direct the chemotaxis of certain leukocyte and lymphocyte subsets during diapedesis (Smith, 1993; reviewed in Bacon and Schall, 1996). The same activating factors responsible for step two in leukocyte endothelial cell recognition may, in some instances, be derived from surrounding tissues and be able to provide chemoattractant gradients for step 4 (diapedesis). In other instances, step 2 activating signals may be localized to the vascular surface, such that subsequent diapedesis may require distinct chemo-attractant gradients to allow migration of attached leukocytes through and away from the vascular wall. Moreover, the biology of chemoattractant receptors permits additional means of separating the signals involved in step 2 vs. step 4 (Campbell, J. Cell Biol., in press). Adhesion-triggering requires near simultaneous occupancy of large numbers of surface receptors, and thus is favored at high receptor density and agonist concentration; in contrast, chemotaxis is thought to depend on *differential receptor occupancy* at the leading and trailing edges of the lymphocyte, and thus is optimal in gradients near the Kd and is in fact *inhibited* at high agonist concentrations. Chemotactic responses also tend to be highly sensitive, and thus do not share the step 2 requirement for high levels of receptor expression. Independent control of steps 2 and 4 is expected to increase dramatically the potential for diversity and specificity in the regulation of recruitment. Finally, it is interesting to consider that the specificity of diapedesis may be controlled not only by subset-specific chemoattractants and receptors, but possibly also by the tissue-selective expression of adhesive extracellular matrix substrates, providing selective "traction" for subsets of leukocytes displaying appropriate counterreceptors.

MUCOSAL LYMPHOCYTE-ENDOTHELIAL DECISION CASCADES FOR HOMING TO PEYER'S PATCHES AND TO INTESTINAL LAMINA PROPRIA

The best characterized examples of lymphocyte homing involve recruitment to the mucosal lymphoid organs, Peyer's patches (PP), through high endothelial venules (HEV), and to the intestinal lamina propria through post-capillary venules (LP) (Bargatze, 1995; Berlin, 1995). Video epifluorescence microscopy and inhibitory antibodies have been used to dissect the roles of adhesion receptors in lymphocyte interactions with these venules *in situ.* These *in vivo* studies reveal distinct sequential roles for participating lymphocyte adhesion receptor/HEV ligand pairs, thus defining the nature of the mucosal combinatorial adhesion cascade in lymphocyte homing to PP-HEV and, surprisingly, showing that recruitment to LP (at least through step 3 arrest) occurs as the consequence of a single receptor-ligand adhesion event.

A key role for $\alpha 4\beta 7$ and MAdCAM-1 in mucosal homing

Mucosal homing receptor – Identification of $\alpha 4\beta 7$

As first suggested by Holzmann and Weissman (Holzmann, 1989; Kilshaw, 1991; Hu, 1992) $\alpha 4\beta 7$ integrin acts as the central lymphocyte adhesion receptor defining the Peyer's patch and gut lamina propria tissue selective homing phenotypes. The anti-$\alpha 4$ MAb R1–2, generated against a mucosal selective binding lymphoma, blocked binding both of lymphomas and of normal lymphocytes to PP HEV in the Stamper-Woodruff frozen section assay (Stamper and Woodruff, 1976). Unfortunately, the R1–2 MAb was not effective in preventing *in vivo* lymphocyte homing to PP.

More recently, using a cross section of anti-$\alpha 4$, $\beta 7$, and anti-$\alpha 4\beta 7$ MAbs, successfully demonstrated *in vivo* that $\alpha 4\beta 7$ had a central role in the homing of lymph node lymphocytes and blasts to mucosal sites (Hamann, 1994). Interestingly, this study also showed that both LFA1 and L-selectin were also significant players in lymphocyte homing to PP. Previously, Berlin *et al.* (Berlin, 1993) demonstrated *in vitro* $\alpha 4\beta 7$ dependent binding to the mucosal addressin MAdCAM-1, establishing $\alpha 4\beta 7$ as the lymphocyte homing receptor for MAdCAM-1.

The role of $\alpha4\beta7$ was confirmed and further defined in *in situ* experiments using video microscopic analysis to study, in real time, the adhesion of lymphocytes to PP-HEV *in vivo* (Bargatze, 1995). These studies of lymph node cell (primarily naive lymphocytes) homing revealed that $\alpha4\beta7$ slowed rolling initiated by L-selectin and that its participation was required for complete arrest in PP. On the other hand, it was also found that when $\alpha4\beta7$ was expressed at high levels (e.g. on TK1 lymphoma cells) or in a high affinity form (e.g., on experimentally preactivated lymphocytes), it could mediate lymphocyte arrest directly with no requirement for L-selectin or LFA1 participation. Arrest of these $\alpha4\beta7^{hi}$ cells in PP-HEV was not sensitive to pertussis toxin. These findings demonstrated that recruitment could take the form of a multistep cascade for naive lymphocytes or be facilitated by a single, integrin $\alpha4$, when presented in a high affinity form on the surface of an activated or $\alpha4\beta7^{hi}$ cell.

Importantly, $\alpha4\beta7$ is differentially expressed on naive B and T cells ($\alpha4\beta7^{low}$), and on subsets of memory T cells that are $\alpha4\beta7^{hi}$ MAdCAM-1 binding. Presumably mucosal memory cells or $\alpha4\beta7^{-}$ unable to bind MAdCAM-1, comprise non mucosal memory (Erle, 1994; Schweighoffer, 1993; L. Rott, J. Immunol., in press). This differential expression may play a critical role in segregating mucosal vs non-mucosal immunity.

Mucosal Addressin – MAdCAM-1

The mucosal addressin, first identified and functionally characterized by Streeter *et al.* (1988), was shown to support *in vivo* homing of normal lymphocytes and to PP-HEV. Animals treated with anti-MAdCAM-1 MAb MECA-367 and infused with labeled lymphocytes showed a >80% reduction of homing to PP, clearly defining a key role for MAdCAM-1 in PP lymphocyte recruitment. In situ studies confirm that MAdCAM-1 is critically important in $\alpha4\beta7$–dependent interactions in PP and LP venules (Bargatze, 1995; Berlin, 1995). Cloning of MAdCAM-1 revealed several key domains including two N-terminal Ig domains homologous to VCAM-1 and ICAM-1 responsible for the binding of $\alpha4\beta7$ (Briskin, 1993; Briskin, 1996). A mucin-like domain may serve to extend the $\alpha4\beta7$-binding domain away from the membrane for better presentation and, as discussed below, can also be modified by HEV to present L-selectin-binding carbohydrate ligands. Mouse but not human MAdCAM-1 also has a proximal IgA-like immunoglobulin domain of uncertain function (Briskin, 1993; Briskin, in press).

L-selectin and Carbohydrate Mediated MAdCAM-1 Adhesion

The MEL-14 antigen, L-selectin, was originally characterized as supporting recruitment of lymphocytes to peripheral nodes and in fact was originally named the peripheral node homing receptor (Gallatin, 1983). Its significance has been broadly expanded since this first description and it is now known to be functionally expressed by nearly all leukocyte types (Lewinsohn, 1987; reviewed in: Butcher and Picker, 1996; and Tedder, 1995).

Based on the earlier results implicating its preferential role in lymph node homing, it was initially surprising to find that L-selectin (as indicated by MEL-14 blockade) allowed significant interaction of peripheral node selective homing lymphomas in PP-HEV (Bargatze, 1990). In the time since that study Hamann (1991 and 1994) showed a significant dependent of lymphocyte homing to PP on L-selectin participation but, a role for L-selectin in an adhesion cascade had not been defined. Looking *in situ*, using fluorochrome labeled mesenteric node lymphocytes it was possible to observe rolling behaviors of lymphocytes (Bargatze, 1993) that closely resembled those of L-selectin neutrophil-dependent rolling *in vivo* (von Andrian, 1991; Ley, 1991). Using L-selectin blocking MAb (MEL-14), the *in situ* analysis allowed identification of L-selectin as the key molecule in initiating rolling in PP HEV. Blocking of L-selectin resulted in a marked reduction of lymphocyte-HEV rolling and greatly increased the number of non-interactive cells passing through venules. Without L-selectin participation, adhesion is predominantly achieved through microvillous expressed $\alpha4\beta7$, a pathway that is inefficient for $\alpha4\beta7^{lo}$ naive cells but that may be critical to $\alpha4\beta7^{hi}$ memory cell interactions and for homing of immunoblasts expressing preactivated $\alpha4\beta7$ (Berlin 1995, Bargatze 1995).

It was also discovered that L-selectin, consistant with its other adhesion roles, acts as a C-type lectin in its interactions with PP HEV. Moreover, MAdCAM-1 on PP and mesenteric node HEV bears *N*-glycanase-resistant sialic acid-containing carbohydrates that act as the predominant prominent L-selectin ligands (Berg, 1993). L-selectin binding carbohydrate modifications decorate MAdCAM-1 expressed by HEV *in vivo*, but are absent from MAdCAM-1 expressed by cultured bEnd.3 endothelioma cells or by intestinal lamina propria venules under normal circumstances. In situ analyses show that anti-MAdCAM-1 mAbs in fact, inhibit L-selectin dependent rolling in PP HEV, an effect dependent on crosslinking (Bargatze, 1995). Thus, MAdCAM-1 is the dominant PP HEV ligand for both $\alpha4\beta7$ and L-selectin and the only L-selectin ligand cloned to date shown to have a physiological role *in vivo*. The requirement of L-selectin for efficient initiation of naive lymphocytes rolling ensures their targeting to PP, and their exclusion from the intestinal lamina propria which lacks L-selectin ligands.

Lymphocyte Rolling, Contributions of L-selectin and $\alpha4\beta7$

Viewed *in situ* lymphocytes entering a PP HEV can initiate either rolling or sticking interactions. Examination of the behavior of lymphocytes pretreated with the anti-L-selectin MAb MEL-14 or treated with MEL-14 after infusion show a dramatic 85% drop in the frequency of rolling (Bargatze, 1995). Under these conditions there is no significant alteration in the frequency of immediately sticking cells, however the relative percent of cells exhibiting intermediate behaviors (rolling followed by sticking) also drops to near background. This demonstrates that $\alpha4\beta7$ on normal (predominantly naive) unactivated lymphocytes is poor at initiating interactions at the high velocities of the blood flow. If the experiment is reversed and $\alpha4\beta7$ adhesion is blocked the observed result is a loss of sticking and significant increase in rolling cell numbers with a concomitant tripling in the rolling velocity. Thus, the normal unactivated lymphocyte population first uses L-selectin to gain access to the endothelial cell surface and subsequently $\alpha4\beta7$ engages MAdCAM-1, dramatically slowing cell rolling velocity.

In experiments with lymphocytes expressing preactivated integrins, interactions with HEV are dramatically shifted to a shortening of the duration of rolling event (2–3 cell diameters) before sticking. Blocking with anti-L-selectin or anti-LFA1 MAb has no effect on the activated cell behavior. Treatment with anti-$\alpha4\beta7$ in contrast, completely blocks activated lymphocyte arrest. These analyses show that unactivated or partially activated $\alpha4\beta7$ (incapable of arresting cells) actively participates in the slowing of lymphocyte-L-selectin rolling, and that activated $\alpha 4\beta 7$ can act to directly attach and arrest cells from the blood.

Pertussis Toxin Sensitive Events in Lymphocyte Adhesion

In our initial *in situ* experiments, conducted to examine the adhesion of lymphocytes to PP HEV, we discovered that the arrest of rolling lymphocytes was prevented by pretreatment with the wild type pertussis toxin (PTX) (Bargatze, 1993). Cells that normally would show a high frequency of rolling and arrest instead rolled in an intermitant rolling and stopping manner, passing through the HEV without arrest. Control experiments using a modified toxin lacking the ADP-ribosylation properties of the wild type toxin failed to block arrest suggesting a direct role for $G\alpha_i$-protein signaling in triggering lymphocyte arrest. Treatment of cells with phorbol esters restored and enhanced the arrest of PTX-treated cells. From these observations we concluded that a rapid signaling event, now estimated to occur as fast as 1–3 seconds, was responsible for alteration of the binding potential of secondary adhesion supporting receptors on the lymphocyte surface.

In our subsequent *in situ* MAb blocking study we evaluated pertussis toxin for its effects on lymphocyte arrest in the context of molecular regulation adhesion steps (Bargatze, 1995). There we found that pertussis toxin acted to prevent the full realization of the adhesion potential of $\alpha 4\beta 7$, and that PMA or Mn^{++} activated lymphocytes were unaffected by pertussis toxin treatment, sticking efficiently to HEV via MAdCAM-1. These data suggest that $\alpha 4\beta 7$ is the end point in the G-protein signaling cascade leading to PP-HEV lymphocyte arrest. This interpretation of the data is somewhat complicated by the surprising result that a MAb directed against LFA1 can also prevent the conversion in lymphocytes behavior from rolling to arrest, but does not prevent the arrest of preactivated lymphocytes.

Contribution of LFA1 to Mucosal Lymphocyte Homing

Interestingly, as stated above anti-LFA1 treated lymphocytes are blocked from arrest in PP-HEV as revealed in *in situ* adhesion analysis (Bargatze, 1995). The simplest interpretation of this observation is that activation of LFA1 is required (along with $\alpha 4\beta 7$) for adhesion-strengthening following lymphocyte dependant arrest. However, experimentally activated lymphocytes that are treated with anti-LFA1 MAb are not prevented from adhering from the blood flow. Thus, fully activated $\alpha 4\beta 7$ can bind lymphocytes to HEV independently of LFA1. Moreover, on normal lymph node cells the effect of blocking LFA1 function closely parallels the effect of pertussis toxin treatment. Thus an alternative hypothesis is that LFA1 may play a role in regulating the activation state of other lymphocyte integrins, through an as of yet undefined mechanism. Cumulatively, these results show a near absolute dependence on $\alpha 4\beta 7$ participation for PP arrest but also demonstrate an important accessory role for LFA1 in the recruitment of unactivated lymphocytes.

These *in situ* studies also demonstrated the inability of lymphocyte LFA1 to initiate functional contact with endothelium under physiologic *in vivo* conditions: Anti-$\alpha 4\beta 7$ treatment completely prevents engagement of lymphocyte LFA1 in PP-HEV.

Particularly surprising is the inability of LFA1 to participate even when lymphocyte attachment and rolling is initiated through L-selectin. In this setting, the relatively high velocity of rolling via L-selectin alone (i.e., when $\alpha4\beta7$ is blocked) may present reduced opportunities for *in situ* triggering of secondary adhesion mechanisms or may simply be too rapid to permit engagement of LFA1 because of a slower binding on rate for LFA1 than that for $\alpha4$ integrins of an interaction. The differences in requirements for LFA1 vs. $\alpha4$ integrin involvement in lymphocyte-EC interactions are striking, and may reflect in part the specialized concentration of $\alpha4\beta7$ integrins and the relative exclusion of $\beta2$ integrins from microvillous, sites of initial cell contact under flow (Berlin, 1995; Erlandsen, 1993).

Summary of the Current Understanding of Mucosal Homing

In summary, most transfused lymph node cells (LNC) interact with PP-HEV in a series of sequential but overlapping adhesion events: L-selectin >$\alpha4\beta7$ initiates contact point interaction, L-selectin and $\alpha4\beta7$ both participate in rolling, and a $G\alpha_i$-protein mediated activation event triggers sticking and arrest that requires both $\alpha4\beta7$ and LFA1. Rolling via L-selectin alone does not permit efficient engagement of LFA1 in PP, thus requiring participation of $\alpha4\beta7$ (which dramatically reduces rolling velocity) as a "bridging" element. Thus, naive lymphocytes interact with HEV through a full multi-step adhesion cascade. In contrast, preactivated LNC or $\alpha4\beta7^{hi}$ TK1 lymphoma cells (models of gut-homing immunoblasts or memory cells) require only $\alpha4\beta7$ for arrest. These results (Bargatze, 1995) confirm the hypothesized involvement of multi-step decision processes in lymphocyte-endothelial cell recognition *in vivo*, but also emphasize the potential for direct interactions through preactivated $\alpha4$ integrins alone. Thus, the dominant PP HEV ligand for $\alpha4\beta7$ are also for L-selectin is the mucosal vascular addressin MAdCAM-1.

In the small intestinal lamina propria, venules express MAdCAM-1 but lack ligand presentation for L-selectin. In this site, lymphocyte arrest appears to involve primarily if not exclusively $\alpha4\beta7$. Most naive lymphocytes, which express only low levels of $\alpha4\beta7$ (and whose interaction with PP-HEV is largely initiated through L-selectin) do not interact detectably with lamina propria venules. In contrast, preactivated LNC or $\alpha4\beta7^{hi}$ cells adhere well. These *in situ* observations offer an explanation for the ability of gut-homing immunoblasts and memory cells but not naive cells to traffic to the extralymphoid intestinal wall; and for the relative preference of naive cells to home to Peyer's patches.

MUCOSAL VERSUS OTHER LYMPHOCYTE HOMING SPECIFICITIES

Short-term *in vivo* homing studies have confirmed that lymphocyte recruitment to inflamed skin is inhibited significantly by anti-E-selectin, $\alpha4$ integrin and LFA1 antibodies but not anti-$\beta7$. By the analogy of lymphocyte homing to mucosal tissues, the homing of CLA^+ lymphocytes to skin may involve a primary CLA:E-selectin interaction, perhaps with rolling slowed by $\alpha4\beta1$ engagement; an activating event that remains to be characterized; and activation-dependent arrest involving $\alpha4\beta1$:VCAM-1 and LFA1 (Picker, 1992, 1994). Lymphoblasts expressing preactivated $\alpha4\beta1$ may be able to interact independently of CLA and E-selectin. Of course, neutrophils can also bind via sLex and related carbohydrates to the E-selectin in chronically inflamed skin, but may be excluded from extravasation in this setting by the absence of appropriate activating factors (step 2), by their lack of $\alpha 4$ integrin expression or the absence of appropriate $\beta 2$ integrin ligands (step 3), or by a deficiency of neutrophil-specific chemoattractants (step 4).

In considering these models, one must also remember that quantitative as well as qualitative aspects of receptor expression can be critically important. L-selectin, for example, functions as a dominant lymphocyte homing receptor in naive lymphocyte trafficking to peripheral lymph nodes, whose HEV express extraordinarily high levels of the peripheral node addressin (PNAd) ligand for the L-selectin (Berg, 1991). In our view PNAd, which is defined as those peripheral node-HEV components involved in physiologic L-selectin binding and is likely composed of multiple distinct glycoproteins capable of presenting L-selectin-binding carbohydrates. These may include but are clearly not limited to CD34 and MAdCAM-1. GlyCAM-1, originally thought to be an important HEV ligand, appears to be a secreted molecule; and there is no detectable alteration in homing in CD34-deficient mice (Lasky, 1992; Lasky, Blood, in press; Suzuki and Mak, Blood, in press). Anti-L-selectin antibodies are highly effective at inhibiting lymphocyte homing to lymph nodes. One can imagine that L-selectin mediated adhesion in lymph nodes may be sufficient not only for initiating lymphocyte rolling, but perhaps significant slowing through higher ligand density also mediating at least the early phases of lymphocyte arrest, thus greatly facilitating the transition to stable activation-dependent attachment via either $\alpha 4\beta 1$ or LFA1 integrins (which may be able to fulfill parallel roles in some instances, as suggested by recent studies of T. Issekutz (1991)). Alternatively, VAP1 or other molecules may participate as bridging elements in PLN (Salmi and Jalkanen, 1992; Salmi and Jalkanen, J. Exp. Med., in press). In contrast, in Peyer's patches a very loose L-selectin-mediated tethering interaction (involving carbohydrate presentation by MAdCAM-1) facilitates initiation of rolling by naive lymphocytes, still maintains the requirement for $\alpha 4\beta 7$ participation in the adhesion process. Thus the levels of receptor and ligand expression can be very important determinants of the contribution of particular adhesion pathways to the selectivity of lymphocyte trafficking in distinct tissue sites.

CONCLUSIONS

Lymphocyte Recruitment And Implications For Therapeutic Intervention in Leukocyte Trafficking

The multistep model implies that one can interfere with leukocyte recruitment in sites of inflammation by blocking any one of the 4 steps involved; rolling, signaling arrest, diapedesis. In relatively acute clinical syndromes such as reperfusion injury, the selectivity of the therapy may be less important than its overall effectiveness, such that even targeting of adhesion or activating pathways involved very broadly in leukocyte trafficking (e.g., LFA1:ICAM-1, L-selectin) would not be constrained. For many indications, however, it may be more desirable to inhibit particular lymphocyte or leukocyte homing interaction as selectively as possible. Although this may not always be feasible, in some instances it is clear that any particular adhesion receptor ligand pair, and in all likelihood any particular activating factors and/or chemoattractants, will prove to dominate the specificity of a particular homing event of a leukocyte subset and/or tissue-specific fashion. Perhaps the best example at present is the $\alpha4\beta7$:MAdCAM-1 interaction, which appears highly specific for leukocyte trafficking to mucosal sites and represents an attractive target for suppression of inflammation in chronic inflammatory bowel disease and other immune-mediated pathologies of the gastrointestinal tract.

ACKNOWLEDGEMENTS

Portions of the text were adapted and updated from earlier reviews by the authors. Our studies were supported by NIH grants AI37832, the Stanford Digestive Disease Center under DK38707, and GM37734, and an award from the Department of Veterans Affairs. We thank Susan Grossman for assistance with manuscript preparation.

REFERENCES

Arfors, K.E., Lundberg, C., Lindbom L., Lundberg, K., Beatty, P.G. and Harlan, J.M. (1987). A Monoclonal Antibody to the Membrane Glycoprotein Complex CD18 Inhibits Poly morphonuclear Leukocyte Accululation and Plasma Leakage. *In Vivo. Blood*, **69**, 338–340.

Bacon, K.B. and Schall, T.J. (1996). Chemokines as mediators of allergic inflammation. *Int. Arch. Allergy. Immunol.*, **109**, 97–109.

Bargatze, R. F., Streeter, P.R. and Butcher, E.C. (1990). Expression of low levels of peripheral lymph node-associated vascular addressin in mucosal lymphoid tissues: Possible relevance to the dissemination of passaged AKR lymphomas *J. Cell Biochem.*, **42**, 219–227.

Bargatze, R.F. and Butcher, E.C. (1993). Rapid G protein-regulated activation event involved in lymphocyte binding to high endothelial venules. *J. Exp. Med.*, **178**, 367–372.

Bargatze, R.F., Jutila, M.A. and Butcher, E.C. (1995). Distinct roles for L- selectin and integrins $\alpha4\beta7$ and LFA1 in lymphocyte homing to Peyer's patch-high endothelium *in situ*: the multi-step model confirmed and refined. *Immunity*, **3**, 99–108.

Berg, E.L., Yoshino, T., Rott, L.S., Robinson, M.K., Warnock, R.A., Kishimoto, T.K., Picker L.J. and Butcher E.C. (1991). The cutaneous lymphocyte antigen is a skin lymphocyte homing receptor for the vascular lectin endothelial cell-leukocyte adhesion molecule-1. *J. Exp. Med.*, **174**, 1461–1466.

Berg, E.L., Robinson, M.K., Warnock, R.A. and Butcher, E.C. (1991). The human peripheral lymph node vasular addressin is a ligand for LECAM-1, the peripheral lymph node homing receptor. *J. Cell Biol.*, **114**, 343–349.

Berg, E.L., McEvoy, L.M., Berlin, C., Bargatze, R.F. and Butcher, E.C. (1993). L-selectin-mediated lymphocyte rolling on MAdCAM-1. *Nature*, 366, 695–698.

Berlin, C., Berg, E.L., Briskin, M.J., Andrew, D.P., Kilshaw, P.J., Holzman, B., Weissman, I.L., Hamann, A. and Butcher, E.C. (1993). $\alpha4\beta7$ integrin mediates lymphocyte binding to mucosal vascular addressin MAdCAM-1. *Cell*, **74**, 185–195.

Berlin, C., Bargatze, R.F., Campbell, J.J., von Andrian, U.H., Szabo, M.C., Hasslen, S.R., Nelson, R.D., Berg, E.L., Erlandsen, S.L. and Butcher, E.C. (1995). $\alpha4$ integrins mediate lymphocyte attachment and rolling under physiological flow. *Cell*, **80**, 413–433.

Briskin, M.J., McEvoy, L.M. and E. C. Butcher. (1993). MAdCAM-1 displays homology to immunoglobulin and mucin-like adhesion receptors and to IgA1. *Nature*, **363**, 461–464.

Briskin, M.J., Rott, L. and Butcher, E.C. (1996). Structural requirements for mucosal vascular addressin binding to its lymphocyte receptor alpha 4 beta 7. Common themes among integrin-Ig family interactions. *J. Immunol.*, **156**, 719–726.

Butcher, E.C. (1986). The regulation of lymphocyte traffic. In: *Current Topics in Microbiology and Immunology*, **128**, 85–122

Butcher, E.C. (1991). Leukocyte-endothelial cell recognition: Three (or more). steps to specificity and diversity. *Cell*, **67**, 1033–1036.

Butcher, E.C. and Picker, L.J. (1996). Lymphocyte homing and homeostasis. *Science*, **272**, 60–66.

Carlos, T.M., Harlan, J.M. (1994). Leukocyte-endothelial adhesion molecules. *Blood*, **84**, 2068–2101.

Erlandsen, S.L., Hasslen, S.R. and Nelson, R.D. (1993). Detection and spatial distribution of the beta 2 integrin (Mac1). and L-selectin (LECAM-1). adherence receptors on human neutrophils by high-resolution field emission SEM. *J. Histochem. Cytochem.*, **41**(3), 327–333.

Erle, D.J., Briskin, M.J., Buthcher, E.C., Garcia-Pardo, A, Lazarovits, A.I. and Tidswell, M. (1994). Expression and function of the MAdCAM-1 receptor, integrin $\alpha 4\beta 7$, on human leukocytes. *J. Immunol.*, **153**(2), 517– 528).

Fennie, C., Gillet, N., Watson, S.R., an Rosen S.D. (1992). An endothelial ligand for L-selectin is a mucin-like molecule. *Cell*, **69**, 927–938.

Gallatin, W.M., Weissman, I.L. and Butcher. E.C. (1983). A cell surface molecule involved in organ-specific homing of lymphocytes. *Nature.*, **304**, 30–34.

Hamann, A., Jablonski-Westrich, D., Duijvestijn, A., Butcher, E. C., Baisch, H., Harder R. and Thiele, H-G. (1988). Evidence for an accessory role of LFA1 in lymphocyte-high endothelium interaction during homing. *J. Immunol.*, **140**, 693–699.

Hamann, A., Jablonski-Westrich, D., Jonas, P. and Thiele, H-G. (1991). Homing receptors reexamined: mouse LECAM-1 (MEL-14 antigen). is involved in lymphocyte migration into gut-associated lymphoid tissue. *Eur. J. Immun.*, **21**, 2925–2929.

Hamann, A., Andrew, D.P., Jablonski-Westrich, D., Holzmann, B. and Butcher, E.C. (1994). The role of α4-integrins in lymphocyte homing to mucosal tissues *in vivo. J. Immunol.*, **52**, 3282–3293.

Holzmann, B., McIntyre, B.W. and Weissman, I.L. (1989). Identification of a murine Peyer's patch-specific lymphocyte homing receptoras an integrin molecule with an α chain homologous to human VLA-4α. *Cell*, **56**, 37–46.

Hu, M.C., Crowe, D.T., Weissman, I.L. and Holzmann, B. (1992). Cloning and expression of mouse integrin bp (b7); a functional role-specific lymphocyte homing. *Proc. Natl. Acad. Sci. USA*, **89**, 8254–8258.

Imhof, B.A., Dunon, D. (1995). Leukocyte migration and adhesion. *Adv. Immunol.*, **58**, 345–416.

Issekutz, T. B. (1991). Inhibition of *in vivo* lymphocyte migration to inflammation and homing to lymphoid tissues by the TA-2 monoclonal antibody. A likely role for VLA-4 *in vivo. J. Immunol.*, **147**, 4178–4184.

Kelvin, D.J., Michiel, D.F., Johnston, J.A., Lloyd, A.R., Sprenger, H., Oppenheim, J.J. and Wang, J-M. (1993). Chemokines and serpentines: The molecular biology of chemokine receptors. *J. Leukocyte Biol.*, **54**, 604–612.

Kilshaw, P.J. and Murant, S.J. (1991). Expression and regulation of $\beta 7$ (βp). integrins on mouse lymphocytes: relevance to the mucosal immune system. *Eur. J. Immunol.*, **21**, 2591–2597.

Lasky, L.A., Singer, M.S., Dowbenko, D., Imai, Y., Henzel, W.J., Grimley, C., Fennie, C., Gillett, N., Watson, S.R. and Rosen S.D. (1992). An endothelial ligand for L-selectin is a novel mucin-like molecule. *Cell*, **69**, 927–938.

Laudanna, C., Campbell, J.J. and Butcher, E.C. (1996). Role of Rho in chemoattractant-activated leukocyte adhesion through integrins. *Science.*, **271**, 981–983.

Lawrence, M.B. and Springer, T.A. (1991). Leukocytes roll on a selectin at physiological flow rates: Distinction from and prerequisite for adhesion through integrins. *Cell*, **65**, 859–873.

Lewinsohn, D.M., Bargatze, R.F. and Butcher, E.C. (1987). Leukocyte- endothelial cell recognition: evidence of a common molecular mechanism shared by neutrophils, lymphocytes and other leukocytes. *J. Immunol.*, **138**, 4313–4321.

Ley, K., Gaehtgens, P., Fennie, C., Singer, M.S. Lasky, L.A. and Rosen S.D. (1991). Lectin-like adhesion molecule-1 mediates leukocyte rolling in mesenteric venules *in vivo. Blood*, 77, 2553–2555.

Mackay, C.R. (1993). Homing of naive, memory and effector lymphocytes. *Curr. Opin.Immunol.*, **5**, 423–427.

McEver R.P., Moore, K.L. and Cummings R.D. (1995). Leukocyte trafficking mediated by selectin-carbohydrate interactions. *J. Biol. Chem.*, **270**, 11025–11028.

Oppenheim, J.J., Zachariae, C.O., Mukaida N. and Matsushima. K. (1991). Properties of the novel proinflammatory supergene "intercrine" cytokine family. *Annu. Rev. Immunol.*, **9**, 617–648.

Picker, L.J., Butcher, E.C. (1992). Physiological and Molecular Mechanisms of Lymphocyte Homing. *Annual Rev. Immunol.*, **10**, 561–591.

Picker, LJ. (1994). Control of lymphocyte homing. *Curr. Opin. Immunol.*, **6**, 394–406.

Salmi, M. and Jalkanen, S. (1992). A 90 kilodalton endothelial molecule mediating lymphocyte binding in humans. *Science*, 257, 1407–1409.

Schall, T.J. (1991). Biology of the RANTES/SIS cytokine family. *Cytokine*, **3**, 165–183.

Schall, T.J., Bacon, K.B. (1994). Chemokines, leukocyte trafficking and inflammation. *Curr. Opin. Immunol.*, **6**, 865–873.

Schweighoffer, T., Tanaka, Y., Tidswell, M., Erle, D.J., Horgan, K.J., Luce, G.E., Lazarovits, A.I., Buck, D. and Shaw, S. (1993). Selective expression of integrin alpha 4 beta7 on a subset of human CD4+ memory T cells with hallmarks of gut-trophism. *J. Immunol.*, **152**(2), 717–729.

Shimizu, Y., Newman, W., Tanaka, Y. and Shaw, S. (1992). Lymphocyte interactions with endothelial cells. *Immunol. Today*, **13**, 106–112.

Smith, CW. (1993). Transendothelial migration. *Curr. Top. Microbiol. Immunol.*, **184**, 201–212.

Springer, TA. (1994). Traffic signals for lymphocyte recirculation and leukocyte emigration: the multistep paradigm. *Cell*, **76**, 301–314.

Stamper, H.B. and Woodruff J.J. (1976). Lymphocyte homing into lymph nodes: *in vitro* demonstration of the selective affinity of recirculating lymphocytes for high endothelial venules. *J. Exp. Med.*, **144**, 828– 833.

Streeter, P.S., Berg, E.L., Rouse, B.T.N., Bargatze, R.F. and Butcher E.C. (1988). A tissue specific endothelial cell molecule involved in lymphocyte homing. *Nature*, **331**, 41–46.

Tanaka, Y., Adams, D.H., Hubscher, S., Hirano, H., Siebenlist, U. and Shaw S. (1993). T-cell adhesion induced by proteoglycan-immobilized cytokine MIP-1b. *Nature*, **361**, 79–82.

Tedder, T.F., Steeber, D.A., Chen, A. and Engel, P. (1995). The selectins: ' vascular adhesion molecules. *FASEB J.*, **9**(10), 866–873.

Vestweber, D. (1993). The selectins and their ligands. *Curr. Top. Microbiol. Immunol.*, **184**, 65–75.

von Andrian, U.H., J.D. Chambers, L.M. McEvoy, R.F. Bargatze, K.E. Arfors, E.C. Butcher. (1991). Two step model of leukocyte-endothelial cell interaction in inflammation: Distinct roles for LECAM-1 and the leukocyte beta 2 integrins *in vivo. Proc. Natl. Acad. Sci.*, **88**, 7538– 7542.

von Andrian, U.H., Hasslen, S.R., Nelson, R.D., Erlandsen, S.L. and Butcher, E.C. (1995). A central role for microvillous receptor presentation in leukocyte adhesion under flow. *Cell*, **82**(6), 989–999.

4. ADHESION MOLECULES AND THE REGULATION OF THE MIGRATION OF LYMPHOCYTES

NANCY OPPENHEIMER-MARKS and PETER E. LIPSKY

Department of Internal Medicine, The University of Texas Southwestern Medical School, Dallas, Texas, USA

INTRODUCTION

The transendothelial migration of T lymphocytes is an essential step in a number of physiologic and pathologic processes that involve perivascular accumulation of cells in lymphoid and nonlymphoid tissues. As part of their essential function in immune surveillance, T cells repeatedly migrate through post-capillary venules as they continuously recirculate through lymphoid tissues and when they are selectively recruited to extravasate into non-lymphoid sites as well.

Regardless of the migration pathway, the mechanism of transendothelial migration generally involves the concerted activities of specific cell surface adhesion receptors, as well as the functional activity of particular subsets of T cells and endothelial cells (EC) at specific anatomic sites. T cells and EC possess multiple cell surface adhesion receptors which facilitate T cell extravasation. In addition, the intrinsic motility of T cells as well as an ability to become functionally altered by interactions with cells and extracellular matrix molecules and other activation signals, also play critical roles in allowing specific subsets of T cells to migrate into tissue. In addition, expression of adhesion molecules by EC as well as physiologic events within blood vessels, such as the rate of blood flow and the shear stress exerted, tend to focus transendothelial migration to specific anatomic sites, usually the post-capillary venules.

In general, lymphocyte extravasation can be considered as a multi-step adhesive process that involves initial contact of T cells with the EC surface, the induction or activation of cell surface adhesion receptors to mediate stable, firm binding to the endothelium, and then a deadhesion of the bound T cells, such that they transmigrate through the EC layer. Transmigration of T cells is also a receptor mediated process that involves specific receptor-counterreceptor pairs that may or may not have been involved in the initial binding. The repertoire of adhesion receptor pairs that are involved in transendothelial migration may be relatively small in comparison to those that have the capacity to mediate T cell-EC adhesion. Central to this process is the intrinsic migratory capacity of the extravasating T cells such that they move through the EC layer in a pathway directed by specific adhesion molecules.

T CELL RECIRCULATION VERSUS RECRUITMENT

The specific trafficking pathway T cells follow is dependent, in part, on their differentiation state. Thus, *naive* T cells that have not previously encountered antigen, post-thymically, predominantly recirculate from the blood into secondary lymphoid tissues, whereas antigen experienced *memory* T cells are recruited to extravasate into non-lymphoid tissues (Picker and Siegelman 1993). As part of their immune surveillance pathways, memory T cells also enter lymph node, but they primarily do so through the afferent lymphatics.

Antigen priming of naive T cells occurs in lymph nodes and spleen. Although the receptor-mediated mechanisms mediating T cell trafficking into the spleen are not precisely known, naive T cells extravasate into lymph nodes by migrating through the endothelium of specialized postcapillary venules, known as high endothelial venules (HEV) (Figure 1). These blood vessels are distinguished from others based upon their characteristic cuboidal or columnar morphology. In the 1960s, Gowans and Knight (1964) identified HEV as the vascular bed through which T cells migrate when they enter lymph node.These findings were later confirmed and extended by the studies of Stamper and Woodruff (1976), who developed an *in vitro* assay that permitted the quantitative assessment of lymphocyte binding to HEV in peripheral lymph node. The current view of lymphocyte recirculation into lymph node still maintains HEV as the predominant vascular bed through which naive T cells migrate.

In general, naive T cells will enter into lymphoid tissues by exiting the blood stream through HEV (Figure 1). If not specifically stimulated, they leave the tissue through efferent lymphatics and continue to recirculate through lymphoid tissues. If antigen stimulation occurs, naive T cells will differentiate into memory T cells before leaving the lymphoid organ in the efferent lymphatics. Rather then primarily recirculating through lymph nodes, memory T cells alter their trafficking patterns and migrate through the endothelium of postcapillary venules of non-lymphoid tissues (Figure 1).

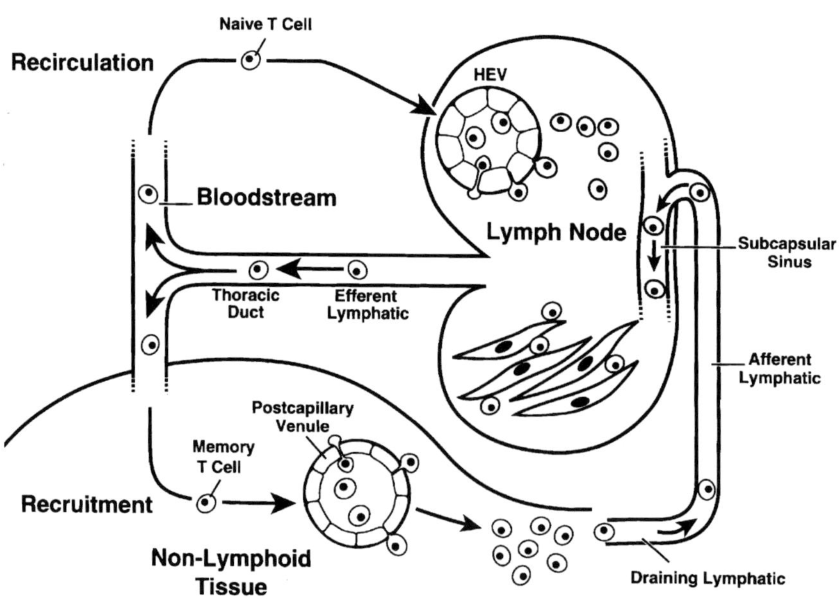

Figure 1 Pathways of naive T cell recirculation into peripheral lymph node and memory T cell recruitment into non-lymphoid tissue.

Table 1 Members of the Selectin Superfamily of Adhesion Receptors That Mediate Leukocyte Interactions With EC.

Receptor	Counter-Receptor	Distribution	Regulation
E-selectin (CD62E)	sialyl Lewis X, L-selectin, CLA, CD15, LFA-1 (CD11a/CD18), CD66, ESL-1	activated endothelial cells	induced by IL-1, TNF-α
L-selectin (CD62L)	E-selectin, P-selectin, MECA79 antigen, CD34, GLYCAM-1, MAdCAM1, charged oligosaccharides	resting leukocytes	decreased by activation stimuli
P-selectin (CD62P)	sialyl Lewis X, L-selectin, CD15 PSGL-1	activated platelets, activated endothelial cells	induced by thrombin, histamine

Inasmuch as it does not appear that all memory T cells accumulate in every non-lymphoid tissue, an endothelial mechanism involving the selective recruitment or retention of particular T cell subsets in tissue appears to operate. Extravasated memory T cells either percolate through the tissue, or they encounter their specific antigen and are retained within the tissue. In either situation, the T cells eventually leave the tissue to enter the afferent lymphatics and they are returned to the draining lymph node. Subsequently, the memory T cells exit the lymph node through the efferent lymphatic, enter the blood stream and circulate until they are recruited to extravasate into particular tissues.

ADHESION RECEPTOR SUPERFAMILIES

Tables 1–3 list the adhesion receptors that are likely to facilitate the extravasation of T cells into perivascular tissues. Adhesion receptors that are members of the selectin, integrin and immunoglobulin superfamilies mediate much of the adhesive interactions that occur during the infiltration of T cells into lymphoid and non- lymphoid tissues, the latter of which will be exemplified in this chapter as sites of chronic inflammation, such as the synovial tissue of patients with rheumatoid arthritis. Additional adhesion receptors that are not members of these superfamilies, such as CD44, also play a role in T cell extravasation. At inflammatory sites, the expression of many receptors on EC is regulated by the actions of proinflammatory cytokines. Each receptor binds at least one counterreceptor expressed by leukocytes, many of which are also functionally regulated by activation stimuli. For example, reagents, such as phorbol esters, which activate protein kinase C as well as engagement of the T cell receptor/CD3 complex, induce changes in adhesion receptor activity without necessarily affecting their expression, such as leukocyte function-associated antigen-1 (LFA-1; CD11a/CD18) and very late antigen-4 (VLA-4; CD49d/CD29), such that they become competent to bind their counterreceptors.

Table 2 Members of the Integrin Superfamily of Adhesion Receptors That Mediate Adhesive Interactions of Leukocytes.

Sub-family	Receptor	Counter-Receptor	Distribution	Regulation
$Beta_1$ (CD29)	VLA-1 (CD49a/CD29)	laminin, collagen	activated T cells, fibroblasts, liver sinusoids, mesangial cells, some cell lines	induced by mitogen, antigen
	VLA-2 (CD49b/CD29)	collagen, laminin	activated T cells, platelets, endothelial cells	increased by mitogen, antigen
	VLA-4 (CD49d/CD29)	VCAM-1, fibronectin	lymphocytes, monocytes, thymus cells, eosinophils, hematopoietic cell lines	functional activation by antigen
	VLA-5 (CD49e/CD29)	fibronectin	lymphocytes, monocytes, thymus cells, endothelial cells, fibroblasts	functional activation by antigen
	VLA-6 (CD49f/CD29)	laminin	lymphocytes, platelets, endothelial cells fibroblats	functional activation by antigen
$Beta_2$ (CD18)	LFA-1 (CD11a/CD18)	ICAM-1 (CD54), ICAM-2 (CD102), ICAM-3 (CD50) E-selectin (CD62E)	all hematopoietic cells except red blood cells	functional activation by antigen, increased expression by various mitogenic stimuli
	Mac-1 (CD11b/CD18)	ICAM-1, fibrinogen, endotoxin, Factor X, iC3b	monocytes, granulocytes, macrophages, large granular lymphocytes, some B and T cells	increased expression by chemotactic factors
	p150/95 (CD11c/CD18)	fibrinogen, iC3b	monocytes, granulocytes, macrophages, large granular lymphocytes, some B cells, hairy cell leukemia cells	increased by TNFα
	$\alpha d\beta 2$	ICAM-3 (CD50)	leukocytes	functional activation by antigen
$Beta_7$	$\alpha 4/\beta 7$	MAdCAM-1, VCAM-1, fibronectin	B and T cells, macrophages	functional activation by antigen
	$\alpha E\beta 7$	E-cadherin	intestinal intraepithelial lymphocytes	induced by TGFβ
	$\alpha ?\beta 7$	?	endothelial cells *in vitro, in vivo*	induced by TNFα, IL-1β

Table 3 Members of the Immunoglobulin Superfamily of Adhesion Receptors That Mediate Leukocyte Interactions With EC.

Receptor	Counter-Receptor	Distribution	Regulation
ICAM-1 (CD54)	LFA-1 (CD11a/CD18), Mac-1 CD11b/CD18), CD43, rhinovirus, P. falciparum	endothelial cells, synovial fibroblasts, synovial macrophages, epithelial cells, germinal center dendritic cells, activated lymphocytes	increased by IL-1, TNF-α, IFN-γ, endotoxin
ICAM-2 (CD102)	LFA-1	endothelial cells, hematopoietic cell lines	constitutive
ICAM-3 (CD50)	LFA-1, $\alpha d\beta 2$	resting leukocytes	constitutive
VCAM-1 (CD106)	VLA-4(CD49d/CD29) $\alpha 4/\beta 7$	endothelial cells, synovial lining cells, germinal center dendritic cells, synovial macrophages, bone marrow fibroblasts	induced by IL-1, TNF-α, IL-4
PECAM-1 (CD31)	$\alpha V\beta 3$, CD31, heparin	endothelial cells, platelets, leukocytes, smooth muscle cells	decreased by activation, increased by TGFβ
MAdCAM-1	$\alpha 4\beta 7$, L-selectin (CD62L)	endothelial cells of mucosal tissue	induced by TNFα IL-1β
LFA-3 (CD58)	CD2	endothelial cells, leukocytes, epithelial cells, some T cells	constitutive

The Selectin Superfamily of Adhesion Receptors

There are three members of the selectin superfamily of adhesion receptors, E-selectin (CD62E), L-selectin (CD62L), and P-selectin (CD62P), each of which appears to be important for the extravasation of lymphocytes into tissue (Table 1). The expression of E-selectin is restricted to EC found at sites of inflammation (Bevilacqua 1993). Normal, unstimulated EC do not express E-selectin, but it is induced by cytokines, including IL-1β and TNFα, and this is dependent on protein synthesis (Pober *et al.*, 1986; Bevilacqua *et al.*, 1987; Bevilacqua 1993). E-selectin was originally thought to be an adhesion receptor strictly for neutrophils, but has since been demonstrated to also mediate the binding of a subset of memory T cells to cytokine activated EC (Bevilacqua *et al.*, 1987; Picker *et al.*, 1991a; Shimizu *et al.*, 1991). P-selectin is a membrane glycoprotein found in secretory granules of platelets and Weibel-Palade bodies of EC (Larsen *et al.*, 1989; McEver *et al.*, 1989). When EC are stimulated by phorbol esters or agonists such as thrombin or histamine, P-selectin is rapidly redistributed to the plasma membrane where it mediates transient binding of a number of cell types, including lymphocytes (McEver *et al.*, 1989; Geng *et al.*, 1990; Kunzendorf *et al.*, 1993). In contrast to E-selectin and P-selectin, which are expressed by EC, L-selectin is found on the surface of resting leukocytes (Lewinsohn *et al.*, 1987). Initially, L- selectin was identified as a homing receptor mediating lymphocyte trafficking into peripheral lymph node, but recent findings suggest that it also may mediate T cell binding to cytokine activated endothelium (Gallatin *et al.*, 1983; Spertini *et al.*, 1991).

All members of the selectin family have a calcium dependent lectin domain at their amino termini, at which most of the binding activity of the molecules occur (Bevilacqua *et al.*, 1989; Johnston *et al.*, 1989; Lasky *et al.* 1989). Next to the lectin domain is a domain homologous to epidermal growth factor (EGF), and then variable numbers of tandem repeats homologous to complement regulatory proteins (CRP). For all selectins, the EGF domains are necessary for recognition of ligand (Kansas *et al.*, 1991; 1994; Pigott *et al.*, 1991). Thus, studies examining the structure-function relationship of these molecules using various anti-selectin mAb or selectin chimeras, have indicated that the EGF domains either contribute to the formation of one or more binding domains that directly interact with ligands, or they are required for the maintenance of the functional conformation of the lectin domain. The CRP domain of L-selectin has been suggested to serve a similar function (Kansas *et al.*, 1991; 1994; Pigott *et al.*, 1991; Watson *et al.*, 1991).

The counterreceptors for each member of the selectin family contain a charged oligosaccharide, with which the lectin domains interact. Thus, both E-selectin and P-selectin bind α(2–3)sialylated, α(1–3)fucosylated lactosaminoglycans, including the sialyly Lewis X (SLe^x) sequence, Neu5Acα2–3Galβ1–4(Fucα1–3)GlcNAcβ-R (Lowe *et al.*, 1990; Phillips *et al.*, 1990; Walz *et al.*, 1990; Zhou *et al.*, 1991). CD15 (Lewis X) has been identified as a counterreceptor for both E-selectin and P-selectin, but it seems to bind to these receptors with an affinity that is much lower than the sialylated form (Larsen *et al.*, 1990; Polley *et al.*, 1991). L-selectin expressed by neutrophils contains the SLe^x sequence and serves as one of the counterreceptors for E-selectin and P-selectin (Picker *et al.*, 1991b; Kishimoto *et al.*, 1991). On T cells, however, L-selectin was reported to lack the SLe^x sequence, and thus not bind to E-selectin. Rather, another cell surface molecule, cutaneous lymphocyte antigen (CLA), that is expressed by a subset of memory T cells, can serve as a counterreceptor for E-selectin (Berg *et al.*, 1991a). In addition to L- selectin and CLA, E-selectin has binding affinity for LFA-1, CD66 and a ligand expressed by the HL-60 cell line, E-selectin ligand (ESL-1) (Kotovouri *et al.*, 1993; Kuijpers *et al, 1992;* Steegmaier *et al.*, 1995). P-selectin also binds other molecules including a recently described mucin-like protein, P-selectin glycoprotein ligand (PSGL-1) (Sako *et al.*, 1993).

Recently, L-selectin counterreceptors that contain charged oligosaccharides have been identified on EC. These include those associated with peripheral lymph node HEV, including CD34 and glycosylation-dependent cell adhesion molecule (GLYCAM-1), as well as the mucosal addressin cell adhesion molecule (MadCAM-1), which is expressed by HEV of mucosal tissues (Lasky *et al.*, 1992; Baumheuter *et al.*, 1993; Jutila, 1994; Berg *et al.*, 1993). In addition, L-selectin appears to mediate lymphocyte binding to unknown structures induced on cytokine activated EC, as well as charged oligosaccharides recognized by a mAb, MECA 79, that was raised against a L-selectin counterreceptor that is expressed by peripheral lymph node HEV (Spertini *et al.* 1991; Streeter *et al.*, 1988; Berg *et al.*, 1991b). MadCAM- 1 and CD34 are examples of adhesion receptors recognized by MECA 79 (Berg *et al.*, 1993; Imai *et al.*, 1991). The MECA 79 antigen is also expressed by endothelium in a number of chronically inflamed tissues, including rheumatoid synovium, thyroid gland, cells of the gastrointestinal tract as well as the islets of the pancreas (Michie *et al.*, 1993; Hanninen *et al.*, 1993).

The selectin adhesion receptors are capable of transducing signals to cells. For L-selectin, this is presumed to occur by interactions of the short cytoplasmic tail with the actin cytoskeleton via α-actinin, since truncation of the cytoplasmic tail blocked this association and leukocyte interactions as well (Pavalko *et al.*, 1995). The cytoplasmic tails of both E-selectin and P-selectin may also interact with the cytoskeleton and deliver intracellular signals inasmuch as ligation of either molecule on EC with specific mAb induces cell rounding as well as a transient rise in calcium (Kaplanski *et al.*, 1994).

The Integrin Superfamily of Adhesion Receptors

Integrin adhesion receptors comprise a large superfamily of molecules that play an adhesive role in many physiologic processes, including embryologic development, hemostasis, thrombosis, wound healing, immune and nonimmune host defense, and tumorgenesis (Albelda and Buck 1990; Hynes 1992). All members of the integrin family are transmembrane glycoproteins composed of two non-covalently linked heterodimers, the α and β subunits (Table 2). Integrins were initially divided into three subfamilies, β1 (CD29) (also known as very late activation antigens, VLA), β2 (CD18) (also known as Leu-CAMs) and β3 (CD61) (also known as cytoadhesins), each having a common β subunit that is capable of associating with a variety of β subunits within a specific group. However, it has since been shown that certain α subunits can associate with more than one β subunit, and that there are at least eight different β subunits. Moreover, several integrins have more than one counterreceptor.

In general, β subunits are highly conserved, and each possesses one transmembrane domain and a short cytoplasmic tail at the carboxy terminus of the molecule (Marcantonio and Hynes 1988; Hogg 1989; Hemler 1990). The extracellular domains of β chains are stabilized by intrachain disulfide bonds. The β chains contain a nonhomologous short carboxy terminus cytoplasmic domain, a single transmembrane domain and a large globular extracellular domain, which contains divalent cation binding regions (Hogg 1989). In the α chains of the β2 subfamily and the α2 member of the β1 subfamily, there is an additional amino acid sequence inserted between the last divalent cation binding site and the amino terminus. This "inserted/interactive", "I" or "A" domain is homologous to repeats within von Willebrand factor, chicken cartilage matrix factor, and a region of complement factor B, and serves as an additional ligand binding site (Hemler 1990; Diamond *et al.*, 1993; Randi and Hogg 1994).

Of particular importance to T cell-EC interactions are the β1 integrin, VLA-4 ($\alpha4\beta1$; CD49dCD29), the β2 integrin, LFA-1 (CD11aCD18), and the β7 integrin, $\alpha4\beta7$, although VLA-1, 2, 3, 5 and 6, $\alpha V\beta3$, as well as $\alpha E\beta7$ play important roles in the interactions of T cells and EC with other cells and with components of ECM (Oppenheimer-Marks and Lipsky 1994; Berlin *et al.*, 1993; Cepek *et al.*, 1993). The β2 family of integrins has three additional members, Mac-1 (CD11bCD18), p150/95 (CD11cCD18) and the recently described αd (Table 2) (Albelda and Buck 1990; Van der Vieren *et al.*, 1995). It is LFA-1, however, that appears to play a prominent role in T cell extravasation into tissue (Oppenheimer-Marks *et al.*, 1990; Oppenheimer-Marks *et al.*, 1991; Kavanaugh *et al.*, 1991). LFA-1, VLA-4 and $\alpha 4\beta 7$ are all expressed by hematopoietic cells, with the exception that neutrophils seem to lack $\alpha 4$ expression, and both $\alpha 4$ and $\beta 7$ are also expressed by non- hematopoietic cells, including EC (Albelda and Buck 1990; Erle *et al.*, 1994; Brezinschek *et al.*, 1996).

Integrin binding of ligand is a divalent cation dependent process that requires the presence of both subunits. Many ligands express specific amino acid sequences, such as RGD and LDV, which are recognized by members of the $\beta 1$, $\beta 3$ and $\beta 7$ families (Albelda and Buck 1990; Hynes 1992). Ligation of integrins induces signal transduction into the cell ("outside-in" signaling). These generally induce the organization of the cytoskeleton as well as the generation of signals that may affect cell differentiation and gene expression (Hynes 1992). In addition, integrins are involved in another type of signaling mechanism that results from ligation of other cell surface molecules. Thus, as a result of binding their specific ligands, intracellular signals are generated which subsequently modulate the ability of integrins to bind their specific ligands ("inside-out" signaling). The mechanisms mediating these signaling events are complex and are not entirely understood.

Integrins are not constitutively active, in that, in general, integrins expressed by resting cells cannot bind their ligands. Cellular events are necessary for them to become competent to bind ligand. For example, ligation of cell surface molecules, including the T cell receptor/CD3 complex, CD2, CD7, CD28, MHC class I, MHC class II or the activation of protein kinase C, induce "inside-out" signals which render LFA-1 competent to bind its counterreceptor(s) (Oppenheimer-Marks *et al.*, 1990; Dustin and Springer 1989; Shimizu *et al.*1990b; 1992; van Kooyk *et al.*, 1989; Spertini *et al.*, 1992; Odum *et al.*, 1992; Hogg *et al.*, 1993). Increased surface expression of LFA-1 is not necessary for this to occur (Hogg *et al.*, 1993). Similar findings have been reported for VLA-4. Thus, for example, when CD31 is ligated by specific anti-CD31 mAb, VLA-4 becomes functionally active(Tanaka *et al.*, 1992). Although the nature of the intracellular signals necessary for the functional activation of integrins are not entirely clear, and they may differ for the various integrins, they seem to involve association of the cytoplasmic domains with each other and with the actin cytoskeleton, the latter of which is mediated by actin binding proteins, including α-actinin, talin, vinculin and paxillin (Burn *et al.*, 1988; Tapley *et al.*, 1989; Sanchez-Mateos *et al.*, 1993; Pardi *et al.*, 1995; Schaller *et al.*, 1995; Lewis and Schwartz 1995;). Specific sequences in the cytoplasmic domains of the α and β chains, which may play important roles in integrin activation, have been identified (Tapley *et al.*, 1989; Dedhar 1994; Reszka *et al.*, 1992; Hibbs *et al.*, 1991). For LFA-1, "inside-out" signaling may not require phosphorylation of the molecule (Hibbs *et al.*, 1991). Tyrosine phosphorylation of the cytoplasmic domain of $\beta 1$ by pp60v-src was found to inhibit ligand binding (Tapley *et al.*, 1989). Structurally, when LFA-1 becomes competent to bind ligand, the molecules appear to aggregate on the cell surface, and new antigenic epitopes are exposed, such that the binding of divalent cations, other than Ca2+, such as Mg2+ or Mn2+ becomes apparent (Dransfield and Hogg 1989; van Kooyk *et al.*, 1991; Dransfield *et al.*, 1992; Landis *et al.*, 1993; Shimizu *et al.*, 1993).

"Outside-in" signaling results when ligand binds to functionally active integrin molecules. Signals are transduced from the integrin into the cell, and this appears to be mediated by actin-containing microfilaments. Thus, the interaction between ligand, integrin and actin microfilaments generates the structural support necessary to maintain adhesion as well as the pathway for signal transduction into the cell. Recent attention has focused on the development of focal adhesion plaques which occur *in vitro* following the ligation of integrins by extracellular matrix molecules (ECM). For example, ligation of $\beta 1$ integrins by ECM causes actin associated cytoskeletal proteins to immobilize integrins, and this may be related to the activity of *rho* GTP binding protein family members (Sastry and Horwitz 1993; Clark and Brugge 1995; Takaishi *et al.*, 1995; Aepfelbacher 1995; Nobes and Hall 1995). Inasmuch as the cytoplasmic domain of integrins do not contain phosphatase or kinase activities, signaling through integrins appears to occur by interactions with initiators of signaling cascades. Thus, many signaling molecules have been identified in adhesion plaques and they have been shown to be specifically activated by ligation of integrins. These include tyrosine kinases (such as $pp60^{c\text{-}src}$, $pp125^{FAK}$ [focal adhesion kinase]), serine/threonine kinases (such as protein kinase C, integrin-linked kinase, MAP kinase, cyclin A kinase), proteins that mediate inositol lipid turnover, G proteins, and ion transporters, including calcium channels and the Na^+/H^+ antiporter (Ingbar *et al.*, 1990; Guan *et al.*, 1991; Schwartz *et al.*, 1991; Torti *et al.*, 1991; Kornberg *et al.*, 1992; Lipfert *et al.*, 1992; Chun and Jacobson 1992; Pelletier *et al.*, 1992; Davis *et al.*, 1993; Cybulsky *et al.*, 1993; Kapron-Bras *et al.*, 1993; McNamee *et al.*, 1993a, 1993b; Schwartz 1993; Vuori *et al.*, 1993; Chen *et al.*, 1994; Schlaepfer *et al.*, 1994; Hannigan *et al.*, 1996). Many of these constituents are also activated by growth factors, the receptors of which are similarly linked to adhesion plaques by the cytoskeleton. Thus, these sites have been suggested to serve as foci of converging signals for cell adhesion, cell differentiation and cell growth.

With regard to the transendothelial migration of lymphocytes, VLA-4 and LFA-1 have been shown to play important roles (Oppenheimer-Marks *et al.*, 1990,1991). Ligation of these molecules by their specific counterreceptors, intercellular adhesion molecule (ICAM-1) and vascular cell adhesion molecule (VCAM-1), respectively, transduces intracellular signals (Damle and Aruffo 1991; van Seventer *et al.*, 1991;1992; Damle *et al.*, 1993; Cabanas and Hogg 1993; Rothlein *et al.*, 1994). Although the precise signaling mechanisms have not been delineated, they appear to involve tyrosine phosphorylation of phospholipase C-$\gamma 1$ and the GTP binding protein, *rho* (Kanner *et al.*, 1993; Tominaga *et al.*, 1993). The cytoplasmic domains of both LFA-1 and VLA-4 colocalize with actin and actin binding proteins (Sanchez-Mateos *et al.*, 1993; Pardi *et al.*, 1995; Burn *et al.*, 1988; Tapley *et al.*, 1989; Sastry and Horwitz 1993; Pardi *et al.*, 1992; Parsey and Lewis 1993). Moreover, T lymphocytes also express $pp125^{FAK}$, and engagement of LFA- 1 or VLA-4 by specific mAb induces tyrosine phosphorylation of a number of polypeptides in lymphocytes, including $pp125^{FAK}$ (Whitney *et al.*, 1993; Maguire *et al.*, 1995; Sato *et al.*, 1995). A recent study has shown that ligation of VLA-4 stimulates "outside-in" signaling, and one of the functional responses is "inside-out" signaling to activate LFA-1 (Bargatze *et al.*, 1995). Ligation of CD44, another adhesion receptor, produces similar results (Koopman *et al.*, 1990). These data are particularly noteworthy in light of the recent findings that VLA-4 is unique among integrins in that it can mediate lymphocyte rolling under physiologic shear (Alon *et al.*, 1995). Similarly, CD44 can mediate lymphocyte rolling in the face of shear stress (Degrendele *et al.*, 1996). Thus, ligation of VLA-4 and/or CD44 may activate LFA-1 dependent adhesion events.

The Immunoglobulin Superfamily of Adhesion Receptors

The immunoglobulin (Ig) superfamily of adhesion receptors has several members that serve adhesive and recognition functions. A number of these receptors are of the C2–type, including ICAM-1, ICAM-2, ICAM-3, VCAM-1, CD31, LFA-3 and MadCAM-1, and they are of particular importance to adhesive interactions between lymphocytes and EC. Other members of the Ig family (C1–type) include the T cell receptor, class I and class II major histocompatibility complex, CD3, CD4 and CD8, all of which are involved in T cell recognition and activation processes and may function to increase cellular adhesion. C2–type Ig adhesion receptors contain a variable number of extracellular Ig-like domains with conserved cysteine residues that form disulfide bonds to stabilize the β sheets of the tertiary structure.

CD31 (platelet-endothelial cell adhesion molecule, PECAM-1) has 6 Ig-like domains, and is expressed by leukocytes, platelets, EC and smooth muscle cells (Muller *et al.*, 1989; Newman *et al.*, 1990) (Table 3). CD31 is highly enriched at intercellular junctions of EC, and plays a critical role in interendothelial cell adhesion and monolayer formation (Albelda *et al.*, 1990; 1991; Muller *et al.* 1989). Several studies have indicated that CD31 on leukocytes and EC also play a key role in leukocyte migration, although two recent reports indicate that CD31(-) T cells comprise the major subset exhibiting a transendothelial migratory capacity through unstimulated EC (Muller *et al.*, 1993; Bird *et al.* 1993; Brezinschek *et al.* 1995). CD31 can dimerize with itself and thereby mediate homotypic adhesions which may involve alignment of the entire CD31 molecule in an antiparallel configuration with CD31 on an apposing cell (Fawcett *et al.*, 1995). In addition, CD31 binds heparin as well as $\alpha V\beta 3$, the latter of which appears to involve the second Ig domain (Watt *et al.*, 1993; Piali *et al.*, 1995). Ligation of CD31 by specific mAb induces "inside-out" signaling resulting in VLA-4 activation on T cells and Mac-1 activation on monocytes and neutrophils (Tanaka *et al.*, 1992; Berman and Muller *et al.*, 1995). The large cytoplasmic tail of CD31 may facilitate signal transduction inasmuch it has been shown to be phosphorylated during activation of platelets and T cells, and to associate with the cytoskeleton of activated platelets (Zehnder *et al.*, 1992). CD31 mRNA and protein are decreased in Jurkat cells following activation, and increased in U937 cells by TGFβ (Zehnder *et al.*, 1992; Lastres *et al.*, 1994).

MadCAM-1 is a recently identified member of the Ig family of adhesion receptors that is expressed by EC of mucosal tissue and sinus lining cells of spleen (Table 3) (Briskin *et al.*, 1993; Kraal *et al.* 1995. The first two domains of MadCAM-1 share strong homology with ICAM-1 and VCAM-1; the first is similar to the first domain of both ICAM-1 and VCAM-1, and the second domain shares homology with the fifth domain of VCAM-1 (Briskin *et al.*, 1993). The third domain of MadCAM-1 is homologous to IgA. MadCAM-1 also contains a serine/threonine rich mucin-like region which may serve to present carbohydrate ligands to cells. There are at least two counterreceptors for MadCAM-1, the integrin $\alpha 4\beta 7$, which also binds VCAM-1 and fibronectin, although these two ligands do not seem to be involved in lymphocyte infiltration into mucosal tissue, and L-selectin, which based upon its lectin-like activity presumably binds the mucin like region (Berg *et al.*, 1993; Berlin *et al.* 1993; Andrew *et al.*, 1994; Hamann *et al.*, 1994). Recently, alternatively spliced MadCAM-1 was identified that lacks the mucin/IgA-homologous exon encoded sequence (Schiffer *et al.*, 1995). This may explain the observation that the MECA79 antigen, which is homologous to this type of carbohydrate region, is not found on all species of MadCAM-1 in Peyer's patch or on an endothelioma cell line, and thus the reason L-selectin does not interact with all species of MadCAM-1. Point mutations of residues within domains 1 and 2 indicate that $\alpha 4\beta 7$ binding to MadCAM-1 is through these domains (Briskin *et al.*, 1996). The regulation and signaling mechanisms of MadCAM-1 have not been completely delineated, although a recent report has indicated that it is induced by TNFα on an endothelioma cell line (Sikorski *et al.*, 1993).

ICAM-1 (CD54) is a heavily glycosylated member of the Ig superfamily. In addition to EC, ICAM-1 is expressed by a variety of cell types including synovial fibroblasts, synovial macrophage type A cells, dermal fibroblasts, epithelial cells, tissue macrophages, germinal center dendritic cells, and activated lymphocytes (Table 3) (Staunton *et al.*, 1988). ICAM-1 is constitutively expressed by many of these cell types, including EC, and its expression increases following exposure of the EC to proinflammatory cytokines, such as IL-1, TNFα, and IFNγ (Dustin and Springer 1988; Renkonen 1989; Voraberger *et al.*, 1991). ICAM-1 possesses 5 Ig domains. It is closely related to ICAM-2 (CD102), which contains 2 Ig domains that are homologous to the two most amino-terminal Ig domains of ICAM-1, and ICAM-3 (CD50), which contains 5 Ig domains that are homologous to the domains with the corresponding positions in ICAM-1 and ICAM-2 (Staunton *et al.*, 1989; Vazeux *et al.*, 1992; Fawcett *et al.*, 1992; de Fougerolles *et al.*, 1993). ICAM-2 is expressed by EC as well as a subpopulation of lymphocytes, monocytes, splenic sinusoids and dendritic cells (Voraberger *et al.*, 1991; Nortamo *et al.*, 1991a; deFougerolles *et al.*, 1991; Hart and Prickett 1993). Unlike ICAM-1, the expression of ICAM-2 is not modulated by cytokines (Nortamo *et al.*, 1991b). ICAM-3, on the other hand, is expressed by resting leukocytes of all lineages but is absent from resting and cytokine activated EC (Cordell *et al.*, 1993).

The ICAM molecules share little, if any, homology in their cytoplasmic domains (de Fougerolles *et al.*, 1993). Whether this is related to their involvement in different signaling mechanisms and/or localization within cells is uncertain at this time. The ICAMs are counterreceptors for members of the $\beta 2$ integrin family of adhesion receptors. Thus, they all bind LFA-1 (Marlin and Springer 1987; Staunton *et al.*, 1989). ICAM-1 also binds MAC-1 as well as CD43, rhinovirus and Plasmodium falciparum infected erythrocytes (Diamond *et al.*, 1990; Berendt *et al.*, 1989; Marlin *et al.*, 1990). ICAM-3 may play an important role in regulating LFA-1 affinity (Campanero *et al.*, 1993). Moreover, ICAM-3 also binds the newly described $\beta 2$ integrin, $\alpha d\beta 2$ (Van der Vieren *et al.*, 1995). All ICAMs are involved in the costimulation of T cell activation (Altmann *et al.*, 1989; van Seventer *et al.*, 1990; 1991; Hernandez-Caselles *et al.*, 1993; deFougerolles *et al.*, 1994; Damle *et al.*, 1992). ICAM-1 mediates many types of cell-cell adhesive interactions. Thus, homotypic T cell adhesion, homotypic B cell adhesion, T cell-accessory cell and T cell-synovial fibroblast interactions involve the activities of ICAM-1 and LFA-1. Particularly noteworthy are the findings from mAb blocking experiments that ICAM-1 on EC plays an important role in the adhesion and transendothelial migration of memory T cells (Oppenheimer-Marks *et al.*, 1991). The role of interactions with ICAM-1 in mediating lymphocyte migration has been documented by the findings that large granular lymphocytes and activated B cells specifically migrate on tissue culture plastic precoated with purified ICAM-1 (Carpen *et al.*, 1991; Dustin *et al.*, 1992).

Ligation of ICAM-1 on brain EC induces $p60^{src}$ tyrosine phosphorylation of the actin binding protein, cortactin (Durieu-Trautmann *et al.*, 1994). Moreover, ligation of ICAM-1 on T cells leads to MHC class II and IL2 receptor expression on B cells, and it appears to dampen the Ca^{2+} increase resulting from crosslinking of surface IgM (Poudrier and Owens 1994; van Horssen *et al.*, 1995). ICAM-1 has been shown to associate with actin containing microfilaments in COS transfected cells (Carpen *et al.*, 1992). The signaling events in EC resulting from ICAM-1 ligation during the transendothelial migration of lymphocytes are unknown. On the other hand, ligation of ICAM-3 on T cells induces "inside-out" signaling such that LFA-1 and VLA-4 become competent to bind ICAM-1 and fibronectin, respectively (Cid *et al.*, 1994). Suggestions have been made that ICAM-3 is involved in the initial phase of an immune response inasmuch as when crosslinked on Jurkat cells, Ca^{2+} levels increase and the activity of the tyrosine kinases, $p56^{lck}$ and $p59^{fyn}$ also increase (Juan *et al.*, 1994). Similar events presumably occur when LFA-1 binds to ICAM-3. The binding of ICAM-3 to LFA-1 appears to occur through the I domain at an epitope that is distinct from the one used to interact with ICAM-1 (Landis *et al.*, 1993; 1994). Moreover, neither ICAM-2 nor ICAM-3 bind to Mac-1, a member of the $\beta2$ integrin family that can bind to ICAM-1 (deFougerolles *et al.*, 1995; Diamond *et al.*, 1991).

VCAM-1 (CD106) is another member of the Ig superfamily of adhesion receptors (Osborn *et al.*1989; Rice *et al.*, 1990). VCAM-1 was initially reported to have 6 Ig domains. However, recent evidence indicates that the major form of VCAM-1 on EC has 7 Ig domains, with the extra domain inserted between domains 3 and 4 (Hession *et al.*, 1991). The inserted domain, which has a peptide sequence homologous to one of the VLA-4 binding sites of fibronectin, is involved in ligand recognition (Osborn *et al.*, 1992; Clements *et al.*, 1994). VCAM-1 is expressed by synovial lining cells, germinal center dendritic cells, bone marrow derived fibroblasts, and macrophages (Koch *et al.*, 1991; Koopman *et al.*, 1991; Morales-Ducret *et al.*, 1991; Ryan *et al.*, 1991). It is variably expressed by EC at sites of chronic inflammation, including rheumatoid synovium (Koch *et al.*, 1991; Morales-Ducret *et al.*, 1991). Activation of human umbilical vein EC (HUVEC) *in vitro* with the cytokines, IL-1β, IL-4 or TNFα, induces VCAM-1 synthesis and expression (Osborn *et al.*, 1992; Iademarco *et al.*, 1995). In contrast, neither IL-1β nor IL-4 effectively induce VCAM-1 on dermal microvascular EC (Swerlick *et al.*, 1992; Petzelbauer *et al.*, 1993). On murine stromal cells and EC, VCAM-1 is expressed as a transmembrane protein and as a glycophosphatidylinositol (GPI) linked molecule, both of which are involved in adhesive interactions (Kumar *et al.*, 1994). LFA-3 is another Ig receptor family member that has a GPI linkage (Dustin *et al.*, 1987). LFA-3, which is expressed by a number of cell types, including EC, binds CD2, an Ig family member that is expressed by T cells and some natural killer cells (Makgobu *et al.*, 1989; Springer 1990). To date, a role for LFA-3 and CD2 in lymphocyte extravasation has not been clearly established.

VLA-4 is a counterreceptor for VCAM-1 that binds to domains 1 and 4 of the VCAM-1 molecule (Osborn *et al.*, 1992). There is a peptide sequence In domain 4 containing the tripeptide, LDV, which VLA-4 recognizes, and which is normally found in a region of the CS (connecting segment)-1 segment within the alternatively spliced IIICS region of fibronectin (Clements *et al.*, 1994). Recent studies indicate that in addition to domain 1, this region of VCAM-1 is important in mediating T cell interactions with EC, inasmuch as CS1 peptides of the IIICS domain inhibited T cell binding to EC as well as to substrates of recombinant VCAM-1 (May *et al.*, 1993;

Makarem *et al.*, 1994). The integrin, $\alpha 4\beta 7$, also binds VCAM-1 and fibronectin, although it has not been established whether it recognizes LDV in VCAM-1 (Ruegg *et al.*, 1992; Andrew *et al.*, 1994). It is likely to bind this region, however, since it binds to this sequence in fibronectin (Ruegg *et al.*, 1992).

Monoclonal antibody blocking studies demonstrated that VCAM-1 can mediate lymphocyte adhesion to cytokine activated EC (Carlos *et al.*, 1990; Rice *et al.*, 1990; Oppenheimer-Marks *et al.*, 1991; Osborn *et al.*, 1992). This was also shown by examination of the adhesive behavior of cells isolated from patients with leukocyte adhesion deficiency (LAD) (Haskard *et al.*, 1989; Schwartz *et al.*, 1990; Kavanaugh *et al.*, 1991; Vennegoor *et al.*, 1992). Unlike normal cells, hematopoietic cells from these individuals do not express members of the $\beta 2$ integrins (Springer 1985). Unlike LAD neutrophils, which do not bind to EC, LAD T cells are competent to EC. Part of the binding to activated EC is mediated by VCAM-1 on the EC and VLA-4 on T cells. Results from other studies indicate that VLA-4/VCAM-1 interactions are important in mediating T cell binding to cytokine activated EC, but they do not appear to play a role in the transendothelial migration of T cells (Oppenheimer-Marks *et al.*, 1991). Since binding is requisite to migration, these data may partially explain the *in vivo* findings in which mAb to VLA-4 blocked the appearance of T cells in sites of inflammation (Issekutz 1991; Issekutz and Issekutz 1991). Like the ICAMs, VCAM-1 binding to its counterreceptor delivers intracellular signals to T cells. Thus, in Jurkat cells, immobilized VCAM-1 stimulated $pp125^{FAK}$ tyrosine phosphorylation (Maguire *et al.*, 1995). The nature of the signaling events within EC occurring when VCAM-1 is ligated has not yet been delineated.

THE ENDOTHELIUM OF LYMPHOID AND NON-LYMPHOID TISSUES

The endothelium plays an essential role in a number of mechanisms involved in the maintenance of hemostasis, inflammatory responses and immune function. Its activity is necessary for the control of vasomotor tone, vascular permeability, regulation of platelet adhesion and activation, modulation of the complement cascade, activation of procoagulant and fibrinolytic pathways, regulation of leukocyte adhesion and transendothelial migration, and immune modulation, by cytokine secretion and cellular interactions (Kavanaugh and Oppenheimer-Marks, 1992). Inherent in the behavior of the endothelium is its ability to maintain a normal resting homeostatic state, and to respond when there is an imbalance. Poised at the interface of the blood and extravascular tissue, the endothelium is acted upon by both environments. During normal lymphocyte recirculation, T cells extravasate into lymph node parenchyma by migrating through the endothelium of HEV. Although the precise mechanisms involved in the induction and maintenance of these specialized venules have not been completely delineated, EC do possess the ability to alter their morphologic and biologic phenotypes in response to environmental stimuli in their microenvironment. Thus, interference with afferent lymphatics of peripheral lymph nodes causes EC to lose their cuboidal shape and their ability to mediate lymphocyte recirculation (Hendricks *et al.*, 1983). Administration of antigen into these nodes caused a reappearance of HEV and reinitiation of lymphocyte extravasation. By contrast, injection of endotoxin into the lymph node had no effect on the development of HEV, nor was extravasation restored. Moreover, flattening of EC in HEV occurs following neonatal thymectomy, chronic drainage of the thoracic duct, treatment with antisera to thymocytes or prolonged treatment with prednisone (Parrott *et al.*, 1966; Goldschneider and McGregor, 1980; Ropke, 1973; Syrjanen, 1978; Miller, 1969). Taken together, these results indicate that the activity of T cells is involved in the development and maintenance of HEV and lymphocyte recirculation. However, HEV are not necessary for lymphocyte extravasation inasmuch as the endothelium of postcapillary venules in non-lymphoid, non-inflamed tissue is largely flat. Moreover, trafficking occurs into peripheral lymph node of nude rats, which lack HEV, although the transit time through flat endothelium is considerably longer than that observed in normal rats (Fossum *et al.*, 1983).

At sites of chronic inflammation, some EC assume morphologic and cytochemical characteristics of HEV (Freemont *et al.*, 1983; Iguchi and Ziff 1986; Freemont 1987). The development of HEV-like blood vessels has been examined in sites of delayed type hypersensitivity reactions in the skin of baboons, and was found to depend on the exposure of the tissue to proinflammatory cytokines (Munro *et al.*, 1989). Thus, concomitant stimulation with TNFα and IFNγ induced the appearance of HEV as well as the adhesion and extravasation of inflammatory cells. Exposure to multiple cytokines in chronically inflamed human tissue is likely to induce HEV in a similar manner. The HEV-like blood vessels in synovial tissue of patients with RA, like those of lymphoid tissue, have been demonstrated to be sites of adhesion of lymphocytes, and thus are suggested to be the sites through which mononuclear cells infiltrate into the inflamed tissue (Oppenheimer-Marks and Ziff 1986; Jalkanen *et al.*, 1986). HEV also may be important in regulating tissue access for only certain subsets of cells. In this regard, HEV have been suggested to be involved in providing routes for extravasation for B cells, and not T cells, into sites of acute inflammation, induced by infectious agents (Meeusen *et al.*, 1991). In rheumatoid synovium, HEV are found in areas containing aggregates of CD4+ T cells suggesting that in this tissue, HEV may play a role in the extravasation of these particular T cell subsets (Iguchi and Ziff 1986; Ziff and Iguchi 1987). Thus, although the HEV-like blood vessels may not be required for all lymphocyte extravasation, in certain circumstances they may be utilized as specialized sites for the infiltration of specific populations of cells. HEV, therefore, may offer a biologic advantage over flat EC by presentation of an optimum architecture for the organization of cell surface adhesion receptors necessary for efficient trafficking, but they do not appear to be absolutely necessary for lymphocyte entry into tissue.

LYMPHOCYTE RECIRCULATION

T cell extravasation into peripheral lymph node is largely dependent on the constitutive activity of T cell expressed L-selectin that binds to HEV. This interaction is favored and detectable under conditions of flow shear, and is responsible for weak interactions of lymphocytes to HEV by binding to charged oligosaccharides expressed on EC surface receptors, including CD34 and GLYCAM-1 (Lasky *et al.*, 1992; Baumheuter *et al.*, 1993; Finger *et al.*, 1996). Recent *in vivo* studies support an essential role for L-selectin in the trafficking of lymphocytes into peripheral lymph node. Thus, administration of an anti-L-selectin mAb to mice inhibited CD4+ naive T lymphocyte trafficking into peripheral lymph node (Bradley *et al.*, 1994). Moreover, lymphocytes of L-selectin deficient mice that completely lack L-selectin expression, do not bind to HEV, nor do they localize to peripheral lymph node (Tedder *et al.*, 1995).

The subsequent sequence of adhesive events mediating tighter T cell-EC interactions and transendothelial migration into peripheral lymph node are not as well understood. An "inside-out" signaling mechanism may activate integrin-mediated binding. To date, it has been demonstrated that a pertussis toxin inhibitable, presumably *g*-protein mediated signaling event is involved in the induction of tight lymphocyte adhesion to the endothelium of peripheral lymph node (Bargatze and Butcher 1993). LFA-1 is considered to play a key role inasmuch as specific mAb to LFA-1 inhibited lymphocyte binding to HEV of peripheral lymph nodes (Pals *et al.*, 1988; Hamann *et al.*, 1988). Recent reports support this and also have demonstrated that $\alpha 4$ integrins can mediate lymphocyte rolling as well as stimulate "inside-out" signaling mechanisms which induce LFA-1 mediated T cell adhesion to HEV of lymphoid tissue (Bargatze *et al.*, 1995; Alon *et al.*, 1995). The mechanisms involved in $\alpha 4$ integrin activation are not clear, but they may involve ligation of CD31 (Tanaka *et al.*, 1992). Other *in vitro* studies have indicated a direct role for $\alpha 4$ integrins in mediating firm adhesion to EC isolated from rat peripheral lymph node HEV (Szekanecz *et al.*, 1992).

To access mucosal lymphoid tissue, a different set of adhesion receptors is used. Thus, the EC express MAdCAM-1 which binds the integrin, $\alpha 4\beta 7$ expressed by T cells (Berlin *et al.*, 1993). $\alpha 4\beta 7$ mediates lymphocyte rolling as well as adhesion under static conditions. Ligation of $\alpha 4\beta 7$, which reduces the rolling velocity of lymphocytes appears to be necessary for the engagement of LFA-1 (Bargatze *et al.*, 1995). CD44 is another adhesion receptor that can facilitate T cell entry into mucosal lymphoid tissue, although the precise mechanism has not yet been determined (Haynes *et al.*, 1989). CD44 is closely related to proteoglycan link protein, and has a number of ligand specificities, including a mucosal tissue specific vascular addressin as well as the extracellular matrix components, hyaluronic acid and collagen (Haynes *et al.*, 1989; Aruffo *et al.*, 1989). Recently, CD44 was found to mediate lymphocyte rolling on substrates of hyaluronic acid (Degrendele *et al.*, 1996). Moreover, the binding of hyaluronic acid to CD44 induces an increase in intracellular calcium and an association of CD44 with the cell cytoskeleton (Bourguignon *et al.*). Ligation of CD44 by a specific anti-CD44 monoclonal antibody induces LFA-1 dependent lymphocyte aggregation (Koopman *et al.*, 1990; Kelleher *et al.*, 1995).

The nature of the adhesion receptors mediating the transendothelial migration of lymphocytes into lymph node have not been studied as extensively as have those mediating T cell-EC binding. Nevertheless, results from *in vitro* experiments examining T cell migration through cultures of rat HEV suggest that $\alpha 4$ integrins may be involved, and have militated against a role for CD31 (Bird *et al.*, 1993; Szekanecz *et al.*, 1992). In vivo studies examining the effects of anti-LFA-1 mAb on lymphocyte trafficking into peripheral lymph node found a decrease in T cell accumulation in the tissue implying a role for this molecule in T cell-EC binding and/or T cell migration (Hamann *et al.*, 1988; 1993).

LYMPHOCYTE RECRUITMENT INTO NON-LYMPHOID TISSUE

To ensure effective immune surveillance, there also are mechanisms mediating the selective recruitment of particular T cell populations into non-lymphoid tissue. Thus, for example, during an inflammatory response, specific subsets of memory T cells accumulate in tissue. The endothelium of chronically inflamed tissue expresses adhesion receptors involved in the adhesion and transendothelial migration of T cells. Many of these receptors, including VCAM-1, E-selectin, and P-selectin, are not expressed by EC of peripheral lymphoid tissue or noninflamed, nonlymphoid tissue (Pober and Cotran 1990). These receptors, which are expressed as a result of exposure of the endothelium to proinflammatory cytokines, such as TNFα and IL-1β, bind to their specific counterreceptors on T cells and may facilitate the selective recruitment of T cells into inflammatory sites.

A number of adhesion receptors are likely to facilitate the infiltration of T cells into sites of inflammation. As is the case for L-selectin, the adhesion receptors CD44 and CD31 do not seem to require an activation step, and thus these constitutively active adhesion receptors potentially could initiate lymphocyte binding to endothelial cells. Many reports have indicated that L-selectin has a prominent role in mediating low affinity interactions of neutrophils that are tethered to the vascular wall (Picker *et al.*, 1991b; Butcher 1991; Lawrence and Springer 1991). It is widely accepted that during acute inflammation, local blood flow through postcapillary venules is not severely impeded, and thus shear force may impose a negative influence on the migration of neutrophils out of the blood into acutely inflamed tissue (Robbins and Cotran 1979). Thus, in response to "inside-out" signaling events derived from the binding of factors emerging from the endothelium and/or perivascular tissue, neutrophils are stimulated and acquire the capacity to utilize LFA-1 for stabilized adhesion and transendothelial migration (Butcher 1991). It remains to be established, however, whether L-selectin plays a similar role in mediating T cell interactions with the endothelium of non-lymphoid tissue, such as sites of inflammation. If such were the case, then in inflamed tissue L-selectin would most likely mediate tethered binding of lymphocytes early in the evolution of disease, when blood flow diminishes, but is not static as occurs in chronic inflammation. Moreover, L-selectin(+) T cells that are capable of migrating would be the most likely candidates to utilize this molecule during their extravasation. Inasmuch as MECA 79 antigen is expressed at inflammatory sites, and L-selectin binds to TNFα activated EC *in vitro*, a role for L-selectin in mediating T cell interactions at inflammatory sites cannot be ruled out (Michie *et al.*, 1993; Hanninen *et al.*, 1993; Spertini *et al.*, 1992). Moreover, recent studies have shown that activated endothelium recruits a population of L-selectin(+) T cells to migrate that were not present in the subset observed to migrate through unstimulated EC (Brezinschek *et al.*, 1995). Whether these are Th-1–like L- selectin(+) memory T cells or whether they play an essential role in the subsequent evolution of the inflammatory response has not been determined (Kanejane *et al.*, 1996).

Stabilized lymphocyte binding to the endothelium is largely mediated by integrin adhesion receptors. VLA-4 and LFA-1 have been shown to mediate T cell-EC binding under static conditions, although all of the binding cannot be accounted for by the activities of these two molecules (Oppenheimer-Marks *et al.*, 1991). VLA-4 recognition of VCAM-1 is most important in binding T cells to cytokine activated endothelial cells. The role of fibronectin, as well as the CS-1 site in VCAM-1, in mediating T cell interactions with activated endothelium have not yet been determined. Examination of mucosal immunoblast binding to the HEV-like blood vessels in rheumatoid synovium demonstrated a role for VLA-4 but only a minor one for $\alpha4\beta7$ (Salmi *et al.*, 1995). Thus, it is uncertain whether $\alpha4\beta7$ facilitates T cell extravasation into rheumatoid synovium even though this tissue is enriched in $\alpha4\beta7$(+) T cells in comparison to peripheral blood and synovial fluid (Lazarovits and Karsh 1993). It also is unclear whether $\alpha 4\beta 7(+)$ T cells are selectively recruited from the peripheral blood into synovial tissue, or whether local activation events induce the expression of this integrin by extravasated T cells.

LFA-1 has been reported to function during the adhesion of resting and activated T cells to unstimulated EC by binding to ICAM-1 and ICAM-2 (Postigo and Sanchez-Madrid, 1993; Oppenheimer-Marks, 1991). Both, E-selectin and P-selectin, which are expressed by activated EC also appear to play a role in lymphocyte binding (Picker *et al.*, 1991a; Kunzendorf *et al.*, 1993; Shimizu *et al.*, 1991). Thus, a subset of memory T cells adheres to E-selectin. For many of these T cells, this binding is mediated by CLA (Picker *et al.*, 1991a). Whether CLA requires an activation step to bind to E-selectin has not yet been determined. There are other adhesion receptors that could potentially play a role in mediating T cell interactions with the endothelium at sites of inflammation, including vascular adhesion protein-1 (VAP-1), which can mediate lymphocyte binding, and is expressed by the endothelium of many tissues, including normal peripheral lymph node as well as those at sites of chronic inflammation (Salmi *et al.*, 1993). CD2 binding to its counterreceptors CD58 (LFA-3), CD59 or CD48 may also facilitate lymphocyte binding to endothelium. CD2 is expressed by all T cells and plays an accessory role during T cell activation (Postigo and Sanchez-Madrid 1993). EC have been reported to express all three CD2 counterreceptors (Meri *et al.*, 1993; Arvieux *et al.*, 1986). EC found *in vitro* and *in vivo*, including those in rheumatoid synovium, were recently shown to express $\beta 7$, although the significance of this is uncertain at this time (Brezinschek *et al.*, 1996).

TRANSENDOTHELIAL MIGRATION

In contrast to T cell binding to endothelial cells, there are a limited number of adhesion receptors that have been identified that mediate the transendothelial migration of lymphocytes. That transendothelial migration and binding are separable events is indicated by the observation that not all adherent T cells transmigrate through an EC layer *in vivo* or *in vitro*. The two processes appear to be regulated by different intracellular signaling events. For example, agents that elevate cAMP levels in both T cells and EC inhibit transendothelial migration but not T cell-EC binding (Oppenheimer-Marks *et al.*, 1994). The signaling events leading to transendothelial migration of T cells have not been precisely delineated but appear to involve protein kinase C inasmuch as treatment of T cells with phorbol esters stimulates motility and transendothelial migration (Oppenheimer-Marks *et al.*, 1990). Signals that regulate transendothelial migration may be induced by binding of adhesion receptors as both LFA-1 and VLA-4 deliver "outside-in" co-stimulatory signals to T cells (Wacholtz *et al.*, 1989; Davis *et al.*, 1990; Shimizu *et al.*, 1990). Thus, ligation of receptors involved in T cell-EC binding may not only be important in mediating cell-to-cell contact but also in transmitting signals that alter the capacity of the cells to migrate across the endothelium.

A number of *in vitro* experiments have been carried out to identify the adhesion receptors that mediate the transendothelial migration of T cells. Using blocking mAb, it has been demonstrated that the transendothelial migration of T cells bound to EC involves LFA-1 and ICAM-1, but not VLA-4, VCAM-1, LFA-3 or E-selectin

(Oppenheimer-Marks *et al.*, 1991). Importantly, this was noted even with IL-1β activated EC, in which the initial interaction was mediated by VLA-4/VCAM-1 interactions. In addition to LFA-1/ICAM-1, the homing receptor CD44, was also found to play a role in the transendothelial migration of T cells, although the counterreceptor for CD44 on cultured EC that CD44 binds has not yet been identified (Oppenheimer-Marks *et al.*, 1990). Thus, although a number of adhesion receptors mediate initial adhesive events between T cells and EC, they do not necessarily serve a similar function during transendothelial migration. The role of various adhesion molecules in T cell recruitment into inflammatory sites has also been examined *in vivo*. In these studies, selective inhibition of T cell infiltration into inflammatory sites was demonstrated using specific anti-VLA-4 and anti-LFA-1 monoclonal antibodies. However, these studies could not distinguish between effects of the antibodies on binding from those on migration.

There is compelling evidence to support a role for ICAM-1 in lymphocyte infiltration into rheumatoid synovium. Thus, administration of a specific monoclonal antibody against ICAM-1 to rheumatoid arthritis patients with early and active disease had a marked modifying effect on disease activity, causing lymphocytosis, as well as an inhibition of skin test reactivity, presumably by altering lymphocyte trafficking into tissue (Kavanaugh *et al.*, 1994). Another study that examined the role of $\beta 2$ integrins in leukocyte infiltration into the synovial tissue in an antigen induced arthritis model in the rabbit found that a monoclonal antibody against $\beta 2$ also profoundly effected disease outcome and markedly reduced leukocyte infiltration into the inflamed tissue (Jasin *et al.*, 1992). Similar results have been noted in studies using a monoclonal antibody to ICAM-1 in the rat model of adjuvant induced arthritis (Iigo *et al.*, 1991). These findings are in contrast to those recently reported in which an anti-LFA-1 monoclonal antibody had no effect on lymphocyte trafficking into the synovial tissue in the rat model of adjuvant arthritis (Issekutz and Issekutz 1993). The reasons for the conflicting data are not clear at this time but may be related to antibody effects on the trafficking of peripheral blood lymphocytes versus those isolated from spleen.

The available data are consistent with a model in which adhesion receptors exert overlapping functions to facilitate the entrance of lymphocytes into inflamed tissues. In addition, it is likely that the extravasation of lymphocytes into inflammatory sites is facilitated by the elaboration of mediators that possess chemotactic activity. Chemokines, such as MCP-1, RANTES, IL-8, MIP-1α and MIP-1β specifically induce the chemotaxis of specific populations of T cells (Prieschl *et al.*, 1995). Signals transmitted by these mediators may serve several functions, including the activation of lymphocyte adhesion receptors to facilitate their binding and transendothelial migration, as well as the directed movement into and within the tissue of specific populations of T cells. Recent reports support this and demonstrate a role for MIP-1α and MCP-1 in the transendothelial migration of lymphocytes (Brezinschek *et al.*, 1995; Cai *et al.*, 1996). RANTES, on the other hand, was found to mediate the migration of lymphocytes only in the absence of an EC monolayer, and therefore may play a more important role in regulating localization of T cells within the tissue after transendothelial migration (Brezinschek *et al.*, 1995). Since the ligation of the receptor for MIP-1β by the chemokine was found to stimulate the activation of VLA-4, a role for this cytokine in facilitating initial contact of T cells with the endothelium must be considered (Tanaka *et al.*, 1993).

CONCLUSIONS

Transendothelial migration is an intrinsic property of distinct T cell subpopulations, which exhibit specific phenotypic characteristics. The activities of a variety of adhesion receptors play a central role in the normal recirculation of T cells into lymphoid tissues as well as their selective recruitment into non-lymphoid tissues . Adhesion molecules exert pleiotropic effects during T cell extravasation into tissue, including the mediation of lymphocyte interactions with EC and extracellular matrix molecules, transendothelial migration, as well as the delivery of co-stimulatory signals involved in activation of these cells. The orchestration of the entry of specific subsets of T cells into perivascular tissues is central to effective immune surveillance, whereas dysregulation may play an important role in the perpetuation of chronic inflammatory diseases.

REFERENCES

Aepfelbacher, M. (1995). ADP-ribosylation of Rho enhances adhesion of U937 cells to fibronectin via the alpha 5 beta 1 integrin receptor. *FEBS Letters*, **363**, 78–80.

Albelda, S.M. and Buck, C.A. (1990). Integrins and other cell adhesion molecules. *FASEB J.*, **4**, 2868–2880.

Albelda, S.M., Oliver, P.D., Romer, L.H. and Buck, C.A. (1990). EndoCAM: A novel endothelial cell-cell adhesion molecule. *J. Cell Biol.*, **110**, 1227–1237.

Albelda, S.M., Muller, W.A., Buck, C.A. and Newman, P.J. (1991). Molecular and cellular properties of PECAM-1 (endoCAM/CD31): A novel vascular cell-cell adhesion molecule. *J. Cell Biol.*, **114**, 1059–1068.

Alon, R., Kassner, P.D., Carr, M.W., Finger, E.B., Hemler, M.E. and Springer, T.A. (1995). The integrin VLA-4 supports tethering and rolling in flow on VCAM-1. *J. Cell Biol.*, **128**, 1243–1253.

Altmann, D.M., Hogg, N., Trowsdale, J. and Wilkinson D. (1989). Cotransfection of ICAM-1 and HLA-DR reconstitutes human antigen-presenting cell function in mouse L cells. *Nature*, **338**, 512–514.

Andrew, D.P., Berlin, C., Honda, S., Yoshino, T., Hamann, A., Holzmann, B., Kilshaw, P.J. and Butcher, E.C. (1994). Distinct but overlapping epitopes are involved in $\alpha 4\beta 7$-mediated adhesion to vascular cell adhesion molecule-1, mucosal addressin-1, fibronectin and lymphocyte aggregation. *J. Immunol.*, **153**, 3847–3861.

Aruffo, A., Stamenkovic, I., Melnick, M., Underhill. ?. and Seed, B. (1989). CD44 is the principal cell surface receptor for hyaluronate. *Cell*, **61**, 1303–1313.

Arvieux, J., Willis, A.C. and Williams, A.F. (19??). MRC OX-45 antigen: A leukocyte/endothelium rat membrane glycoprotein of 45000 molecular weight. *Mol. Immunol.*, **23**, 983–990.

Bargatze, R.F. and Butcher, E.C. (1993). Rapid G protein regulated activation event involved in lymphocyte binding to high endothelial venules. *J. Exp. Med.*, **178**, 367–372.

Bargatze, R.F., Jutila, M.A. and Butcher, E.C. (1995). Distinct roles of L-selectin and integrins alpha 4 beta 7 and LFA-1 in lymphocyte homing to Peyer's patch-HEV in situ: The multistep model confirmed and refined. *Immunity*, **3**, 99–108.

Baumheter, S., Singer, M.S., Henzel, W., Hemmerich, S., Renz, M., Rosen, S.D. and Lasky L.A. (1993). Binding of L-selectin to the vascular sialomucin CD34. *Science*, **262**, 436–438.

Berendt, A.R., Simmons, D.L., Tansey, J., Newbold, C.I. and Marsh, K. (1989). Intercellular adhesion molecule-1 is an endothelial cell adhesion receptor for *Plasmodium falciparum*. *Nature*, **341**, 57–59.

Berg, E.L., Yoshino, T., Rott, L.S., Robinson, M.K., Warnock, R.A., Kishimoto, T.K., Picker, L.J. and Butcher, E.C. (1991a). The cutaneous lymphocyte antigen is a skin lymphocyte homing receptor for the vascular lectin endothelial cell-leukocyte adhesion molecule. *J. Exp. Med.*, **174**, 1461–1466.

Berg, E.L., Robinson, M.K., Warnock, R.A. and Butcher, E.C. (1991b). The human peripheral lymph node vascular addressin is a ligand for LECAM-1, the peripheral lymph node homing receptor. *J. Cell Biol.*, **114**, 343–349.

Berg, E.L., McEvoy, L.M., Berlin, C., Bargatze, R.F. and Butcher, E.C. (1993). L-selectin mediated lymphocyte rolling on MAdCAM-1. *Nature*, **366**, 695–698.

Berlin, C., Berg, E.L., Briskin, M.J., Andrew, D.P., Kilshaw, P.J., Holzmann, B., Weissman, I.L., Hamann, A. and Butcher, E.C. (1993). $\alpha 4\beta 7$ integrin mediates lymphocyte binding to the mucosal vascular addressin MAdCAM-1. *Cell*, **74**, 185–195.

Berman, M.E. and Muller, W.A. (1995). Ligation of platelet/endothelial cell adhesion molecule 1 (PECAM/CD31) on monocytes and neutrophils increases binding capacity of leukocyte CR3 (CD11b/CD18). *J. Immunol.*, **154**, 299–307.

Bevilacqua, M.P., Pober, J.S., Mendrick, D.L., Cotran, R.S. and Gimbrone, Jr. M.A. (1987). Identification of an inducible endothelial-leukocyte adhesion molecule. *Proc. Natl. Acad. Sci.*, **84**, 9238–9242.

Bevilacqua, M.P., Stengelin, S., Gimbrone, Jr. M.A. and Seed, B. (1989). Endothelial leukocyte adhesion molecule-1: An inducible receptor for neutrophils related to complement regulatory proteins and lectins. *Science*, **243**, 1160–1165.

Bevilaqua, M.P. (1993). Endothelial-leukocyte adhesion molecules. *Annu. Rev. Immunol*, **11**, 767–804.

Bird, I.N., Spragg, J.H., Ager, A. and Matthews, N. (1993). Studies of lymphocyte transendothelial migration: Analysis of migrated cell phenotypes with regard to CD31 (PECAM-1), CD45RA and CD45RO. *Immunol.*, **80**, 553–560.

Bourguignon, L.W., Lokeshwar, V.B., Chen, X. and Kerrick, W.G. (1993). Hyaluronic acid induced lymphocyte signal transduction and HA receptor (GP85/CD44) cytoskeletal interaction. *J. Immunol.*, **151**, 6634–6644.

Bradley, L.M., Watson, S.R. and Swain, S.L. (1994). Entry of naive CD4 T cells into peripheral lymph nodes requires L-selectin. *J. Exp. Med.*, **180**, 2401–2406.

Brezinschek, R.I., Lipsky, P.E., Galea, P., Vita, R. and Oppenheimer-Marks, N. (1995). Phenotypic characterization of CD4+ T cells that exhibit a transendothelial migratory capacity. *J. Immunol.*, **154**, 3062–3077.

Brezinschek, R.I., Brezinschek, H.P., Lazarovits, A.I., Lipsky, P.E. and Oppenheimer-Marks, N. (1996). Expression of the $\beta 7$ inetgrin by human endothelial cells. *Amer. J. Pathol.*, **149**, 1651–1660.

Briskin, M.J., McEvoy, L.M. and Butcher, E.C. (1993). MAdCAM-1 has homology to immunoglobulin and mucin-like adhesion receptors and to IgA. *Nature*, **363**, 461–464.

Briskin, M.J., Rott, L. and Butcher, E.C. (1996). Structural requirements for mucosal vascular addressin binding to its lymphocyte receptor $\alpha 4\beta 7$. *J. Immunol.*, **156**, 719–726.

Burn, P., Kupfer, A. and Singer, S.J. (1988). Dynamic membrane-cytoskeletal interactions: Specific association of integrin and talin arises in vivo after phorbol ester treatment of peripheral blood lymphocytes. *Proc. Natl. Acad. Sci.*, **85**, 497–501.

Butcher, E.C. (1991). Leukocyte-endothelial cell recognition: Three (or more) steps for specificity and diversity. *Cell*, **67**, 1033–1036.

Cabanas, C. and Hogg, N. (1993). Ligand intercellular adhesion molecule1 has a necessary role in activation of integrin lymphocyte function-associated molecule 1. *Proc. Natl. Acad. Sci.*, **90**, 5838–5842.

Cai, J.P., Hudson, S., Ye, M.W. and Chin, Y.H. (1996). The intracellular signaling pathways involved in MCP-1 stimulated T cell migration across microvascular endothelium. *Cell. Immunol.*, **167**, 269–275.

Campanero, M., del Pozo, M.A., Arroyo, A.G., Sanchez-Mateos, P., Hernandez-Caselles, T., Craig, A., Pulido, R. and Sanchez-Madrid, F. (1993). ICAM-3 interacts with LFA-1 and regulates the LFA-1/ICAM-1 cell adhesion pathway. *J. Cell Biol.*, **123**, 1007–1016.

Carlos, T.M., Schwartz, B.R., Kovach, N.L., Yee, E., Rosso, M., Osborn, L., Chi-Rosso, G., Newman, B., Lobb, R. and Harlan, J.M. (1990). Vascular cell adhesion molecule-1 mediates lymphocyte adherence to cytokine-activated cultured human endothelial cells. *Blood*, **76**, 965–970.

Carpen, O., Dustin, M.L., Springer, T.A., Swafford, J.A., Beckett, L.A. and Caulfield, J.P. (1991). Motility and ultrastructure of large granular lymphocytes on lipid bilayers reconstituted with adhesion receptors LFA-1, ICAM-1 and two isoforms of LFA-3. *J. Cell Biol.*, **115**, 861–871.

Carpen, O., Pallai, P., Staunton, D.E. and Springer, T.A. (1992). Association of intercellular adhesion molecule-1 (ICAM-1) with actin-containing cytoskeleton and α-actinin. *J. Cell Biol.*, **118**, 1223–1234.

Cepek, K.L., Prker, C.M., Madara, J.L. and Brenner, M.B. (1993). Integrin $\alpha^E\beta7$ mediates adhesion of T lymphocytes to epithelial cells. *J. Immunol.*, **150**, 3459–3470.

Chen, Q., Kinch, M.S., Lin, T.H., Burridge, K., Juliano, R.L. (1994). Integrin-mediated cell adhesion activates mitogen-activated protein kinases. *J. Biol. Chem.*, **269**, 26602–26605.

Chun, J.S. and Jacobson B.S. (1992). Requirement for diacylglycerol and protein kinase C in HELA cell-substratum adhesion and their feedback, amplification of arachidonic acid production for optimum cell spreading. *Mol. Biol. Cell*, 4, 271–281.

Cid, M.C., Esparza, J., Juan, M., Miralles, A., Ordi, J., Vilella, R., Urbano-Marquez, A., Gaya, A., Vives, J. and Yague, J. (1994). Signaling through CD50 (ICAM-3) stimulates T lymphocyte binding to human umbilical vein endothelial cells and extracellular matrix proteins via an increase in beta 1 and beta 2 integrin function. *Eur. J. Immunol.*, **24**, 1377–1382.

Clark, E.A. and Brugge, J.S. (1995). Integrins and signal transduction pathways: The road taken. *Science*, **268**, 233–239.

Clements, J.M., Newham, P., Shepard, M., Gilbert, R., Dudgeon, T.J., Needham, L.A., Edwards, R.M., Berry, L., Brass, A. and Humphries, M.J. (1994). Identification of a key integrin-binding sequence in VCAM-1 homologous to the LDV active site in fibronectin. *J. Cell Sci.*, **107**, 2127–2135.

Cordell, J.L., Pulford, K., Turley, H., Jones, M., Micklem, K., Doussis, I.A., Tyler, X., Mayne, K., Gatter, K.C. and Masonn, D.Y. (1993). Cellular distribution of human leucocyte adhesion molecule ICAM-3. *J. Clin. Pathol.*, 47, 143–147.

Cybulsky, A.V., Carbonetto, S., Cyr, M.D., McTavish, A.J., Huang, Q. (1993). Extracellular matrix stimulated phospholipase activation is mediated by beta 1 integrin. *Am. J. Physiol.*, **264**, C323–C332.

Damle, N.K. and Aruffo, A. (1991). Vascular cell adhesion molecule 1 induces T-cell antigen receptor-dependent activation of CD4+ T lymphocytes. *Proc. Natl. Acad. Sci.*, **88**, 6403–6407.

Damle, N.K., Klussman, K., Linsley, P.S., Aruffo, A. and Ledbetter, J.A. (1992). Differential regulatory effects of intercellular adhesion molecule-1 on costimulation by the CD28 counter-receptor B7. *J. Immunol.*, **149**, 2541–2548.

Damle, N.K., Klussman, K., Leytze, G., Ochs, H.D., Aruffo, A., Linsley, P.S. and Ledbetter, J.A. (1993). Costimulation via vascular cell adhesion molecule-1 induces in T cells increased responsiveness to the CD28 counter-receptor B7. *Cell. Immunol.*, **148**, 144–156.

Davis, C.M., Danehower, S.C., Laurenza, A. and Molony, J.L. (1993). Identification of the role of the vitronectin receptor and protein kinase C in the induction of endothelial cell vascular formation. *J. Cell. Biochem.*, **51**; 206–218.

Davis, L.S., Oppenheimer-Marks, N., Bednarczyk, J.L., McIntyre, B.W. and Lipsky, P.E. (1990). Fibronectin promotes proliferation of naive and memory T cells by signaling through both the VLA-4 and VLA-5 integrin molecules. *J. Immunol.*, **145**, 785–793.

de Fourgerolles, A.R., Stacker, S.A., Schwarting, R. and Springer, T.A. (1991). Characterization of ICAM-2 and evidence for a third counter-receptor for LFA-1. *Eur. J. Immunol.*, **174**, 253–267.

de Fourgerolles, A.R., Klickstein, L.B. and Springer, T.A. (1993). Cloning and expression of intercellular adhesion molecule 3 reveals strong homology to other immunoglobulin family counter-receptors for lymphocyte function-associated antigen-1. *J. Exp. Med.*, 177, 1187–1192.

de Fourgerolles, A.R., Qin, X. and Springer, T.A. (1994). Characterization of the function of intercellular adhesion molecule (ICAM)-3 and comparison with ICAM-1 and ICAM-2 in immune responses. *J. Exp. Med.*, **179**, 619–629.

de Fourgerolles, A.R., Diamond, M.S. and Springer, T.A. (1995). Heterogeneous glycosylation of ICAM-3 and lack of interaction with Mac-1 and p150,95. *Eur. J. Immunol.*, **25**, 1008–1012.

Dedhar, S. (1994). Novel functions for calreticulin: Interaction with integrins and modulation of gene expression? *TIBS*, **19**, 269–271.

Degrendele, H., Estess, P., Picker, L.J. and Siegelman M.H. (1996). CD44 and its ligand hyaluronate mediate rolling under physiologic flow- A novel lymphocyte-endothelial cell primary adhesion pathway. *J. Exp. Med.*, **183**, 1119–1130.

Diamond, M.S., Staunton, D.E., deFougerolles, A.R., Stacker, S.A., Garcia-Aguilar, J., Hibbs, M.L. and Springer, T.A. (1990). ICAM-1 (CD54): A counter-receptor for Mac-1 (CD11b/CD18). *J. Cell Biol.*, **111**, 3129–3139.

Diamond, M.S., Garcia-Aguilar, J., Bickford, J.K., Corbi, A.L. and Springer, T.A. (1993). The I domain is a major recognition site on the leukocyte integrin Mac-1 (CD11b/CD18) for four distinct adhesion ligands. *J. Cell Biol.*, **120**, 1031–1043.

Dransfield, I. and Hogg, N. (1989). Regulated expression of Mg2+ binding epitope on leukocyte integrin α subunits. *EMBO J.*, **8**, 3759–3765.

Dransfield, I., Cabanas, C., Craig, A. and Hogg, N. (1992). Divalent cation regulation of the function of the leukocyte integrin LFA-1. *J. Cell Biol.*, **116**, 219–226.

Durieu-Trautmann, O., Chaverot, N., Cazaubon, S., Strosberg, A.D. and Couraud, P.O. (1994). Intercellular adhesion molecule 1 activation induces tyrosine phosphorylation of the cytoskeleton-associated protein cortactin in brain microvessel endothelial cells. *J. Biol. Chem.*, **269**, 12536–12540.

Dustin, M.L., Selvaraj, P., Mattaliano, R.J. and Springer, T.A. (1987). Anchoring mechanisms for LFA-3 cell adhesion glycoprotein at membrane surface. *Nature*, **329**, 846–850.

Dustin, M.L. and Springer, T.A. (1988). Lymphocyte function-associated antigen-1 (LFA-1) interaction with intercellular adhesion molecule-1 (ICAM-1) is one of at least three mechanisms for lymphocyte adhesion to cultured endothelial cells. *J. Cell Biol.*, **107**, 321–331.

Dustin, M.L. and Springer, T.A. (1989). T-cell receptor cross-linking transiently stimulates adhesiveness through LFA-1. *Nature*, **341**, 619–623.

Dustin, M.L., Carpen, O. and Springer, T.A. (1992). Regulation of locomotion and cell-cell contact area by the LFA-1 and ICAM-1 adhesion receptors. *J. Immunol.*, **148**, 2654–2663.

Erle, D.J., Briskin, M.J., Butcher, E.C., Garcia-Pardo, A., Lazarovits, A.I. and Tidswell, M. (1994). Expression and function of the MadCAM-1 receptor, integrin $\alpha4\beta7$, on human leukocytes. *J. Immunol.*, **153**, 517–528.

Fawcett, J., Holness, C.L.L., Needham, L.A., Turley, H., Gatter, K.C., Mason, D.Y. and Simmons D.L. (1992). Molecular cloning of ICAM-3, a third ligand for LFA-1, constitutively expressed on resting leukocytes. *Nature*, **360**, 481–484.

Fawcett, J., Buckley, C., Holness, C.L., Bird, I.N., Spragg, J.H., Saunders, J., Harris, A. and Simmons D.L. (1995). Mapping the homotypic binding sites on CD31 and the role of CD31 adhesin in the formation of interendothelial cell contacts. *J. Cell Biol.*, **128**, 1229–1241.

Finger, E.B., Purl, K.D., Alon, R., Lawrence, M.B., von Andrian, U.H. and Springer, T.A. (1996). Adhesion through L-selectin requires a threshold hydrodynamic shear. *Nature*, **379**, 266–269.

Fossum, S., Smith, M.E., Ford, W.L. (1983). The recirculation of T and B lymphocytes in the athymic, nude rat. *Scand. J. Immunol.*, **17**, 551–557.

Freemont, A.J. (1987). Molecules controlling lymphocyte-endothelial interactions in lymph nodes are produced in vessels in inflamed synovium. *Ann. Rheum. Dis.*, **4**, 924–928.

Freemont, A.J., Jones, C.J.P., Bromley, M. and Andrews, P. (1983). Changes in vascular endothelium related to lymphocyte collections in diseased synovia. *Arthritis Rheum.*, **26**, 1427–1433.

Gallatin, W.M., Weissman, I.L. and Butcher, E.C. (1983). A cell-surface molecule involved in organ-specific homing of lymphocytes. *Nature*, **304**, 30–34.

Geng, J-G., Bevilacqua, M.P., Moore, K.L., McIntyre, T.M., Prescott, S.M., Kim, J.M., Bliss, G.A., Zimmerman, G.A. and McEver, R.P. (1990). Rapid neutrophil adhesion to activated endothelium mediated by GMP-140. *Nature*, **343**, 757–760.

Gowans, J.L. and Knight, E.J. (1964). The route of recirculation of lymphocytes in the rat. *Proc. R. Soc. Lond. [Biol.]*, **159**, 257–282.

Goldschneider, I. and McGregor, D.D. (1968). Migration of lymphocytes and thymocytes in the rat. I. The route of migration from blood to spleen and lymph nodes. *J. Exp. Med.*, **127**, 155–168.

Guan, J.L., Trevithick, J.E. and Hynes, R.O. (1991). Integrin interaction induces tyrosine phosphorylation of a 120 kDa protein. *Cell Regul.*, **2**, 951–964.

Hamann, A., Jablonski-Westrich, D., Duijvestijn, A., Butcher, E.C., Baisch, H., Harder, R. and Thiele, H.G. (1988). Evidence for an accessory role of LFA-1 in lymphocyte-high endothelium interaction during homing. *J. Immunol.*, **140**, 693–699.

Hamann, A. and Jablonski-Westrich, D. (1993). Integrins and L-selectin in lymphocyte-endothelium interactions and homing into gut associated tissue. In C. Sorg, D.J. Ruiter (eds.), *The Vascular Endothelium in Inflammation.* Behring Institute Mitteilungen, Die Medizinische, FRG, pp. 30–35.

Hamann, A., Andrew, D.P., Jablonski-Westrich, D., Holzmann, B. and Butcher, E.C. (1994). Role of alpha 4-integrins in lymphocyte homing to mucosal tissues in vivo. *J. Immunol.*, **152**, 3282–3293.

Hannigan, G.E., Leeung-Hagesteijn, C., Fitz-Gibbon, L., Coppolino, M.G., Radeva G., Filmus J., Bell J.C., Dedhar S. (1996). Regulation of cell adhesion and anchorage dependent growth by a new β1 inetgrin linked protein kinase. *Nature*, **379**, 91–96.

Hanninen, A., Taylor, C., Streeter, P.R., Stark, L.S., Sarte, J.M., Shiizuru, J.A., Simell, O. and Michie, S.A. (1993). Vascular addressins are i nduced on islet vessels during insulitis in nonobese diabetic mice and are involved in lymphoid cell binding to islet endothelium. *J. Clin. Invest.*, **92**, 2509–2515.

Haskard, D.O., Strobel, S., Thornhill, M., Pitzalis, C., Levinsky, R.J. (1989). Mechanisms of lymphocyte adhesion to endothelial cells: studies using a LFA-1 deficient cell line. *Immunol.*, **66**, 111–116.

Haynes, B.F., Telen, M.J., Hale, L.P. and Denning, S.M. (1989). CD44–A molecule involved in leukocyte adherence and T cell activation. *Immunol. Today*, **10**, 423–428.

Hart, D.N.J. and Prickett, T.C.R. (1993). Intercellular adhesion molecule-2 (ICAM-2) expression on human dendritic cells. *Cell Immunol.*, **148**, 447–454.

Hemler, M.E. (1990). VLA proteins in the integrin family. Structures, functions and their role on leukocytes. *Annu. Rev. Immunol.*, **8**, 365–400.

Hendriks, H.R. and Eestermans, I.L. (1983). Disappearance and reappearance of high endothelial venules and immigrating lymphocytes in lymph nodes deprived of afferent lymphatic vessels: A possible regulatory role of macrophages in lymphocyte migration. *Eur. J. Immunol.*, **13**, 663–669.

Hession, C., Tizard, R., Vassallo, C., Schiffer, S.B., Goff, D., Moy, P., Chi-Rosso, G., Luhowskyj, S., Lobb, R. and Osborn, L. (1991). Cloning of an alternate form of vascular cell adhesion molecule-1 (VCAM1). *J. Biol. Chem.*, **266**, 6682–6685.

Hernandez-Caselles, T., Rubio, G., Campanero, M.R., del Pozo, M.A., Muro, M., Sanchez-Madrid, F. and Aparicio, P. (1993). *Eur. J. Immunol.*, **23**, 2799–2806.

Hibbs, M.L., Jakes, S., Stacker, S.A., Wallace, R.W. and Springer, T.A. (1991). The cytoplasmic domain of the integrin lymphocyte function-associated antigen 1β subunit: Sites required for binding to inetrcellular adhesion molecule 1 and the phorbol ester-stimulated phosphorylation site. *J. Exp. Med.*, **174**, 1227–1238.

Hogg, N. (1989). The leukocyte integrins *Immunol. Today*, **10**, 111–114.

Hogg, N., Harvey, J., Cabanas, C. and Landis, R.C. (1993). Control of leukocyte integrin activation. *Amer. Rev. Resp. Dis.*, **148**, S55–S59

Hynes, R.O. (1992). Integrins: Versatility, modulation and signaling in cell adhesion. *Cell*, **69**, 11–25.

Iademarco, M.F., Barks, J.L. and Dean D.C. (1995). Regulation of vascular cell adhesion molecule-1 expression by IL-4 and TNFβ in cultured endothelial cells. *J. Clin. Invest.*, **95**, 264–271.

Iguchi, T. and Ziff, M. (1986). Electron microscopic study of rheumatoid synovial vasculature. *J. Clin. Invest.*, 77, 355–361.

Iigo, Y., Takashi, T., Tamatani, T., Miyasaka, M., Higashida, T., Yagita, H., Okumura, K. and Tsukada, W. (1991). ICAM-1 dependent pathway is critically involved in the pathogenesis of adjuvant arthritis in rats. *J. Immunol.*, **147**, 4167–4171.

Imai, Y., Singer, M.S., Fennie, C., Lasky, L.A. and Rosen, S.D. (1991). Identification of a carbohydrate-based endothelial ligand for a lymphocyte homing receptor. *J. Cell Biol.*, **113**, 1213–1221.

Ingbar, D.E., Prusty, D., Frangione, J., Cragoe, E.J., Jr. Lechene, C. and Schwartz, M. (1990). Control of intarcellular pH and growth by fibronectin in capillary endothelial cells. *J. Cell Biol.*, **110**, 1803–1812.

Issekutz, T.B. (1991). Inhibition of in vivo lymphocyte migration to inflammation and homing to lymphoid tissues by the TA-2 monoclonal antibody. A likely role for VLA-4 in vivo. *J. Immunol.*, **147**, 4178–4184.

Issekutz, T.B. and Issekutz, A.C. (1991). T lymphocyte migration to arthritic joints and dermal inflammation in the rat: Differing migration patterns and the involvement of VLA-4. *Clin. Immunol. Immunopathol.*, **61**, 436–447.

Jalkanen, S., Steere, A.C., Fox, R.I. and Butcher, E.C. (1986). A distinct endothelial cell recognition system that controls lymphocyte traffic into inflamed synovium. *Science*, **233**, 556–558.

Jasin, H.E., Lightfoot, E., Davis, L.S., Rothlein, R., Faanes, R.B. and Lipsky, P.E. (1992). Amelioration of antigen-induced arthritis in rabbits treated with monoclonal antibodies to leukocyte adhesion molecules. *Arthritis Rheum.*, **35**, 541–549.

Johnston, G.I., Cook, R.G. and McEver, R.P. (1989). Cloning of GMP-140, a granule membrane protein of platelets and endothelium: Sequence similarity to proteins involved in cell adhesion and inflammation. *Cell*, **56**, 1033–1044.

Juan, M., Vinas, O., Pino-Otin, M.R., Places, L., Martinez-Caceres, E., Barcelo, J.J., Miralles, A., Vilella, R., de la Fuente, M.A., Vives, J., Yague, J. and Gaya, A. (1994). CD50 (intercellular adhesion molecule 3) stimulation induces calcium mobilization and tyrosine phosphorylation through p59fyn and p56lck in Jurkat T cell line. *J. Exp. Med.*, **179**, 1747–1756.

Jutila, M.A. (1994). Function and regulation of leukocyte homing receptors. *J. Leuk. Biol.*, **55**, 133–140.

Kanejane, H., Kasahara, Y., Niida, Y., Yachie, A., Sugii, S., Takatsu, K., Tanigichi, N. and Miyawaki, T. (1996). Expression of L-selectin (CD62L) discriminates Th1–and Th2–like cytokine-producing memory CD4+ T cells. *Immunol.*, **87**, 186–190.

Kanner, S.B., Grosmire, L.S., Ledbetter, J.A. and Damle, N.K. (1993). β2-integrin LFA-1 signaling through phospholipase C-γ_1 activation. *Proc. Natl. Acad. Sci.*, **90**, 7099–7103.

Kansas, G.S., Spertini, O., Stoolman, L.M. and Tedder, T.F. (1991). Molecular mapping of functional domains of the leukocyte receptor for endothelium, LAM-1. *J. Cell Biol.*, **114**, 351–358.

Kansas, G.S., Saunders, K.B., Ley, K., Zakrzewicz, A., Gibson, R.M., Furie, B.C., Furie, B. and Tedder, T.F. (1994). A role for the epidermal growth factor-like domain of P-selectin in ligand recognition and cell adhesion. *J. Cell Biol.*, **124**, 609–618.

Kaplanski, G., Farnarier, C., Benoliel, A.M., Foa, C., Kaplanski, S., Bongrand, ?. (1994). A novel role for E- and P-selectins: shape control of endothelial cell monolayers. *J. Cell Sci.*, **107**, 2449–2457.

Kapron-Bras, C., Fitz-Gibbon, L., Jeevaratnam, P., Wilkins, J., Dedhar, S. (1993). Stimulation of tyrosine phosphorylation and accumulation of GTP-bound p21ras upon antibody mediated $\alpha 2\beta 1$ integrin activation in T lymphoblastic cells. *J. Biol. Chem.*, **268**, 20701–20704.

Kavanaugh, A.F., Lightfoot, F., Lipsky, P.E. and Oppenheimer-Marks, N. (1991). Role of CD11/CD18 in adhesion and transendothelial migration of T cells. Analysis utilizing CD18–deficient T cell clones. *J. Immunol.*, **146**, 4149–4156.

Kavanaugh, A.F. and Oppenheimer-Marks, N. (1992). The role of the endothelium in the pathogenesis of vasculitis. In E.C. LeRoy, (ed.), *Systemic Vasculitis: The Biological Basis* Marcel Dekker, Inc, New York, pp 27–48.

Kavanaugh, A.F., Davis, L.S., Nichols, L.A., Norris, S.H., Rothlein, R., Scharschmidt, L.A. and Lipsky, P.E. (1994). Treatment of refractory rheumatoid arthritis with a monoclonal antibody to intercellular adhesion molecule-1 (ICAM-1). *Arthritis Rheum.*, **37**, 992–999.

Kelleher, D., Murphy, A., Feighery, C. and Casey, E.B. (1995). Leukocyte function-associated antigen-1 (LFA-1) and CD44 are signaling molecules for cytoskeleton-dependent morphological changes in activated T cells. *J. Leuk. Biol.*, **58**, 539–546.

Kishimoto, T.K., Warnock, R.A., Jutila, M.A., Butcher, E.C., Lane, C., Anderson, D.C. and Smith, C.W. (1991). Antibodies against human neutrophil LECAM-1 (LAM-1/Leu-8/Dreg 56 antigen) and endothelial cell ELAM-1 inhibit a common CD18-independent adhesion pathway *in vitro*. *Blood*, **78**, 805–811.

Koch, A.E., Burrows, J.C., Haines, G.K., T.M., Harlan, J.M. and Leibovich, S.J. (1991). Immunolocalization of endothelial and leukocyte adhesion molecules in human rheumatoid and osteoarthritic synovial tissue. *Lab. Invest.*, **64**, 313–320.

Koopman, G., van Kooyk, Y., de Graaff, M., Meyer, C.J.L.M., Figdor, C.G. and Pals, S.T. (1990). Triggering of the CD44 antigen on T lymphocytes promotes T cell adhesion through the LFA-1 pathway. *J. Immunol.*, **145**, 3589–3593.

Koopman, G., Parmentier, H.K., Schuurman, H-J., Newman, W., Meijer, C.J.L.M. and Pals, S.T. (1991). Adhesion of human B cells to follicular dendritic cells involves both the lymphocyte function-associated antigen 1/intercellular adhesion molecule a and very late antigen 4/vascular cell adhesion molecule 1 pathways. *J. Exp. Med.*, **173**, 1297–1304.

Kornberg, L., Earp, H.S., Parsons, J.T., Schaller, M. and Juliano, R.L. (1992). Cell adhesion or integrin clustering increases phosphorylation of a focal adhesion-associated tyrosine kinase. *J. Biol. Chem.*, **267**, 23439–23442.

Kotovouri, P., Tontti, E., Pigott, R., Shepard, M., Kiso, M., Hasegawa, A., Renkonen, R., Nortamo, P., Altieri, D.C. and Gahmberg, C.G. (1993). The vascular E-selectin binds to the leukocyte integrins CD11/CD18. *Glycobiology*, **3**, 131–136.

Kraal, G., Schornagel, K., Streeter, P.R., Holzman, B. and Butcher, E.C. (1995). Expression of the mucosal vascular addressin, MadCAM-1, on sinus lining cells in the spleens. *Am. J. Path.*, **147**, 763–771.

Kuijpers, T.W., Hoogerwerf, M., van der Laan, L.J.W., Nagel, G., van der Schoot, C.E., Grunert, F. and Roos, D. (1992). CD66 nonspecific crossreacting antigens are involved in neutrophil adherence to cytokine-activated endothelial cells. *J. Cell Biol.*, **118**, 457–466.

Kumar, A.G., Dai, X.Y., Kozak, C.A., Mms, M.P., Gotto, A.M. and Ballantyne, C.M. (1994). Murine VCAM-1. Molecular cloning, mapping and analysis of a truncated form. *J. Immunol.*, **153**, 4088–4098.

Kunzendorf, U., Notter, M., Hock, H., Distler, A., Diamanstein, T. and Walz, G. (1993). T cells bind to the endothelial adhesion molecule GMP-140 (P-selectin). *Transplantation*, **56**, 1213–1217.

Landis, R.C., Bennett, R.I. and Hogg, N. (1993). A novel LFA-1 activation epitope maps to the I domain. *J. Cell Biol.*, **120**, 1519–1527.

Landis, R.C., McDowall, A., Holness, C.L., Littler, A.J., Simmons, D.L. and Hogg, N. (1994). Involvement of the "I" domain of LFA-1 in selective binding to ligands ICAM-1 and ICAM-3. *J. Cell Biol.*, **126**, 529–537.

Larsen, E., Celi, A., Gilbert, G.E., Furie, B.C., Erban, J.K., Bonfanti,R., Wagner, D.D. and Furie, B. (1989). PADGEM protein: A receptor that mediates the interaction of activated platelets with neutrophils and monocytes. Cell 59, 305–312.

Larsen, E., Palabrica, T., Sajer, S., Gilbert, G.E., Wagner, D.D., Furie, B.C. and Furie, B. (1990). PADGEM-dependent adhesion of platelets to monocytes and neutrophils is mediated by a lineage-specific carbohydrate, LNF III (CD15). *Cell*, **63**, 467–474.

Lasky, L.A., Singer, M.S., Yednock, T.A., Dowbenko, D., Fennie, C., Rodriguez, H., Nguyen, T., Stachel, S. and Rosen, S.D. (1989). Cloning of a lymphocyte homing receptor reveals a lectin domain. *Cell*, **56**, 1045–1055.

Lasky, L.A., Singer, M.S., Dowbenko, D., Imai, Y., Henzel, W.J., Grimley, C., Fennie, C., Gillett, N., Watson, S.R. and Rosen, S.D. (1992). An endothelial ligand for L-selectin is a novel mucin-like molecule. *Cell*, **69**, 927–938.

Lastres, P., Almendro, N., Bellon, T., Lopez-Guerrero, J.A., Eritja, R., Bernabeu, C. (1994). Functional regulation of platelet/endothelial cell adhesion molecule-1 by TGF-$\beta 1$ in promonocytic U-937 cells. *J. Immunol.*, **153**, 4206–4218.

Lawrence, M.B. and Springer, T.A. (1991). Leukocytes roll on a selectin at physiologic flow rates: Distinction from and prerequisite for adhesion through integrins. Cell 65, 859–873.

Lazarovits, A.I. and Karsh, J. (1993). Differential expression in rheumatoid synovium and synovial fluid of $\alpha 4\beta 7$ integrin: A novel receptor for fibronectin and vascular cell adhesion molecule-1. *J. Immunol.*, **151**, 6482–6489.

Lewinsohn, D.M., Bargatze, R.F. and Butcher, E.C. (1987). Leukocyte-endothelial cell recognition: Evidence of a common molecular mechanism shared by neutrophils,lymphocytes and other leukocytes. *J. Immunol.*, **138**, 4313–4321.

Lewis, J.M. and Schwartz, M.A. (1995). Mapping in vivo associations of cytoplasmic proteins with integrin beta 1 cytoplasmic domain mutants. *Molec. Biol. Cell*, **6**, 151–160.

Lipfert, L., Haimovich, B., Schaller, M.D., Cobb, B.S., Parsons, J.T. and Brugge, J.S. (1992). Integrin-dependent phosphorylation and activation of the protein tyrosine kinase pp125FAK in platelets. *J. Cell Biol.*, **119**, 905–912.

Lowe, J.B., Stoolman, L.M., Nair, R.P., Larsen, R.D., Berhend, T.L. and Marks, R.M. (1990). ELAM-1–dependent cell adhesion to vascular endothelium determined by a transfected human fucosyltransferase cDNA. *Cell*, **63**, 475–484.

Maguire, J.E., Danahey, K.M., Burkly, L.C. and van Seventer, G.A. (1995). T cell receptor- and beta 1 integrin-mediated signals synergize to induce tyrosine phosphorylation of focal adhesion kinase (pp125FAK) in human T cells. *J. Exp. Med.*, **182**, 2079–2090.

Makerem, R., Newham, P., Askari, J.A., Green, L.J., Clements, J., Edwards, M., Humphries, M.J. and Mould, A.P. (1994). Competitive binding of vascular cell adhesion molecule-1 and the HepII/IIICS domain of fibronectin to the integrin $\alpha 4\beta 4$. *J. Biol. Chem.*, **269**, 4005–4011.

Makgoba, M.W., Sanders, M.E. and Shaw, S. (1989). The CD2–LFA-3 and LFA-1–ICAM pathways: Relevance to T-cell recognition. *Immunol. Today*, **10**, 417–422.

Marcantonio, E.E. and Hynes, R.O. (1988). Antibodies to the conserved cytoplasmic domain of the integrin $\beta 1$ subunit react with proteins in vertebrates, invertebrates and fungi. *J. Cell Bio.*, **106**, 1765–1772.

Marlin, S.D. and Springer, T.A. (1987). Purified intercellular adhesion molecule-1 (ICAM-1) is a ligand for lymphocyte function-associated antigen-1 (LFA-1). *Cell*, **51**, 813–819.

Marlin, S.D., Staunton, D.E., Springer, T.A., Stratowa, C., Sommergruber, W. and Merluzzi, V.J. (1990). A soluble form of intercellular adhesion molecule-1 inhibits rhinovirus infection. *Nature*, **344**, 70–72.

May, M.J., Entwistle, G., Humphries, M.J. and Ager, A. (1991). VCAM-1 is a CS1 peptide-inhibitable adhesion molecule expressed by lymph node high endothelium. *J. Cell Sci.*, **106**, 109–119.

McEver, R.P., Beckstead, J.H., Moore, K.L., Marshall-Carlson, L. and Bainton, D.F. (1989). GMP-140, a platelet α-granule membrane protein, is also synthesized by vascular endothelial cells and is localized in weibel-palade bodies. *J. Clin. Invest.*, **84**, 92–99.

McNamee, H.P., Ingber, D.E. and Schwartz, M.A. (1993). Adhesion to fibronectin stimulates inositol lipid synthesis and enhances PDGF-induced inositol lipid breakdown. *J. Cell Biol.*, **121**, 673–678.

Meeusen, E., Lee, C-S. and Brandon, M. (1991). Differential migration of T and B cells during an acute inflammatory response. *Eur. J. Immunol.*, **21**, 2269–2272.

Meri, S., Mattila, P. and Renkonen, R. (1993). Regulation of CD59 expression on the human endothelial cell line E.A.hy926. *Eur. J. Immunol.*, **23**, 2511.

Michie, S.A., Streeter, P.R., Bolt, P.A., Butcher, E.C. and Picker, L.J. (1993). The human peripheral lymph node vascular addressin. An inducible endothelial antigen involved in lymphocyte homing. *Am. J. Pathol.*, **143**, 1688–1698.

Miller, 3d J.J. (1969). Studies of the phylogeny and ontogeny of the specialized lymphatic tissue venules. *Lab. Invest.*, **21**, 484–490.

Morales-Ducret, C.R.J., Wayner, E., Elices, M.J., Alvaro-Garcia, J.M., Zvaifler, N.J. and Firestein, G.S. (1992). Alpha 4/beta1 integrin (VLA-4) ligands in arthritis. Vascula cell adhesion molecule-1 expression in synovium and on fibroblast-like synoviocytes. *J. Immunol.*, **21**, 1424–1431.

Muller, W.A., Ratti, C.M., McDonnell, S.L. and Cohn, Z.A. (1989). A human endothelial cell-restricted extrernally disposed plasmalemmal protein enriched in intracellular junctions. *J. Exp. Med.*, **170**, 399–414.

Muller, W.A., Weigel, S.A., Deng, X. and Phillip, D.M. (1993). PECAM-1 is required for transendothelial migration of leukocytes. *J. Exp. Med.*, **178**, 449–460.

Muller, W.A. (1995). The role of PECAM-1 (CD31) in leukocyte emigration: studies *in vitro* and *in vivo*. *J. Leuk. Biol.*, **57**, 523–528.

Munro, J.M., Pober, J.S. and Cotran, R.S. (1989). Tumor necrosis factor and interferon-γ induce distinct patterns of endothelial activation and associated leukocyte accumulation in skin of *Papio Anubis*. *Am. J. Path.*, **135**, 121–133.

Newman, P.J., Berndt, M.C., Gorsky, J., White, G.C., Paddock, L.S. and Muller, W.A. (1990). PECAM-1 (CD31): Cloning and relation to adhesion molecules of the immunoglobulin gene superfamily. *Science*, **247**, 1219–1222.

Nobles, C.D. and Hall A. (1995). Rho, rac and cdc42 GTPases regulate the assembly of multimolecular focal complexes associated with actin stress fibers lamellipodia and filopodia. *Cell*, **81**, 53–62.

Nortamo, P., Salcedo, R., Timonen, T., Patarroyo, M., Gahmberg, C.G. (1991a). A monoclonal antibody to the human leukocyte adhesion molecule inetrcellular adhesion molecule-2. Cellular distribution and molecular characterization of the antigen. *J. Immunol.*, **146**, 2530–2535.

Nortamo, P., Li, R., Renkonen, R., Timenon, T., Prieto, J., Patarroyo, M. and Gahmberg, C.G. (1991b). The expression of human intercellular adhesion molecule-2 is refractory to inflammatory cytokines. *Eur. j. Immunol.*, **21**, 2629–2632.

Odum, N., Yoshizumi, H., Okamoto, Y., Kamikawaji, N., Kimura, A., Nishimura, Y. and Sasazuki, T. (1992). Signal transduction by HLA class II molecules in human T cells: Induction of LFA-1 dependent and independent adhesion. *Human Immunol.*, **35**, 71–84.

Oppenheimer-Marks, N. and Ziff, M. (1986). Binding of normal human mononuclear cells to blood vessels in rheumatoid arthritis synovial membrane. *Arthritis Rheum.*, **29**, 789–792.

Oppenheimer-Marks, N., Davis, L.S. and Lipsky, P.E. (1990). Human T lymphocyte adhesion to endothelial cells and transendothelial migration. Alteration of receptor use relates to the activation status of both the T cell and the endothelial cell. *J. Immunol.*, **145**, 140–148.

Oppenheimer-Marks, N., Davis, L.S., Bogue, D.T., Ramberg, J. and Lipsky, P.E. (1991). Differential utilization of ICAM-1 and VCAM-1 during the adhesion and transendothelial migration of human T lymphocytes. *J. Immunol.*, **147**, 2913–2921.

Oppenheimer-Marks, N. and Lipsky, P.E. (1994). Transendothelial migration of T cells in chronic inflammation. *The Immunologist*, **2**, 58–64.

Osborn, L., Hession, C., Tizard, R., Vassallo, C., Luhowskyj, S., Chi-Rosso, G. and Lobb, R. (1989). Direct expression cloning of vascular cell adhesion molecule 1, a cytokine-induced endothelial protein that binds to lymphocytes. *Cell*, **59**, 1203–1211.

Osborn, L., Vassallo, C. and Benjamin, C.D. (1992). Activated endothelium binds lymphocytes through a novel binding site in the alternatively spliced domain of vascular cell adhesion molecule-1. *J. Exp. Med.*, **176**, 99–107.

Pals, S.T., Den Otter, A., Miedema, F., Kabel, P., Keizer, G.D., Scheper, R.J. and Meijer, C.J.L.M. (1988). Evidence that leukocyte function-associated antigen-1 is involved in recirculation and homing of human lymphocytes via high endothelial venules. *J. Immunol.*, **137**, 1851–1853.

Pardi, R., Inverardi, L., Rugarli, C. and Bender, J.R. (1992). Antigen-receptor complex stimulation triggers protein kinase C-dependent CD11a/CD18–cytoskeleton association in T lymphocytes. *J. Cell Biol.*, **116**, 1211–1220.

Pardi, R., Bassi, G., Inverardi, L., Rovida, E. and Bender, J.R. (1995). Conserved regions in the cytoplasmic domains of the leukocye integrin $\alpha L\beta 2$ are involved in endoplasmic reticulin retention, dimerization and cytoskeletal association. *J. Immunol.*, **155**, 1252–1263.

Parrott, D.V., De Sousa, M.A. and East, J. (1966). Thymus-dependent areas in the lymphoid organs of neonatally thymectomized mice. *J. Exp. Med.*, **123**, 191–204.

Parsey, M.V. and Lewis, G.K. (1993). Actin polymerization and pseudopod reorganization accompany anti-CD3-induced growth arrest of Jurkat T cells. *J. Immunol.*, **151**, 1881–1893.

Pavalko, F.M., Walker, D.M., Graham, L., Goheen, M., Doerschuk C.M. and Kansas, G.S. (1995). The cytoplasmic domain of L-selectin interacts with cytoskeletal proteins via α-actinin: Receptor positioning in microvilli does not require interaction with α-actinin. *J. Cell Biol.*, **129**, 1155–1164.

Pelletier, A.J., Bodary, S.C. and Levinson, A.D. (1992). Signal transduction by the platelet integrin alpha Iib beta 3: Induction of calcium oscillations required for protein tyrosine phosphorylation and ligand induced spreading of stably transformed cells. *Mol. Biol. Cell*, **3**, 989–998.

Petzelbauer, P., Bender, J.R., Wilson, J. and Pober, J.S. (1993). Heterogeneity of dermal microvascular endothelial cell antigen expression and cytokine responsiveness in situ and in cell culture. *J. Immunol.*, **151**, 5062–5072.

Phillips, M.L., Nudelman, E., Gaeta, F.C.A., Perez, M., Singhal, A.K., Hakomori, S-I. and Paulson J.C. (1990). ELAM-1 mediates cell adhesion by recognition of a carbohydrate ligand, sialyl-LeX. *Science*, **250**, 1130–1132.

Piali, L., Hammel, P., Uherek, C., Bachmann, F., Gisler, R.H., Dunon, D. and Imhof, B.A. (1995). CD31/PECAM-1 is a ligand for $\alpha v\beta 3$ integrin involved in adhesion of leukocytes to endothelium. *J. Cell Biol.*, **130**, 451–460.

Picker, L.J., Kishimoto, T.K., Smith, W.C., Warnock, R.A. and Butcher, E.C. (1991a). ELAM-1 is an adhesion molecule for skin homing T cells. *Nature*, **349**, 796–799.

Picker, L.J., Warnock, R.A., Burns, A.R., Doerschuk, C.M., Berg, E.L. and Butcher, E.C. (1991b). The neutrophil selectin LECAM-1 presents carbohydrate ligands to the vascular selectins ELAM-1 and GMP-140. *Cell*, **66**, 921–933.

Picker, L.J. and Siegelman, M.H. (1993) Lymphoid tissues and organs. In Paul W.E. (Ed). *Fundamental Immunology*3rd edn.New York: Raven Press. pp. 145–197.

Pigott, R., Needham, L.A., Edwards, R.M., Walker, C. and Power, C. (1991). Structural and functional studies of the endothelial activation antigen endothelial leukocyte adhesion molecule-1 using a panel of monoclonal antibodies. *J. Immunol.*, **147**, 130–135.

Pober, J.S., Bevilacqua, M.P., Mendrick, D.L., Lapierre, L.A., Fiers, W. and Gimbrone, Jr. M.A. (1986). Two distinct monokines, interleukin 1 and tumor necrosis factor, each independently induce biosynthesis and transient expression of the same antigen on the surface of cultured human vascular endothelial cells. *J. Immunol.*, **136**, 1680–1687.

Pober, J.S. and Cotran, R.S. (1990). The role of endothelial cells in inflammation. *Transplantation*, **50**, 537–544.

Polley, M.J., Phillips, M.L., Wayner, E., Nudelman, E., Singhal, A.K., Hakomori, S-I. and Paulson J.C. (1991). CD62 and endothelial cell-leukocyte adhesion molecule 1 (ELAM-1) recognize the same carbohydrate ligand sialyl-Lewis x. *Proc. Natl. Acad. Sci.*, **88**, 6224–6228.

Postigo, A.A. and Sanchez-Madrid, F. (1993). Adhesion and homing molecules. *Transplant. Proc.*, **25**, 65–69.

Poudrier, J. and Owens, T. (1994). CD54/intercellular adhesion molecules 1 and major histocompatability complex II signaling induces B cells toexpress interleukin 2 receptors and complements help provided through CD40 ligation. *J. Exp. Med.*, **179**, 1417–1427.

Prieschl, E.E., Kulmburg, P.A. and Baumruker, T. (1995). The nomenclature of chemokines. *Int. Arch. Allergy Immunol.*, **107**, 475–483.

Randi, A.M. and Hogg, N. (1994) I domain of B_2 integrin lymphocyte function- associated antigen-1 contains a binding site for ligand intercellular adhesion molecule-1. *J. Biol. Chem.*, **269**, 12395–12398.

Renkonen, R. (1989). Regulation of intercellular adhesion molecule-1 expression on endothelial cells with correlation to lymphocyte-endothelial binding. *Scand. J. Immunol.*, **29**, 717–721.

Reszka, A.A., Hayashi, Y. and Horwitz, A.F. (1992). Identification of amino acid seuqneces in the inetgrin $\beta 1$ cytoplasmic domian implicated in cytoskeletal association. *J. Cell Biol.*, **117**, 1321–1330.

Rice, G.E., Munro, J.M. and Bevilacqua, M.P. (1990). Inducible cell adhesion molecule 110 (INCAM-110) is an endothelial receptor for lymphocytes. A CD11/CD18–independent adhesion mechanism. *J. Exp. Med.*, **171**, 1369–1374.

Robbins, S.L. and Cotran, R.S. (1979). *Pathological Basis of Disease 2nd edition.*W.B. Saunders and Co., Philadelphia.

Ropke, C. (1973). Affects of antithymocyte serum and antilymphocyte serum on the postnatal development of the thymolymphatic system in BALB-C mice. *J. Reticuloendo. Soc.*, **13**, 78–89.

Rothlein, R., Kishimoto, T.K. and Mainolfi, E. (1994). Cross-linking of ICAM-1 induces co-signaling of an oxidative burst from mononuclear leukocytes *J. Immunol.*, **152**, 2488–2495.

Ruegg, C., Postigo, A.A., Sikorski, E.E., Butcher, E.C., Pytela, R. and Erle, D.J. (1992). *J. Cell Biol.*, **117**, 179–189.

Ryan, D.H., Nuccie, B.L., Abboud, C.N. and Winslow, J.M. (1991). Vascular cell adhesion molecule-1 and the integrin VLA-4 mediate adhesion of human B cell precursors to cultured bone marrow adherent cells. *J. Clin. Invest.*, **88**, 995–1004.

Sako, D., Chang, X.J., Barone, K.M., Vachino, G., White, H.M., Shaw, G., Veldman, G.M., Bean, K.M., Ahern, T.J., Furie, B., Cumming, D.A. and Larsen, G.R. (1993). Expression cloning of a functional glycoprotein ligand for P-selectin. *Cell*, **75**, 1179–1186.

Salmi, M., Kalimo, K. and Jalkanen, S. (1993). Induction and function of vascular adhesion protein-1 at sites of inflammation. *J. Exp. Med.*, **178**, 2255–2260.

Salmi, M., Andrew, D.P., Butcher, E.C. and Jalkanen S. (1995). Dual binding capacity of mucosal immunoblasts to mucosal and synovial endothelium in humans: Dissection of the molecular mechanisms. *J. Exp. Med.*, **181**, 137–149.

Sanchez-Mateos, P., Campanero, M.R., Balboa, M.A. and Sanchez-Madrid, F. (1993). Co-clustering of $\beta 1$ integrins, cytoskeletal proteins and tyrosine- phosphorylated substrates during integrin-mediated leukocyte aggregation. *J. Immunol.*, **151**, 3817–3828.

Sastry, S.K. and Horwitz, A.F. (1993). Integrin cytoplasmic domains: mediators of cytoskeletal linkages and extra- and intracellular initiated transmembrane signaling. *Curr. Opin. Cell Biol.*, **5**, 819–831.

Sato, T, Tachibana, K., Nojima, Y., D'Avirro, N. and Morimoto, C. (1995). Role of the VLA-4 molecule in T cell costimulation. Identification of the tyrosine phosphorylation pattern induced by the ligation of VLA-4. *J. Immunol.*, **155**, 2938–2947.

Schaller, M.D., Otey, C.A., Hildebrand, J.D. and Parsons, J.T. (1995). Focal adhesion kinase and paxillin bind to peptides mimicking β integrin cytoplasmic domains. *J. Cell Biol.*, **130**, 1181–1187.

Schiffer, S.G., Day, E., Latanision, S.M., Tizard, R. and Osborn, L. (1995). An alternatively spliced mRNA encoding functional domains of murine MadCAM-1. *Biochem. Biophys. Res. Commun.*, **216**, 170–176.

Schlaepfer, D.D., Hanks, S.K., Hunter, T. and van der Geer, P. (1994). Integrin mediated signal transductiotn linked to Ras pathway by GRB2 binding to focal adhesion kinase. *Nature*, **372**, 786–789.

Schwartz, B.R., Wayner, E.A., Carlos, T.M., Ochs, H.D. and Harlan, J.M. (1990). Identification of surface proteins mediating adherence to CD11/CD18–deficient lymphoblastoid cells to cultured human endothelium. *J. Clin. Invest.*, **85**, 2019–2022.

Schwartz, M.A. (1993). Spreading of human endothelial cells on fibronectin or vitronectin triggers elevation of intracellular free calcium. *J. Cell Biol.*, **120**, 1003–1010.

Schwartz, M.A., Lechene, C. and Ingber, D.E. (1991). Insoluble fibronectin activates the Na+/H+ antiporter by clustering and immobilizing integrin alpha 5 beta 1, independent of cell shape. *Proc. Natl. Acad. Sci.*, **88**, 7849–7853.

Shimizu, Y., van Seventer, G.A., Horgan, K.J. and Shaw S. (1990a). Costimulation of proliferative responses of resting CD4+ T cells by the interaction of VLA-4 and VLA-5 with fibronectin or VLA-6 with laminin. *J. Immunol.*, **145**, 59–67.

Shimizu, Y., van Seventer, G.A., Horgan, K.J. and Shaw, S. (1990b). Regulated expression and binding of three VLA ($\beta 1$) integrin receptors on T cells. *Nature*, **345**, 250–253.

Shimizu, Y., Shaw, S., Graber, N., Gopal, T.V., Horgan, K.J., van Seventer, G.A. and Newman, W. (1991). Activation-independent binding of human memory T cells to adhesion molecule ELAM-1. *Nature*, **349**, 799–802.

Shimizu, Y., van Seventer, G.A., Ennis, E., Newman, W., Horgan, K.J. and Shaw, S. (1992). Crosslinking of the T cell-specific accessory molecules CD7 and CD28 modulates T cell adhesion. *J. Exp. Med.*, **175**, 577–582.

Shimizu, Y. and Mobley, J.L. (1993). Distinct divalent cation requirements for integrin mediated CD4+ T lymphocyte adhesion to ICAM-1, fibronectin, VCAM-1 and invasin. *J. Immunol.*, **151**, 4106–4115.

Sikorski, E.E., Hallmann, R., Berg, E.L. and Butcher, E.C. (1993). The Peyer's patch high endothelial receptor for lymphocytes, the mucosal vascular addressin, is induced on a murine endothelial cell line by tumor necrosis factor-α and IL-1. *J. Immunol.*, **151**, 5239–5250.

Spertini, O., Luscinskas, F.W., Kansas, G.S., Munro, J.M., Griffin, J.D., Gimbrone, Jr. M.A. and Tedder, T.F. (1991). Leukocyte adhesion molecule-1 (LAM-1, L-selectin) interacts with an inducible endothelial cell ligand to support leukocyte adhesion. *J. Immunol.*, **147**, 2565–2573.

Spertini, F., Chatila, T. and Geha, R.S. (1992). Engagement of MHC class I molecules induces cell adhesion via both LFA-1 dependent and LFA-1 independent pathways. *J. Immunol.*, **148**, 2045–2049.

Springer, T.A. (1985). The LFA-1, Mac-1 glycoprotein family and its deficiency in an inherited disease. *Fed. Proc.*, **44**, 2660–2663.

Springer, T.A. (1990). Adhesion receptors of the immune system. *Nature*, **346**, 425–434.

Stamper, H.B. and Woodruff, J.J. (1976). Lymphocyte homing into lymph nodes: In vitro demonstration of the selective affinity of recirculating lymphocytes for high-endothelial venules. *J. Exp. Med.*, **144**, 828–833.

Staunton, D.E., Marlin, S.D., Stratowa, C., Dustin, M.L. and Springer, T.A. (1988). Primary structure of ICAM-1 demonstrates interaction between members of the immunoglobulin and integrin supergene families. *Cell*, **52**, 925–933.

Staunton, D.E., Dustin, M.L. and Springer, T.A. (1989). Functional cloning of ICAM-2, a cell adhesion ligand for LFA-1 homologous to ICAM-1. *Nature*, **339**, 61–64.

Steegmaier, M., Levinovitz, A., Isenmann, S., Borges, E., Lenter, M., Kocher, H.P., Kleuser, B. and Vestweber, D. (1995). The E-selectin-ligand ESL-1 is a variant of a receptor for fibroblast growth factor. *Nature*, **373**, 615–620.

Streeter, P.R., Rouse, B.T.N. and Butcher, E.C. (1988) Immunohistologic and functional characteriization of a vascular addressin involved in lymphocyte homing into peripheral lymph nodes. *J. Cell Biol.*, **107**, 1853–1862.

Swerlick, R.A., Lee, K.H., Li, L.J., Sepp, N.T., Caughman, S.W. and Lawley, T.J. (1992). Regulation of vascular cell adhesion molecule 1 on human dermal microvascular endothelial cells. *J. Immunol.*, **149**, 698–705.

Syrjanen, K.J. (1978). Post-capillary venules of the lymphatic tissues in mice with special reference to the depletion of T-lymphocyte population. *Exp. Molec. Path.*, **29**, 291–302.

Szekanecz, Z., Humphries, M.J. and Ager, A. (1992). Lymphocyte adhesion to high endothelium is mediated by two $\beta 1$ integrin receptors for fibronectin, $\alpha 4\beta 1$ and $\alpha 5\beta 1$. *J. Cell Sci.*, **101**, 885–894.

Takaishi, K., Sasaki, T., Kameyama, T., Tsukita, S., Tsukita, S. and Takai, Y. (1995). Translocation of activated Rho from the cytoplasm to membrane ruffling area, cell-cell adhesion sites and cleavage furrows. *Oncogene*, **11**, 39–48.

Tanaka, Y., Albelda, S.M., Horgan, K.J., van Seventer, G.A., Shimizu, Y., Newman, W., Hallam, J., Newman, P.J., Buck, C.A. and Shaw, S. (1992). CD31 expressed on distinctive T cell subsets is a preferential amplifier of $\beta 1$ integrin-mediated adhesion. *J. Exp. Med.*, **176**, 245–253.

Tanaka, Y., Adams, D.H., Hubscher, S., Hirano, H., Siebenlist, U. and Shaw, S. (1993). T cell adhesion induced by proteoglycan-immobilized cytokine MIP-1β. *Nature*, **361**, 79–82.

Tapley, P., Horwitz, A., Buck, C., Duggan, K. and Rohrschneider, L. (1989). Integrins isolated from Rous sarcoma virus-transformed chicken embryo fibroblasts. *Oncogene*, **4**, 325–333.

Tedder, T.F., Steeber, D.A. and Pizcueta, P. (1995). L-selectin-deficient mice have impaired leukocyte recruitment into inflammatory sites. *J. Exp. Med.*, **181**, 2259–2264.

Tominaga, T., Sugie, K., Hirata, M., Morii, N., Fukata, J., Uchida, A., Imura, H. and Narumiya, S. (1993). Inhibition of PMA-induced, LFA-1–dependent lymphocyte aggregation by ADP ribosylation of the small molecular weight GTP binding protein. *J. Cell Biol.*, **120**, 1529–1537.

Torti, M., Sinigaglia, F., Ramaschi, G. and Balduini, C. (1991). Platelet glycoprotein Iib-IIIa is associated with 21kDa GTP binding protein. *Biochim. Biophys. Acta*, **1070**, 20–26.

Van der Vieren, M., Trong, H.L., Wood, C.L., Moore, P.F., St. John, T., Staunton, D.E. and Gallitin, W.M. (1995). A novel leukointegrin, $\alpha d\beta 2$, binds preferentially to ICAM-3. *Immunity*, **3**, 683–690.

van Horssen, N., Loman, S., Rijkers, G.T., Boom, S.E. and Bloem, A.C. (1995). Co-ligation of ICAM-1 (CD54) and membrane IgM negatively affects B cell receptor signaling. *Eur. J. Immunol.*, **25**, 154–158.

van Kooyk, Y., van de Wiel-van Kemenade, P., Weder, P., Kuijpers and T.W., Figdor, C.G. (1989). Enhancement of LFA-1 mediated cell adhesion by triggering through CD2 or CD3 on T lymphocytes. *Nature*, **342**, 811–813.

van Kooyk, Y., Weder, P., Hogervorst, F., Verhoeven, A.J., van Seventer, G., te Velde, A.A., Borst, J., Keizer, G.D. and Figdor, C.G. (1991). Activation of LFA-1 through a Ca2+-dependent epitope stimulates lymphocyte adhesion. *J. Cell Biol.*, **112**, 345–354.

van Seventer, G.A., Shimizu, Y., Horgan, K.J. and Shaw, S. (1990). The LFA-1 ligand ICAM-1 provides an important costimulatory signal for T cell receptor-mediated activation of resting T cells. *J. Immunol.*, **144**, 4579–4586.

van Seventer, G.A, Newman, W., Shimizu, Y., Nutman, T.B., Tanaka, Y., Horgan, K.J., Gopal, T.V., Ennis, E., O'Sullivan, D., Grey, H. and Shaw, S. (1991). Analysis of T cell stimulation by superantigen plus major histocompatability complex class II molecules or by CD3 monoclonal antibody: Costimulation by purified adhesion ligands VCAM-1, ICAM-1, but not ELAM-1. *J. Exp. Med.*, **174**, 901–913.

van Seventer, G.A., Bonvini, E., Yamada, H., Conti, A., Stringfellow, S., June, C.H. and Shaw, S. (1992). Costimulation of T cell receptor/CD3–mediated activation of resting human CD4+ T cells by leukocyte function-associated antigen-1 ligand intercellular adhesion molecule-1 involves prolonged inositol phospholipid hydrolysis and sustained increase of intracellular Ca2+ levels. *J. Immunol.*, **149**, 3872–3880.

Vazeux, R., Hoffman, P.A., Tomita, J.K., Dickinson, E.S., Jasman, R.L., St. John, T. and Gallatin, W.M. (1992). Cloning and characterization of a new intercellular adhesion molecule ICAM-R. *Nature*, **360**, 485–488.

Vennegoor, C.J.G.M., van de Wiel-van Kemenade, E., Huijbens, R.J.F., Sanchez-Madrid, F., Melief, C.J.M. and Figdor, C.G. (1992). Role of LFA-1 and VLA-4 in the adhesion of cloned normal and LFA-1 (CD11/CD18)-deficient T cells to cultured endothelial cells. Indication of a new adhesion pathway. *J. Immunol.*, **148**, 1093–1101.

Voraberger, G., Schafer, R. and Stratowa, C. (1991). Cloning of the human gene for intercellular adhesion molecule 1 and analysis of its 5'-regulatory region. Induction by cytokines and phorbol ester. *J. Immunol.*, **147**, 2777–2786.

Vuori, K. and Ruoslahti, E. (1993). Activation of protein kinase C precedes $\alpha 5\beta 1$ integrin mediated cell spreading on fibronectin. *J. Biol. Chem.*, **268**, 21459–21462.

Wacholtz, M.C., Patel, S.S. and Lipsky, P.E., (1989). Leukocyte function-associated antigen-1 is an activation molecule for human T cells. *J. Exp. Med.*, **170**, 431–448.

Walz, G., Aruffo, A., Kolanus, W., Bevilacqua, M. and Seed, B. (1990). Recognition by ELAM-1 of the sialyl-LeX determinant on myeloid and tumor cells. *Science*, **250**, 1132–1135.

Watson, S.R., Imai, Y., Fennie, C., Geoffrey, J., Singer, M., Rosen, S.D. and Lasky, L.A. (1991). The complement binding-like domains of the murine homing receptor facilitate lectin activity. *J. Cell Biol.*, **104**, 725–731.

Watt, S.M., Williamson, J., Genevier, H., Fawcett, J., Simmons, D.L., Hatzfeld, A., Nesbitt, S. and Coombe D.R. (19??). The heparin binding PECAM-1 adhesion molecule is expressed by CD34+ hematopoietic precursor cells with early myeloid and B-lymphoid cell phenotypes. *Blood*, **82**, 2649–2663.

Whitney, G.S., Chan, P.Y., Blake, J., Cosand, W.L., Neubauer, M.G., Aruffo, A. and Kanner, S.B. (1993). Human T and B Lymphocytes express a structurally conserved focal adhesion kinase, pp125FAK. *DNA Cell Biol.*, **12**, 823–830.

Zehnder, J.L., Hirai, K., Shatsky, M., McGregor, J.L., Levitt, L.J., Leung, L.L. (1992). The cell adhesion molecule CD31 is phosphorylated after cell activation. Down regulation of CD31 in activated T lymphocytes. *J. Biol. Chem.*, **267**, 5243–5249.

Zhou, Q., Moore, K.L., Smith, D.F., Varki, A., McEver, R.P. and Cummings, R.D. (1991). The selectin GMP-140 binds to sialylated fucosylated lactosaminoglycans on both myeloid and nonmyeloid cells. *J. Cell Biol.*, **115**, 557–564.

Ziff, M. and Iguchi, T. (1987). Electron microscopic study of rheumatoid synovial vasculature: Intimate relationship between tall endothelium and lymphoid aggregation. *Int. J. Tissue Reactions*, **9**, 273–276.

5. CHEMOKINES AS REGULATORS OF LEUKOCYTE ADHESION AND MIGRATION

DENNIS D. TAUB[1,*] and WILLIAM J. MURPHY[2]

[1] *Clinical Services and* [2] *Intramural Research and Support Programs, SAIC-Frederick, National Cancer Institute-Frederick Cancer Research and Development Center, Frederick, MD 21702–1201, USA*

INTRODUCTION

A key feature in a number of inflammatory conditions, such as infection, hypersensitivity reactions, or autoimmune diseases, is the presence of infiltrating leukocytes. The extravasating leukocytes are critical for host defense, leading to clearance of the inciting factors such as infectious agents. However, it should be appreciated that leukocyte recruitment may also contribute to the pathogenesis of an underlying disease. The maintenance of leukocyte recruitment during inflammation requires communication between infiltrating leukocytes and the endothelium as well as stromal and parenchyma cells (Strieter, 1993; Strieter, 1996a; Ward, 1974). These signals are mediated via the generation of early response cytokines [such as IL-1 and TNFα], the expression of surface adhesion molecules, and the production of chemotactic molecules. In addition to leukocyte trafficking into diseased or damaged tissues, leukocytes also emigrate into normal lymphoid and non-lymphoid organs to fulfill their function in immunosurveillance as well as exiting the circulation at the end of their life span. Thus, a better understanding of the complexities of leukocyte extravasation will greatly assist our ability to orchestrate, regulate, and control various pathological disease states as well as understand normal leukocyte trafficking.

Leukocyte-endothelial cell transmigration is believed to be a multistep process involving four sequential steps: *rolling, triggering, adhesion*, and *migration*. During the last decade, several pairs of ligand-counterligand molecules have been described on the surface of leukocytes and endothelial cells which play a role in cell adhesion (Butcher, 1991; Butcher, 1996; Carlos, 1994; Haynes, 1989; Hynes, 1992; Shimizu. 1992; Springer, 1990; Springer, 1994; Springer, 1995; Tanaka, 1993c). These adhesion molecules, which belong to different structural classes, provide the basis for leukocyte-endothelial cell adhesion as well as leukocyte transmigration across the endothelial barrier and into the target tissue. Under conditions of lateral blood flow, different types of adhesion molecules have been implicated as participants in the different types of leukocyte adhesion and transmigration. This differential use of adhesion molecules has subsequently led to the development of a multistep adhesion paradigm (Butcher, 1991; Butcher, 1996; Carlos, 1994; Haynes, 1989; Hynes, 1992; Shimizu. 1992; Springer, 1990; Springer, 1994; Springer, 1995; Tanaka, 1993c). This concept was originally developed from the studies of neutrophil-endothelial cell adhesion, but it has become clear that it can be extended, with reservations, to interactions of endothelial cells

*Correspondence: Dennis D. Taub, Ph.D., National Institute on Aging, NIH, Gerontology Research Center, 4940 Eastern Avenue, Baltimore, MD., 21224-2780. Tel: (410) 558-8110.

with other leukocytes, including the process of lymphocyte homing into lymphoid organs and normal tissues (Butcher, 1996; Springer, 1995). According to this model, the initial interaction of leukocytes with the endothelium is mediated by "selectins" and their sugar counterligands on the surface of both leukocytes and endothelial cells. This reversible interaction is quite weak and, without additional adhesion steps, cannot lead to leukocyte emigration. Due to the shear stress of blood flow, selectin-mediated adhesion results in leukocyte "rolling" or "tethering" on the endothelial cell surface (Butcher, 1996; Carlos, 1994; Haynes, 1989; Lawrence, 1991; Lo, 1989; Lorant, 1985; Springer, 1994; Springer, 1995). This primary adhesion step slows leukocyte transit within the circulation, thus permitting leukocytes to sample the vascular barrier for proadhesive factors (Abbassi, 1993). During the second adhesion step, leukocyte integrins interact with immunoglobulin-like endothelial cell adhesion molecules which stop leukocyte rolling and promote firm leukocyte adhesion and spreading on the endothelium. This second step is induced by the activation of integrins constitutively expressed on the leukocyte cell surface (Butcher, 1996; Springer, 1995). Many integrin molecules are conformationally inactive and require a proadhesive signal(s) to be "activated" to bind their respective ligands (Butcher, 1996; Springer, 1995). This enhanced adhesiveness appears to be based on a qualitative alteration in the integrin heterodimers that are constitutively expressed on the plasma membrane (Butcher, 1996; Springer, 1994; Springer, 1995; Tanaka, 1993c). This rapid alteration in integrin adhesion has been attributed to an increase in integrin avidity rather than a quantitative increase in integrin heterodimer expression on the leukocyte surface (Vedder, 1988). Various soluble cytokines and chemotactic agents as well as cell-bound factors have been shown to promote leukocyte-endothelial cell adhesion by rapidly altering the adhesive potential of integrin molecules (Tanaka, 1993c; Rot, 1996; Springer, 1994; Springer, 1995). Once activated, integrins bind to their specific endothelial cell surface counterligands, which belong to the immunoglobulin superfamily of adhesion molecules. However, under conditions of lateral blood flow, activated integrins cannot initiate leukocyte binding to the endothelium due to excessive shear forces (Butcher, 1996; Springer, 1995). Besides increasing integrin avidity, leukocyte activation also leads to shape change, an increase in membrane rigidity, and shedding of L-selectin (Arfors, 1987; Butcher, 1996; Springer, 1994; Springer, 1995). All of these factors probably contribute to the inability of activated leukocytes to "roll" on endothelium.

While the adhesive interactions between leukocytes (via L-selectin, $\beta 1$ and $\beta 2$ integrins) and endothelial cells (via P- and E-selectin, VCAM-1, and ICAM-1) are prerequisite events for successful leukocyte extravasation into lymphoid tissue or sites of inflammation, the subsequent steps that lead to diapedesis and migration beyond the vascular compartment are dependent upon the continued expression of $\beta 1$ and $\beta 2$ integrins and the directional movement along a leukocyte-specific chemotactic gradient (Butcher, 1991; Butcher, 1996; Hakkert, 1991; Smith, 1989; Springer, 1994; Springer, 1995). However, these initial adhesion steps can only occur in a certain sequence; if leukocyte integrins are activated in the circulation prior to their initial selectin-mediated interaction with the endothelium, the leukocytes lose their ability to adhere and migrate into the tissue (Butcher, 1996; Osborne, 1990; Rot, 1996; Springer, 1995; Tanaka, 1993c). It has been previously proposed, by analogy with *in vitro* leukocyte migration assays, that leukocyte emigration is driven by concentration gradients of chemoattractants. Recent studies have dramatically increased our knowledge concerning the identification of chemotactic cytokines that mediate leukocyte:endothelial transmigration. Information gained from studying various cytokines and other immunomodulators have generated exciting new concepts regarding lymphocyte recirculation and the genesis of inflammatory responses.

CHEMOATTRACTANTS AND LEUKOCYTE ADHESION

Leukocytes are believed to be directed to sites of inflammation under the influence of chemotactic molecules through a process termed *chemotaxis.* Chemotactic molecules induce the directed migration of cells, which can be observed *in vitro* as the movement of cells in response to increasing concentration gradients of the soluble chemoattractant. Often confused with chemotaxis is the complementary process by which cells move along gradients of substrate bound chemoattractant molecules termed *haptotaxis.* Haptotaxis is believed to be the predominant process by which leukocytes and other cell types migrate into normal as well as compromised tissues. This mechanism allows the exposure to chemoattractants of only those leukocytes that have already established their first adhesive interaction with the endothelium. Chemotaxis and haptotaxis should also be distinguished from *chemokinesis* (the increased random movement of cells) as many inflammatory mediators such as histamine, IL-1, and IFN-γ are predominantly chemokinetic in their ability to facilitate cell movement (Oppeneheim, 1991; Oppenheim, 1996; Rot, 1992a; Rot, 1993; Schall, 1991; Schall, 1994; Taub, 1993a; Taub, 1994b). Chemotactic factors are generated within inflammatory sites and diffuse toward local capillary beds, inducing leukocytes to traverse the endothelium and enter the compromised tissues. Many chemoattractants have been shown to bind to various substates within and beyond the tissue lesion and thus facilitate the haptotactic migration of leukocytes (Rot, 1992a; Rot, 1993; Rot, 1996). The effects of many chemotactic mediators are not solely confined to cell motility, as many of these factors have also been shown to activate leukocyte adhesion, granule release, enzyme release, and cell-mediated cytotoxicity as well as enhanced microbicidal properties when present at higher concentrations. The distinct and differential roles of these various chemotactic mediators in the genesis of inflammatory and immune responses have been extensively investigated.

The leukocyte receptors and endothelial counter-receptors involved in the initial and final steps of this adhesion cascade have been thoroughly described (Butcher, 1996; Hynes, 1992; Springer, 1994; Springer, 1995; Tanaka, 1993c). However, the physiologically relevant molecules involved in the adhesion triggering step remains a controversial issue. Early studies of chemoattractants have identified a number of single defined proteins and protein mixtures as inducers of leukocyte migration and adhesion, including activated serum components, platelet activating factor (e.g. PAF), casein, lysophosphatidyl choline, lactoferrin, anti-IgG, endotoxin, human plasma eicosinoids (e.g.. PGE_2, LTB_4), bacterial-derived peptides (e.g.. f-met-leu-phe, fMLP), mast cell-, neutrophil- and lymphocyte-derived chemoattractant factors, β-thromboglobulin, LPS, cytokines (e.g.. IL-1β, IL-2, IL-3, IL-4, IL-6, IL-10, IL-12, TNFα, IFN-γ, GM-CSF, M-CSF, TGFβ), growth factors (e.g.. HGF), and opioids (Allevena, 1994; Baggiolini, 1994; Bevilacqua, 1985; Brett, 1989; Brevario, 1988; Cavender, 1986; Colditz, 1985; Colditz, 1992; Cramer, 1984; Cybulsky, 1988; El-Naggar, 1981; Fernandez, 1978; Heagy, 1990; Jagels, 1994; Keller, 1977; Keller, 1981; Johnston,

1994, Goetzl, 1981; Linbom, 1982; Marasco, 1984; Marder, 1985; May, 1992; Oseas, 1981; Schiffman, 1975; Shukula, 1992; Shyy, 1993; Smith, 1979; Smith, 1980; Spagnuolo, 1980; Taub, 1994b; Thureson-Klein, 1984). Many of these chemotactic molecules have also been shown to facilitate leukocyte adhesion to endothelial cells and purified adhesive ligands both *in vitro* and *in vivo* (Mantovani, 1992). Hoover and colleagues (1984) have also reported that adherence of human neutrophils to bovine aortic endothelial cells was enhanced by C5a, zymosan-activated human serum, and fMLP. Similarly, neutrophils stimulated with fMLP, C5a, LTB_4, or PAF adhere to human endothelial cells by a CD11b-CD18-dependent mechanism (Arfors, 1987; Bussolino, 1986; Camussi, 1990; Camussi, 1995; Charo, 1985; Charo, 1986; Cramer, 1984; Furie, 1991; Gimbrone, 1984; Hoover, 1978a; Hoover, 1978b; Hoover, 1984; Kishimoto, 1989; Kuijpers, 1991; Lacasse, 1991; Lewis, 1984; Linbom, 1982; Luscinskas, 1989; Mantovani, 1992; Marasco, 1984; Marder, 1985; Miglioris, 1984; Miglioris, 1987; Palmblad, 1981; Shukula, 1992; Thureson-Klein, 1984; Tonnesen, 1984; Yue, 1991; Zimmerman, 1992). This enhanced adhesiveness occurs within seconds and is not mediated by the recruitment of newly expressed integrin molecules to the cell surface. Inhibition of CD11b-CD18 expression does not block neutrophil adhesion to non-activated or cytokine-activated endothelial cells or cell aggregation (Butcher, 1991; Butcher, 1996; Carlos, 1994; Furie, 1991; Kishimoto, 1989; Palmblad, 1981; R. Smith, 1979; Springer, 1994; Springer, 1995). In addition, after activation of neutrophils with each of these chemoattractants, the CD11b-CD18 integrin heterodimer, which is present in subcellular granules of neutrophils, is translocated to the cell surface within minutes of exposure to the chemotactic stimuli (Bainton, 1987; Berger, 1984; Springer, 1994; Springer, 1995). Similar endothelial adhesion results have been observed with fMLP, PAF, and C5a acting on monocytes and macrophages (Camussi, 1990; Hakkert, 1991; Issekutz, 1995; Jagels, 1992; Jagels, 1994; Marder, 1985; Migliorisi, 1984; Miglioroso, 1987; Shiffmann, 1975; Smith, 1980; Vaddi, 1994). Furthermore, the proadhesive effects of these chemoattractants appear to vary depending on the concentration. For example, concentrations of fMLP, C5a, LTB_4, and PAF found to be optimal for cell migration were found to be less than optimal for promoting cell adherence to glass or plastic while high concentrations of these attractants (>100nM), which reduced leukocyte chemotaxis, promoted leukocyte adherence. Smith and coworkers (1979) have demonstrated that high concentrations of fMLP and C5a enhanced human neutrophil adherence to glass while simultaneously reducing cell motility. C5a-dependent transmigration of monocytes across a quiescent human endothelial cell barrier also requires active CD11/CD18 ($\beta 2$) molecules (Charo, 1986; Furie, 1991). However, this requirement was found to only be partial as mAb to CD18 inhibits only 70% of the C5a response. Subsequent studies have demonstrated that the VLA-4 ($\alpha 4/\beta 1$) complex on the surface of the monocytes is also important for C5a-mediated transendothelial migration. PAF- and fMLP-induced monocyte transmigration were also found to operate through both CD18- and CD29-dependent pathways (Furie, 1991). Furthermore, vasoactive intestinal peptide (VIP), C5a, fMLP, and casein are chemotactic for human monocytes and T lymphocytes and, in several studies, enhance unstimulated T cell adhesion to a number of adhesive ligands, including ICAM-1, VCAM-1, and fibronectin (El-Naggar, 1981; Johnston, 1994). Many of these chemotactic factors also exert dramatic effects on circulating white blood cell levels (Jagels, 1992). All of these chemotactic factors induce cell adhesion and migration through specific G-protein-linked seven transmembrane receptors on the leukocyte cell surface (Bokoch, 1995; Horuk, 1994; Kelvin, 1993; Schall, 1994). These receptors are members of the rhodopsin or serpentine receptor superfamily, and have the characteristic seven hydrophobic spanning transmembrane regions. Chemoattractant receptors have been difficult to purify biochemically, but molecular cloning has led to the identification of several receptors with varying degrees of homology. A more detailed discussion of these receptors will be presented later in this chapter.

A number of laboratories have also shown that various inflammatory cytokines such as IL-1β, IL-2, IL-3, IL-4, IL-6, IL-10, IL-12, TNFα, IFN-γ, TGFβ, M-CSF, and GM-CSF serve as physiological triggers of leukocyte adhesion and migration. These molecules are released at sites of inflammation and induce inflammatory infiltrates of leukocytes upon injection *in vivo* (Munro, 1989). TNF-α, IFN-γ, and IL-1 increase the adhesiveness of endothelial cells for neutrophils, monocytes, and lymphocytes and induce the expression of ICAM-1, E-selectin, and VCAM-1 (Bevilacqua, 1985; Brett, 1977; Brevario, 1988; Bussolino, 1986; Cavender, 1986; Cavender, 1991; Furie, 1989; Furie, 1991; Gamble, 1985; Hakkert, 1991; Henning, 1987; Hoover, 1978a; Jagels, 1994; Moser, 1989; Pohlman, 1986; Ulich, 1987; Yu, 1985). Similarly, IL-4 increases lymphocyte adhesion and induces VCAM-1 expression by endothelial cells (Butcher, 1996; Springer, 1995). Injections of IL-1α or TNFα into the skin of mice induces both neutrophil and monocyte infiltration while IFN-γ only induces a predominant monocytic infiltrate (Munro, 1989). In addition, TNF-α and IFN-γ but less so IL-1 induce T lymphocyte recruitment upon injection into the skin of rodents (Munro, 1989). Each of these cytokines have also been shown to possess modest chemotactic/ chemokinetic activity *in vitro* (Taub, 1993a; Taub, 1994b). In addition, these inflammatory cytokines also modify neutrophil, monocyte, and lymphocyte integrin activation and usage in trafficking studies (Ulich, 1987; Yu, 1985). In vivo injections of radiolabelled monocytes has demonstrated a differential role for β_1 and β_2 integrin molecules in mediating monocyte emigration (Issekutz, 1995). Antibodies to CD11a alone or in combination with anti-CD11b or -VLA-4 but not anti-CD11b alone inhibited IL-1α-, TNFα-, IFNγ-, and C5a-induced monocyte accumulation in rats. Anti-VLA-4 mAb significantly inhibited monocyte migration in response to these cytokines but not C5a. These studies suggest that different adhesion activating agents may differentially mediate adhesion and migration through distinct integrin molecules.

While many cytokines modulate cellular adhesion and promote leukocyte recruitment, few have been shown to directly affect integrin avidity. Rapid cytokine triggering of integrin-mediated adhesion has been demonstrated for a number of leukocyte populations including neutrophils, monocytes, and T lymphocytes (Butcher, 1996; Springer, 1994; Springer, 1995; Tanaka, 1993c). Recent studies have identified a family of structurally-related cytokines which selectively promote rapid leukocyte adhesion to endothelial cells and various extracellular matrix proteins and also induce the directional migration of various leukocyte subsets both *in vivo* and *in vitro*. These chemotactic cytokines termed "chemokines" appear to be excellent physiologically-relevant candidates for the rapid triggering of integrin-mediated adhesion. The remainder of this chapter will focus on the role of chemokines as mediators of leukocyte adhesion and migration both *in vivo* and *in vitro* as well as their role in various inflammatory disease states and lymphocyte homing.

CHEMOKINES AND THEIR CHARACTERISTICS

Chemokines are secreted by many different cell types in response to both endogenous and exogenous stimuli and have been shown to elicit the selective accumulation of inflammatory leukocytes *in vivo* (Baggiolini, 1991; Baggiolini, 1994; Matsushima, 1992; Miller, 1992; Oppenheim, 1991; Schall, 1991; Schall, 1994; Taub, 1994b). Since the discovery, cloning, and characterization of interleukin-8 (IL-8) in 1988 (Matsushima, 1988; Yoshimura, 1987), more than 30 unique human and mouse cytokines have been identified as members of the chemokine superfamily of peptides. Chemokines exhibit between 20 to 70% homology in their amino acid sequences and are related, with one exception, by a conserved motif containing four cysteine residues (Figure 1). The chemokine superfamily can be separated into three subfamilies, C-X-C (or α), C-C (or β), and C, based on the presence or absence of an intervening amino acid residue located between the first two of the four conserved cysteines (Baggiolini, 1994; Matsushima, 1992; Miller, 1992; Oppenheim, 1991; Schall, 1991; Schall, 1994; Taub, 1994b). In contrast to the C-X-C and C-C subfamily members, lymphotactin (Ltn), the only C subfamily member, possesses only one pair of NH_2-terminal cysteine amino acid (Kelner, 1994). Chemokine family members show considerable similarity in their amino acid sequence, exhibiting between 20–40% sequence homology (Figure 1). The chemokines in their monomeric form range from 7 to 10 kD, act at concentrations of 10^{-8} to 10^{-10} M, and are characteristically basic heparin-binding proteins. Many chemokines form dimers, trimers, and tetramers upon secretion; however, it remains controversial whether chemokine dimer formation is necessary for optimal ligand interaction with its chemokine-specific receptor. Interestingly, C-X-C chemokines are clustered on human chromosome 4q12–q21 and exhibit between 20 to 50% homology on the amino acid level (Baggiolini, 1994; Oppenheim, 1991; Schall, 1991; Taub, 1994b). C-C chemokines are clustered on human chromosome 17q11–q21 and mouse chromosome 11 and exhibit between 28 to 70% homology on the amino acid level (Baggiolini, 1994; Oppenheim, 1991; Schall, 1991; Taub, 1994b). The C chemokine, Ltn, is located on human chromosome 1 (Kelner, 1995).

The characterizing features of the chemokine subfamilies include their apparent leukocyte specificity (Baggiolini, 1994; Oppenheim, 1991; Schall, 1991; Taub, 1994b). Thus, many of the C-X-C chemokines including interleukin-8 (IL-8), growth related peptide (GROα), macrophage inflammatory protein-2 (MIP-2) or GROβ, GROγ, NAP-2, ENA-78, and GCP-2 induce both neutrophil migration and activation but have no effect on monocytes. However, members of this family are not restricted to interactions with neutrophils as several C-X-C chemokines act on other cell types (e.g.. basophils, T cells, endothelial cells, melanocytes, muscle cells, keratinocytes, etc.). In addition, the C-X-C chemokines, interferon-inducible protein-10 (IP-10) and MIG, have been shown to chemoattract monocytes and T lymphocytes rather than neutrophils (Sarris, 1995; Taub, 1993b; Taub, 1994b) while PF-4 and β-TG attract fibroblasts, but not neutrophils (Senior, 1983). In contrast, C-C chemokines including macrophage inflammatory protein-1α (MIP-1α), MIP-1β, regulated upon activation, normal T cell expressed and secreted (RANTES), macrophage chemotactic protein-1 (MCP-1), MCP-2, MCP-3, I-309, and C10 predominantly chemoattract and activate monocytes/ macrophages and, with the exception of MCP-3, have very little to no effects on neutrophils.

A. C-X-C Chemokines

```
                        1         10         20         30         40         50         60         70
                        .    .    .    .    .    .    .    .    .    .    .    .    .    .    .    .

hIL-8/NAP-1        AVLPRSAKELRCQCIKTYSKPFHPKFIKELRVIESGPHCANT-EIIVKLSD-GRELCLDPKENWVQRVVEKFLKRAENS
hGROα/MGSA            ASVATELRCQCLQTLQGI-HPKNIQSVNVKSPGPHCAQT-EVIATLKN-GRKACLNPASPIVKKIIEKMLNSDKSN
hGROβ/hMIP-2α         APLATELRCQCLQTLQGI-HLKNIQSVKVKSPGPHCAQT-EVIATLKN-GQKACLNPASPMVKKIIEKMLKNGKSN
hGROγ/hMIP-2β         ASVVTELRCQCLQTLQGI-HLKNIQSVNVRSPGPHCAQT-EVIATLKN-GKKACLNPASPMVQKIIEKILNKGSTN
mMIP-2                 VVASELRCQCLKTLPRV-DFKNIQSLSVTPPGPHCAQT-EVIATLKG-GQKVCLDPEAPLVQKIIQKILNKGKAN
mKC                   APIANELRCVCLQTMAGI-HLKNIQSLKVLPSGPHCTQT-EVIATLKN-GREACLDPEAPLVQKIVQKMLKGVPK
hIP-10                VPLSRTVRCTCISISNQPVNPRSLEKLEIIPASQFCPRV-EIIATMKKKGEKRCLNPESKAIKNLL-KAVSKEMSKRSP
mIP-10/mCRG-2         IPLARTVRCNCIHIDDGPVRMRAIGKLEIIPASLSCPRV-EIIATMKKNDEQRCLNPESKTIKNLM-KAFSQKRSKRAP
mMIG                  TLVIRNARCSCISTSRGTIHYKSLKDLKQFAPSPNCNKT-EIIATLKN-GDQTCLDPDDSANVKKLMKEWEKKI..
hPF-4                EAEEDGDLQCLCVKTTSQV-RPRHITSLEVIKAGPHCPTA-QLIATLKN-GRKICLDLQAPLYKKII-KKLLES
hCTAP/β-TG         SLDSDLYAELRCMCIKTTSGI-HPKNIQSLEVIGKGTHCNQV-EVIATLKD-RKICLDPDAPRIKKIVQKKLAGDESAD
hNAP-2                     AELRCMCIKTTSGI-HPKNIQSLEVIGKGTHCNQV-EVIATLKD-GRKICLDPDAPRIKKIVQKKLAGDESAD
hENA-78            GPAAAVLRELRCVCLQTTQGV-HPKMISNLQVFAIGPQCSKV-EVVASLKN-KEICLDPEAPFLKKVIQKILDGGNKEN
mMIP-2                AVVASELRCQCLKTLPRV-DFKNIQSLSVTPPGPHCAQJ-EVIATLKG-GQKVCLDPEAPLVQKIIQKILNGKAN
mSDF-1α             DGKPVSLSYRCPCRFFESHI-ARANVKHLKILNT-PNCALQ-IVARLKNN-NRQVCIDPKLKWIQEYLEKA-LNK
mSDF-1β             DGKPVSLSYRCPCRFFESHI-ARANVKHLKILNT-PNCALQ-IVARLKNN-NRQVCIDPKLKWIQEYLEKA-LNKRLKM
```

B. C-C Chemokines

```
MCP-1/MCAF         QPDAINAPVTC-CYNFTNRKISVQRLASYRRITSS-KCPKE-AVIFKTIV-AKEICADPKQKWVQDSMDHLDKQYQTPKT...
mMCP-1/JE          QPDAVNAPLTC-CYSFTSKMIPMSRLESYKRITSS-RCPKE-AVVFVTKL-KREVCADPKKEWVQTYIKNLDRNQMRSEP...
hMCP-2 (hHC14)    AQPDSVSIPITC-CFNVINRKIPIQRLESYTRITNI-QCPKE-AVIFKTKR-GKEVCADPKERWVRDSMKHLDQIFQNLKP
hMCP-3                    KSTTC-CYRFINKKIPKQRLESYRRTTSS-HCPRE-AVIFK    DKEICADPTQKWVQDFMKHLDKKTQTPKL
hLD78/hMIP-1α      ASLAADTPTAC-CFSYTSRQIPQNFIADYFE-TSS-QCSKP-GVIFLTKR-SRQVCADPSEEWVQKYVSDLELSA
mMIP-1α            APYGADTPTAC-CFSY-SRKIPRQFIVDYFE-TSS-LCSQP-GVIFLTKR-NRQICADSKETWVQETITDLELNA
hACT2 (pAT744)     APMGSDPPTAC-CFSYTARKLPRNFVVDYYE-TSS-LCSQP-AVVFQTKR-SKQVCADPSESWVQEYVYDLEIN
hMIP-1β (hH400)    APMGSDPPTAC-CFSYT-REASSNFVVDYYE-TSS-LCSQP-AVVFQTKR-SKQVCADPSESWVQEYVYDLEIN
MIP-1β             APMGSDPPTSC-CFSYTSRQLHRSFVMDYYE-TSS-LCSKP-AVVFLTKR-GRQICANPSEPWVTEYMSDLEIN
hRANTES              SPYSSDTTPC-CFAYIARPLPRAHIKEYFY-TSG-KCSNP-AVVFVTRK-NRQVCANPEKKWVREYINSLEMS
mRANTES              SPYGSDTTPC-CFAYLSLALPRAHVKEYFY-TSS-KCSNL-AVVFVTRR-NRQVCANPEKKWVQEYINYLEMS
hI-309               KSMQVPFSRC-CFSFAEQEIPLRAILCYRN-TSS-ICSNE-GLIFKLKR-GKEACALDTVGWVQRHRKMLRHCPSKRK
mTCA3                KSMLTVSNSC-CLNTLKKELPLKFIQCYRKMGSS--CPDPPAVVFRLNK-GRESCASTNKTWVQNHLKKVNPC
mfic                 QPDGPNASTC-CYVKKQKIPKRNLKSYRRI-TSS-RCPWE-AVIFKTKK-GMEVCREAHQKWVEEAIAYLDMKTPTPKP
mC10  GLIQEMEKEDRRYNPPII-
                     HQGFQDTSSDC-CFSY-ATQIPCKRFIYYFP-TSG-GCIKP-GIIFISRR-GTQVCADPSDRRVQRCLSTLKQGPRSGNKVIA
```

Figure 1 Amino acid sequences of C-X-C and C-C chemokines (Taub, 1994b).

The biological properties of many of the chemokines have been identified (Baggiolini, 1994; Oppenheim, 1991; Schall, 1991; Taub, 1994b). Most of the C-X-C chemokines, including IL-8, NAP-2, GROα, GROβ, ENA-78, and GCP-2, have been shown to be potent neutrophil but not monocyte chemotactic agents both *in vitro* and *in vivo* (Baggiolini, 1994; Oppenheim, 1991; Schall, 1991; Taub, 1994b). This has been attributed to the presence of the -ELR- residues at position 4–6 near the N-terminus of all of these chemokines (Baggiolini, 1994; Schall, 1994; Taub,

1994b). Studies by Larsen and colleagues (1989) were the first to show that IL-8 induces human T cell migration in *in vitro* chemotaxis assays and attracts some T cells *in vivo*. Subsequently, several investigators have confirmed dose-dependent human T cell migration in response to IL-8, GROα, and MIP-2 under various assay conditions (Bacon, 1993; Bacon, 1995a; Wang, 1995; Xu, 1995). This migration appears to occur through interaction with a promiscuous IL-8 receptor on the T cell surface. Conversely, the C-X-C chemokines, IP-10, MIG, and PF-4, are not chemotactic for neutrophils (Taub, 1993b; Taub, 1994b). In addition, ELR$^+$ C-X-C chemokines also promote the generation of superoxide anion as well as the mobilization intracellular calcium, and the release of enzymes by neutrophil degranulation (Baggiolini, 1994; Schall, 1994; Taub, 1994b). The lack of an ELR motif in the NH_2 terminus of these chemokines, a region shown to be critical for neutrophil migration and interaction with IL-8 receptors, has been shown to be responsible for the lack of activity on neutrophils (Baggiolini, 1994). In addition to inducing their migration, MCP-1 and other C-C chemokines can stimulate monocytes/ macrophages to produce superoxide anion, nitric oxide, and release stored enzymes (Baggiolini, 1994; Oppenheim, 1991; Schall, 1991; Schall, 1994; Taub, 1994b). C-C chemokines also mobilize intracellular calcium and increase intracellular levels of cAMP (Baggiolini, 1994; Oppenheim, 1991; Schall, 1991; Schall, 1994; Taub, 1994b). In addition, both C-C and C chemokines have also been shown to induce T cell subset migration *in vitro* (Schall, 1994; Taub, 1994b). Initial studies by Schall (1990) demonstrated that RANTES preferentially induces CD4+ memory T cell chemotaxis *in vitro*. This migration was dose-dependent and pertussis toxin sensitive. Subsequently, Tanaka (1993a), Taub (1993a; 1993b; 1994b; 1995b), Schall (1993), Carr (1994), Loetscher (1994), and others (Berger, 1996; Gilat, 1995; del Pozo, 1995) have described that the ability of various C-C chemokines, including MIP-1α, MIP-1β, RANTES, MCP-1, MCP-2, MCP-3, C10, fic, TCA-3, I-309, to facilitate human T cell and T cell subset migration *in vitro*. In addition, MIP-1α as well as several other C-C chemokines are potent chemoattractants for human natural killer cells (Maghazachi, 1994; Taub, 1995c). Schall and co-worker (1993) have also reported MIP-1α to be a B cell chemoattractant. Although eosinophils, basophils, and mast cells are developmentally more closely linked to neutrophils, they are primarily chemoattracted and activated by the C-C chemokines (Taub, 1994b). Several of the C-C chemokines namely MCP-1 and RANTES as well as MIP-1α have potent chemoattractant activity on basophils and mast cells (Alam, 1992; Bischoff, 1993; Kuna, 1992a; Kuna, 1992b; Kuna, 1993; Taub, 1995a). MCP-1 and RANTES rapidly induce histamine release by basophils (Alam, 1992; Bischoff, 1993; Kuna, 1992a; Kuna, 1992b; Kuna, 1993). RANTES, MIP-1α, and eotaxin are also potent chemoattractants for eosinophil (Alam, 1993; Ebisawa, 1994; Jose, 1991; Jose, 1994; Rot, 1992b; Schroder, 1996). MCP-1 and RANTES as well as MIP-1α have been shown to induce mast cell migration *in vitro* (Taub, 1995b). In addition, MIP-1α has also been shown to induce histamine release from peritoneal exudate cells, believed to be derived from mast cells (Alam, 1992). Some of the C-X-C and C-C chemokines have been found to chemoattract and induce various biological responses in a number of non-leukocytic cell populations including smooth muscle cells, fibroblasts, endothelial cells, epithelial cells, keratinocytes, and melanocytes (Baggiolini, 1994; Oppenheim, 1991; Schall, 1991; Taub, 1994b).

Many different cell types have been shown to produce chemokines in response to a wide variety of stimuli (Table 1). In addition to leukocytes, endothelial cells, epithelial

Table 1 Stimulants and cellular sources of C-X-C and C-C chemokines (Baggiolini, 1994; Schall, 1991; Taub, 1993c; Taub, 1994b).

Chemokine	Stimuli	Cellular Sources
A. C-X-C Chemokines		
IL-8, GROα, ENA-78, GCP-2	Mitogens (PHA, ConA, LPS) IL-1α, IL-1β, TNFα, IL-3, IFN-γ Anti-CD3 mAb, phorbol esters Viruses, Bacteria	Monocytes/macrophages, T cells (Th1 and Th2), Neutrophils, Keratinocytes, Endothelial cells, Hepatocytes, Chondrocytes, Glioblastoma cells, Astrocytoma cells, NK cells, Melanomas, Fibroblasts, Tumor cell lines, Synovial cells, Osteosarcoma, Epithelial cells
hPF-4	Platelet activators	Platelets
hIP-10	IFNγ, LPS, TNF, anti-CD3 mAb	Monocytes/macrophages, Fibroblasts, Endothelial cells, Keratinocytes, T cells (Th1 and Th2), thymocytes
hβ-TG/ CTAPIII/ NAP-2	Platelet activators	Monocytes/macrophages, Platelets
B. C-C Chemokines		
MCP-1,MCP-2, MCP-3	IL-1, TNF, PDGF, EGF, IFN-γ PHA, LPS, Hydroxyurea, Silica poly (rI):poly (rC) viruses, Listeria anti-CD3, Ag activation	Monocytes/macrophages, endothelial cells, fibroblasts, keratinocytes, T cells
MIP-1α	IL-2, TNF, IL-7 PHA, PMA, anti-CD3, ConA, LPS, Ag activation IgE/Fcε crosslinking	T cells, monocytes, mast cells
MIP-1β	IL-2, TNF PHA, PMA, anti-CD3, Con A, LPS, Ag activation	T cells, monocytes
RANTES	PHA, PMA, anti-CD3,LPS Ag activation	T cells, platelets
I-309/TCA-3	PHA, Anti-CD3, Ag activation IgE/Fcε crosslinking	T cells, mast cells

cells, fibroblasts, synovial cells, alveolar macrophages, keratinocytes, mesangial cells, mesothelial cells, melanocytes, chondrocytes, astrocytes, hepatocytes, and smooth muscle cells can produce both C-X-C and C-C chemokines. Platelets and mast cells are unique in that they release stored chemokines from their granules (Kameyoshi, 1992). Chemokine production occurs in response to either exogenous or endogenous stimuli including lectins, LPS, urate crystals, silicon, phorbol esters, eicosinoids, antigenic stimulation, anti-CD3 mAb, injurious stimuli, infectious agents, and various endogenous inflammatory cytokines (Baggiolini, 1994; Oppenheim, 1991; Schall, 1991; Taub, 1994b). The production of chemokines by both immune and non-immune cells supports the contention that these cytokines may play a pivotal role in orchestrating inflammation and tissue repair.

Chemokine activities are not only restricted to chemotaxis, multiple other pro-inflammatory effects have been described. They have been shown to induce granulocyte, mast cell, and lymphocyte degranulation; increased expression of cell adhesion molecules and enhance adhesion to endothelial cells and extracellular matrix proteins; enhance the cytostatic effects of neutrophils and monocytes on pathogens such as *Candida albicans*; enhance the cytolytic activity against CTL and NK cell target cells; costimulate T cell activation and proliferation; and enhance antigen presenting cell functions (Baggiolini, 1994; Oppenheim, 1991; Schall, 1991; Taub, 1994b). Furthermore, IL-8 as well as several C-C chemokines have been shown to regulate the production of specific isotypes of immunoglobulin by human B cells in combination with anti-μ mAb and IL-4 and/or CD40L (personal observations). Several C-X-C and C-C chemokines have also been implicated as positive (e.g.. IL-8) and negative (e.g.. PF-4, IP-10, and MIG) regulators of angiogenesis and hematopoiesis (Angiolillo, 1995; Cook, 1996; Koch, 1992; Maione, 1990; Strieter, 1996a; Strieter, 1996b; Taub, 1994b). Many of these biological properties are believed to contribute to their role in inflammation and tissue repair as well as the generation of antigen-specific immune responses.

Recently, a model has been proposed that chemokines, rather than remaining in solution, are preferentially immobilized through low affinity binding to heparin-bearing proteoglycans on the endothelial cell surface or to extracellular matrix proteins within the tissues (Rot, 1992a; Rot, 1993; Rot, 1996; Tanaka, 1993c). Proteoglycans are proteins which have been post-translationally modified by the addition of glycosaminoglycan (GAG) side chains at serine residues. GAGs are long polysaccharide chains composed of disaccharide subunits and are subdivided into structural subfamilies: chondroitin sulfate, keratin sulfate, heparan sulfate, dermatan sulfate, and hyaluronic acid (Rot, 1996; Tanaka, 1993c). There are a wide variety of proteoglycans, some of which are integral membrane proteins while others can be found in the extracellular matrix (Tanaka, 1993c). Many endothelial cells express and secrete both heparan sulfate proteoglycans. Chemokines are basic proteins which possess glycosaminoglycan (GAG) – binding sites and bind heparin through their carboxyl termini (Baggiolini, 1994; Rot, 1996; Schall, 1994; Taub, 1994b). Similarly, the cytokines and growth factors, IL-7, TGF-β, GM-CSF, bFGF, and HGF, possess GAG binding sites (Tanaka, 1993c). This binding is believed to be important in cytokine localization as well as their biological activity. These GAG-bindings sites also permit chemokines to bind to proteoglycans with a low to intermediate affinity. In this way, the integrins would only be activated only on those leukocytes that have already established selectin-mediated adhesion. Bound chemokines promote the haptotaxis of leukocytes by triggering adhesion of leukocyte subsets to vascular cell adhesion molecules on endothelial cells, thus facilitating the extravasation of leukocytes from the blood stream (Luster, 1995; Rot, 1993; Rot, 1996; Tanaka, 1993c; Webb, 1993). In addition, chemokines bound to the extracellular matrix proteins within the inflammatory environment or that form a gradient toward the site of inflammation can promote the arrest, recruitment, and localization of migrating and chemokine-responsive immune cells at sites rich in chemokine-producing cells. Thus, the chemotactic and haptotactic properties of chemokines may direct both the exit of leukocytes from the vessels and their migration through the extracellular matrix toward extravascular tissues. This "chemokine presentation" model is supported by several studies demonstrating the presence of IL-8 on the endothelial cells of postcapillary venules and small veins and MIP-1β on the high endothelial venules (HEV) of lymph nodes (Rot, 1996; Tanaka, 1993c). As endothelial cells are known to express proteoglycans on their cell surface, chemokine binding through their GAG sites would permit effective chemokine presentation to rolling leukocytes. It is not clear which endothelial proteoglycans are physiologically relevant to the presentation of chemokines. However, recent studies by Shaw (Tanaka, 1993c) suggest that the inducible endothelial-bound heparin proteoglycan, CD44, may be an important chemokine presentation molecule in vivo. This molecule carries heparan sulfate and chondroitin sulfate side chains which may potentially bind cytokines with GAG-binding sites (Haynes, 1989). However, not all GAGs result in increased leukocyte adhesive responses suggesting that chemokines may bind to various GAGs in a distinct fashion or may differentially alter chemokine structure (Rot, 1996; Tanaka, 1993c). It is likely that the affinity of different endothelial GAGs may vary for different chemokines. This would add an additional level of specificity where the characteristic for vascular- or organ-specific endothelial proteoglycans would determine which chemokine is capable of binding to the endothelium and being presented to leukocytes. In addition, several extracellular matrix proteins (e.g.. fibronectin) have been shown to bind chemokines as well as other cytokines and may serve as additional chemokine presenting structures within the tissue environment. Thus, this form of cytokine presentation may be an important mechanism by which many inflammatory and growth factor signals stimulate leukocyte biological responses in a localized and concentrated form.

Various inflammatory lesions contain disparate proportions of various leukocyte subtypes, often characteristic for the specific lesion. Additionally, various lymphocyte subsets selectively traffic to different lymphoid organs. This selective extravasation most likely involves selective endothelial adhesion via distinct selectin-, addressin-, and integrin-mediated pathways. As chemokines selectively recruit distinct leukocytic subsets both *in vivo* and *in vitro,* their role in the cascade of leukocyte-endothelial cell adhesion may regulate and control the diversity and selectivity of cell recruitment. Several properties of the chemokines enable them function in the selective regulation of leukocyte recruitment into inflammatory tissues: (1) the ability of leukocytic and non-leukocytic cells to rapidly transcribe and produce C-X-C and C-C chemokines upon stimulation with various inflammatory stimuli; (2) the large number of chemokines with distinct and overlapping target cell specificities; (3) the localization of the chemokines on the surface of the endothelium via GAG and/or extracellular matrix molecules, thereby avoiding dilution by flowing blood; (4) the rapid induction of increased adhesive state of leukocytes; and (5) the capacity to form a gradient with the highest chemokine concentration originating at the local site of the inflammatory injury. This capacity to form a gradient is critical as is illustrated by the fact that the intravenous administration of high doses of chemokine which counteracts the formation of a gradient inhibits leukocyte adhesion and transmigration through the endothelial barrier and promotes leukocytosis (??). The inhibition of leukocyte emigration by high plasma chemokine levels is believed to be due to the lack of a tissue-localized chemokine gradient as well as the desensitization of circulating leukocytes to additional chemokine signals. Therefore, the concentration, availability, and distribution of chemokines in vivo all appear to be important factors in promoting cellular adhesion and emigration.

CHEMOKINE RECEPTORS

While a review of chemokine signaling pathways and their receptors is beyond the scope of this chapter, a brief introduction to chemokine receptors is crucial to an understanding of the overlapping features between the many members of the chemokine subfamilies. Like C5a, fMLP, and PAF receptors, the chemokine receptors belong to the superfamily of G protein-coupled seven transmembrane receptors (Amatruda, 1993; Horuk, 1994; Kelvin, 1993; P. Murphy, 1994; Schall, 1994; Wells, 1996). There are similarities in the tertiary structure of chemokine receptors and other members of the G protein-coupled receptor family, yet chemokine receptors initiate unique and specific cellular activities only in response to selected chemokines. To date, about eight human and murine C-X-C and C-C chemokine receptors have been identified and cloned. There are two distinct IL-8 receptors, IL-8RA and IL-8RB, have been cloned (P. Murphy, 1994). IL-8RA only binds IL-8 with high affinity while IL-8RB binds all ELR$^+$ C-X-C chemokines such as IL-8, GROα, GROβ, ENA-78, and GCP-2, but not C-C chemokines. Similarly, the C-C receptors are both promiscuous and specific in their binding properties. For example, Neote (1993a) and Gao (1993) have identified a cDNA from a neutrophil library (C-C CKR1) that could bind MIP-1α with high affinity (approximately 6.5 nM). Ligation of this receptor by MIP-1α and RANTES induces Ca^{++} mobilization in transfected fibroblasts. This receptor has also been shown to bind MCP-3 but not MIP-1β or IL-8. Thus, this receptor is specific for some C-C chemokines but not others. Subsequently, several other C-C chemokine receptors have been cloned that also appear to possess overlapping specificities (P. Murphy, 1994, Oppenheim, 1996). Further characterization of the binding and signaling specificities of these transfected receptors may elucidate their role in leukocyte emigration *in vivo.* More recently, the Duffy receptor, which appears to be a unique chemokine receptor which exists on the surface of human red blood cells, has been shown to be highly promiscuous as it binds both α and β chemokines (Darbonne, 1991; Horuk, 1994; Horuk, 1996; Neote, 1993b; Oppenheim, 1996). While it is believed that red blood cells are not directly affected by chemokine binding, it is believed that this receptor acts as a sink to transport excess chemokine leaking into the peripheral circulation. Further characterization of this receptor may provide insight into the role of this receptor in the systemic regulation of cell migration into inflammatory sites.

Soluble chemokines leaking into the circulation may also be regulated by the presence of Duffy receptors on the cell surface of erythrocytes and certain endothelial cell populations within Duffy-positive individuals (Hadley, 1994; Horuk, 1994; Horuk, 1996). This promiscuous chemokine receptor, which is believed to act as a "sink" for excessive chemokines running off into the circulation, may assist in the maintenance of low levels of chemokines in plasma to maximize the sensitivity of circulating leukocytes to changes in the concentration of chemokines in the tissues (Horuk, 1996). In subjects undergoing experimental IL-1α therapy for cancer, IL-8 was found to bind to red cells *in vivo* and this binding corresponded with a decrease in soluble IL-8 in the plasma. In addition, Duffy-bound IL-8 has been shown to be biologically inactive (Horuk, 1996). If this function is physiologically important, differences in the clinical course of Duffy-positive versus Duffy-negative individuals should be observed during septic responses. However, more detailed clinical and biological studies are required to define the exact role of these receptors *in vivo.*

Signal transduction through these receptors is believed to operate through G proteins, phosphoinositol hydrolysis, arachidonic acid metabolism, production of inositol 1,4,5-triphosphate (IP3), the activation of protein kinase C and serine/threonine kinases, the activation of phosphoinositol 3-kinase (PI-3-kinase), and the rapid elevation of diacylglycerol and cytosolic Ca^{++} levels (Baggiolini, 1994; Bokoch, 1995; Matsushima, 1992; Oppenheim, 1991; Schall, 1991; Schall, 1994; Sozzani, 1991; Taub, 1994b; Turner, 1995). Chemokine-receptor interactions initiate a characteristic pattern of responses including shape change, chemotaxis, degranulation, and respiratory burst. The effector function induced by chemokine agonists depends on the type of G protein that couples to the receptor. All chemokine receptors characterized so far contain one or more of the consensus sequences for Gi protein activation. The first indication that chemokine receptors are coupled to G proteins came from the observations that pretreatment with Bordetella pertussis toxin (PT), which inactivates Gi-type G-proteins by ADP-ribosylation) renders leukocytes unresponsive to chemokines (Amatruda, 1993; Bokoch, 1995; Spangrude, 1985; Wu, 1993). Subsequent studies have demonstrated that chemokines induce GTPase activity of the G protein α subunit (Horuk, 1994; Horuk, 1996, Murphy, 1994; Schall, 1994). The coupling of a G protein to these receptors is believed to be essential for the high affinity binding of both C-X-C and C-C chemokines. Thus, both G protein signaling and protein phosphorylation are crucial to the signal transduction by these ligands. However, the physiological role of each of these signaling pathways remains to be determined.

The cellular responses to chemokines are strictly regulated by a "desensitization" process that is a characteristic for all G protein-coupled receptors (Horuk, 1994; Schall, 1994; Wu, 1993). Prior exposure of a receptor to an activating ligand rapidly results in a state of hyporesponsiveness. Desensitization can be either homologous or heterologous. *Homologous desensitization* is the chemokine-specific desensitization typically induced by exposure to high concentrations of ligand for a short time period and is believed to result from the activation of G protein-coupled receptor kinases. *Heterologous desensitization* may accompany specific receptor desensitization and consists of a reversible loss of responsiveness to multiple chemokine ligands. Heterologous desensitization probably results from phosphorylation of shared signal transducers such as protein kinase A and C and results in the uncoupling of the G protein from the ligated receptor (Oppenheim, 1996). This desensitization phenomenon probably also accounts for the attenuation of leukocyte migration in response to high concentrations of chemokines yielding the typical chemokine "bell-shaped" dose response curves in *in vitro* chemotaxis assays. Desensitization presumably serves to down regulate chemokine responses. Desensitization may actually have a physiological role by preventing the remobilization and departure of leukocytes upon entering sites of inflammation where the chemokine concentrations are believed to be maximal. However, while shutting down chemotaxis, other chemokine-mediated leukocyte functions are not altered by high ligand exposure (Oppenheim, 1996).

CHEMOKINES PROMOTE LEUKOCYTE ADHESION

Increases in neutrophil:endothelial adhesion has been described for fMLP, C5a, PAF, LTB_4, and the C-X-C chemokine, IL-8 (Carveth, 1989; Detmers, 1990). Each of these ligands stimulates neutrophil adhesion to the CAMs on endothelial cells as well as purified extracellular matrix molecules (e.g.. fibrinogen and collagen), the complement protein C3bi, serum, glass, and plastic *in vitro* (Brevario, 1988; Carveth, 1989; Cramer, 1984; Detmers, 1990). Similar to numerous reports examining neutrophil extravasation (Springer, 1994; Springer, 1995; Tanaka, 1993a), IL-8-induced adhesion of human and rodent neutrophils to endothelial cells requires the presence of β_2 integrin molecules on the neutrophil and CAMs on the endothelial cell surface (Carveth, 1989; Detmers, 1990). Carveth (1989), Detmers (1990), Gimbrone (1989), and Huber (1991) have reported that adherence of neutrophils to endothelial cells or their transmigration across an endothelial cell monolayer in a blood vessel construct was enhanced by IL-8 at high concentrations and that this proadhesive/promigratory effect is CD11-CD18-dependent. IL-8-mediated adhesion occurs within seconds of receptor ligation and does not require the recruitment of newly expressed integrin molecules to the cell surface. Neutrophils stimulated by IL-8 undergo dramatic shape changes, rapidly shed L-selectin (LECAM-1) from their cell surface, and express increased levels of β_2 integrins. All of these alterations are essential for efficient neutrophil emigration into inflamed tissues. Neutrophils have been shown to undergo morphological changes and shed L-selectin just prior to the development of firm adhesion to the vascular endothelium (Carlos, 1994; Luscinskas, 1989). Stimulation of neutrophils solely with IL-8 or another activating agent prior to their initial contact with endothelial cells prevents neutrophil selectin-mediated rolling as well as the subsequent firm adhesion and emigration into the inflammatory tissues (Butcher, 1996; Luscinskas, 1989; Schall, 1994; Smith, 1991; Springer, 1994; Springer, 1995; Taub, 1994b). L-selectin facilitates neutrophil rolling and contains the carbohydrate epitope sialo-Lewis X (the counterligand for P- and E-selectin found on the endothelial cell surface) (Berg, 1992). Thus, the shedding of L-selectin after exposure to the proadhesive IL-8 signal appears to hinder additional selectin-mediated neutrophil rolling perhaps promoting integrin-mediated adhesion to the endothelial cell surface. IL-8 bound to CD44 and/or other heparan proteoglycans on the endothelial cell surface is believed to present a concentrated directed signal to rolling neutrophils to facilitate firm integrin-mediated adhesion to vascular endothelium (Rot, 1996). While the ability of other C-X-C chemokines to exert proadhesive effects on neutrophils has not been examined, the number of overlapping activities between IL-8 and the other ELR^+ C-X-C chemokines would suggest similar effects on L-selectin shedding as well as β_2 integrin expression.

An increase in β_2 integrin expression, in particular CD11b-CD18 (Mac-1), on the neutrophil cell surface has also been observed post stimulation with IL-8 (Carveth, 1989). This increase appears to correlate with the IL-8-mediated release of specific granules (Carveth, 1989; Detmers, 1990; Springer, 1994). The CD11b-CD18 integrin heterodimer is present in subcellular granules of neutrophils and is rapidly translocated to the cell surface within minutes of exposure to chemotactic stimuli (Bainton, 1987; Berger, 1984; Detmers, 1990). Chemoattractant stimulation of neutrophils results in a tenfold increase in cell surface Mac-1 through the fusion of secretory granules with the plasma membrane. However, this granular fusion is not sufficient or necessary for the increased adhesiveness to endothelial cells as the addition of inhibitors of neutrophil degranulation to adhesion cultures fails to block IL-8-mediated neutrophil binding to endothelial cell monolayers. While an increase in CD11–CD18 expression on neutrophils may assist in a firmer adhesion to endothelial cells, IL-8 is believed to induce rapid avidity alterations in the constitutively expressed integrin molecules on the neutrophil cell surface. Conformational alterations in CD11–CD18 complex are believed to be associated with increased adhesiveness and has been demonstrated by the use of monoclonal antibodies which recognize only the "active" form of various integrin molecules. Springer and colleagues have found that 10% of the Mac-1 molecules expressed by chemoattractant-activated neutrophils possess the integrin activation epitope (Springer, 1994; Springer, 1995). Furthermore, this mAb also blocked neutrophil binding to ICAM-1 and fibrinogen suggesting that chemoattractant-induced binding is mediated by this subpopulation of activated Mac-1 molecules. Therefore, an avidity change in the β_2 integrin molecules on the neutrophil cell surface permits rolling neutrophils to strongly bind to the ICAMs expressed on the endothelial cell surface within seconds of chemokine exposure. Furthermore, the adhesiveness of the CD11a-CD18 (LFA-1) and CD11c-CD18 (p150,95) complexes on neutrophils are also activated by exposure to IL-8 as well as fMLP and C5a (Furie, 1991; Springer, 1994; Springer, 1995). Chemokine-induced adhesion to endothelial cells via integrin molecules is a transient event occurring within seconds of exposure and stopping minutes later. The transient nature of this integrin activation also provides a possible mechanism for the deadhesion and the retraction of leukocytes trailing edge from the adhesion substrate during cell migration.

Conversely, IL-8 has also been shown under certain circumstances to antagonize neutrophil adhesion to cytokine-activated endothelial cells (Gimbrone, 1989). Supernatants derived from activated mononuclear cells as well as TNFα-stimulated endothelial cells inhibit neutrophil binding to IL-1α-, TNFα-, and IFN-γ-activated endothelial cell monolayers (Gimbrone, 1989). This soluble inhibitory factor was termed *leukocyte adhesion inhibitor* (LAI) and was subsequently determined to be identical in sequence to IL-8 (Gimbrone, 1989). The inhibitory effect of soluble IL-8 (and possibly other chemokines) is probably due to leukocyte shape changes, an increase in membrane rigidity, redistribution of counterligands for endothelial selectins, and/or the loss of L-selectin (Baggiolini, 1994; Springer, 1995; Taub, 1994b). Thus, it appears that chemokines immobilized by the endothelium but not those in solution can mediate selectin- and integrin-mediated adhesion assuring a controlled and efficient extravasation of leukocytes into tissues. The inhibitory actions of soluble chemokines on leukocyte adhesion may have an important role in regulation of leukocyte trafficking. Neutrophils encountering soluble chemokines before their initial adhesion to endothelium would shed L-selectin and undergo neutrophil shape changes preventing neutrophil integrin-mediated adhesion and emigration. Thus, excessive production of soluble IL-8 (or other chemokines) by activated endothelial cells may prevent additional leukocyte extravasation limiting the number of cells entering an inflammatory lesion.

Increases in leukocyte:endothelial adhesion have also been observed with the C-C chemokines, MCP-1, MIP-1α, MIP-1β, TCA-3, and RANTES, and monocytes and basophils (Bacon, 1994; Devi, 1995; Issekutz, 1995; Jiang, 1992; Luo, 1994). Similar to inflammation-induced monocyte trafficking *in vivo* (Butcher, 1991; Meerschaert, 1994; Springer, 1994; Springer, 1995), chemokine-induced adhesion of human and rodent monocytes to unactivated or activated endothelial cells require the presence of β_2 integrin leukocyte surface molecules (Jiang, 1994). Similar to IL-8 on neutrophils, *in vitro* stimulation of human monocytes with MCP-1 induced the expression of the β_2 integrins, in particular CD11b and CD11c, but with little or no change in the levels of CD11a, VLA-4, or LAM-1 (Jiang, 1994; Vaddi, 1994). Similarly, MIP-1α, RANTES, and MCP-1 were also found to modulate CD11b and CD11c expression as well as the common β_2 chain, CD18 (Vaddi, 1994). This enhanced expression was dose-dependent and followed a distinct time course with peak levels at 4 hr and baseline levels by 24 hr. While the increased expression on integrin molecules on monocytes may assist in a more firm cell adhesion, it is believed that, like IL-8, C-C chemokines bound to the inflamed vascular endothelial barrier induce rapid avidity alterations in the constitutively expressed integrin molecules on the monocyte cell surface. This increase in integrin avidity alterations would permit rolling monocytes to strongly bind to VCAM-1 and the ICAMs expressed on the endothelial cell surface within seconds of chemokine activation. MIP-1α and RANTES has also been shown to induce human monocyte transendothelial migration requiring the presence of β_2 but not β_1 integrin molecules on unactivated endothelial cells (Chuluyan, 1995). In addition, activation of the endothelial cell monolayers with IL-1β, LPS, or TNFα supports both β_2- and VLA-4-dependent migration in response to C5a, PAF, MIP-1α, and RANTES (Chuluyan, 1995). In contrast to MIP-1α and RANTES, MCP-1-induced monocyte transmigration across an endothelial cell monolayer appears to be inhibited upon activation of endothelial cells with these inflammatory cytokines suggesting a differential role for the C-C chemokines in facilitating monocyte adhesion/migration (Chuluyan, 1995). Subsequent studies (Brown, 1994; Chuluyan, 1995; Rollins, 1990; Rollins, 1991; Sica, 1990b) have revealed that IL-1β-, IL-4-, TNFα-, and IFN-γ-activated endothelial cells express and secrete significant amounts of MCP-1. However, these results may not totally explain the inhibitory effects of MCP-1 as cytokine-, LPS-, adhesion-, and platelet-activated endothelial cells also produce MIP-1α, RANTES, and IL-8 (Brown, 1994; Kaplanski, 1993; Lukacs, 1994; Sica, 1990a; Strieter, 1989). Analogous to the IL-8/LAI effects described above, soluble MCP-1 as well as other C-C chemokines produced by cytokine-activated endothelial cells may limit monocyte adhesion and emigration into inflammatory tissues.

In summary, leukocyte entry into inflammatory tissues involves multiple activation and adhesion steps. The initial activation of a local vascular endothelial barrier by inflammatory cytokines such as IL-1 and TNFβ seems to be a prerequisite for leukocyte emigration into inflamed tissues. These cytokines upregulate selectin, ICAM, and VCAM-1 expression on the endothelial cell surface providing the adhesive substrates for leukocyte rolling as well as firm adhesion to the endothelium (Butcher, 1996; Springer, 1994; Springer, 1995). In addition, these cytokines also induce the production of IL-8 as well as C-C chemokines by endothelial cells and extravascular cells (Baggiolini, 1994; Oppenheim, 1991; Schall, 1991; Taub, 1994b). These endothelial-derived chemokines are believed to be presented to leukocytes as immobilized ligands on the luminal endothelial cell surface of blood vessels via heparan proteoglycans (Rot, 1996). Chemokines produced by extravascular cells are believed to be actively transported through the cytoplasm of the endothelial cells to the endothelium luminal surface (Rot, 1996). This localization of chemokines at the luminal surface has been shown to be accompanied by the transmigration of leukocytes (Rot, 1996). Thus, under physiological conditions, endothelial-bound chemokines, which promote integrin-mediated leukocyte adhesion to purified adhesion ligands and endothelial cell monolayers *in vitro* (Rot, 1992a; Rot, 1993; Rot, 1996), should promote leukocyte adhesion and transmigration into inflammed tissues. Immobilized chemokine presentation should allow for a more selective recruitment of leukocytes from the circulation as well as a more intense chemokine signal through the chemokine receptors on adhering leukocytes. In this way, adherent leukocytes are then permitted to directionally transmigrate in response to proteoglycan/matrix-bound chemokine concentration gradient in the extravascular tissues. Excessive soluble chemokines produced by endothelial or extravascular cells may serve to downregulate additional leukocyte extravasation and thus control the number of leukocytes infiltrating an inflammatory lesion.

EFFECTS OF C-C CHEMOKINES ON T CELL SUBSETS

Several recent studies have examined the ability of C-X-C, C-C, and C chemokines to modulate T cell migration and adhesion (Baggiolini, 1994; Schall, 1991; Taub, 1994b). Most of the C-C chemokines have been shown to induce human T lymphocyte and NK cell migration both *in vitro* and in vivo (Carr, 1994; Schall, 1990; Schall, 1993; Loetscher, 1994; Maghazachi, 1994; Murphy, 1994; Tanaka, 1993a; Taub, 1993a; Taub, 1993b; Taub, 1995d; Taub, 1995c; Taub, 1995d). Initial studies by Larsen (1989) and Schall (1990) were the first to demonstrate the ability of the chemokines IL-8 and RANTES, respectively, to facilitate human T lymphocyte migration *in vitro*. Subsequent studies by Tanaka (1993a) and Taub (1993a) demonstrated that purified MIP-1β induces human T cells *in vitro* in a dose-dependent manner. While Shaw and colleagues (Tanaka, 1993a) failed to observe T cell migration in response to purified MIP-1α, Taub and coworkers (1993a) have demonstrated that MIP-1α, MIP-1β, and RANTES preferentially induce the directional migration of preactivated human CD4+ and CD8+ as well as the CD45RA+ (naive) and CD29+ (memory) T cells depending on the chemokine utilized and the activation state of the T cell. These investigators also found that the *in vitro* methodology used to purify T cells greatly affected the migratory ability of the purified populations (Taub, 1995d). T cells purified using prolonged purification protocols involving multiple incubations at 37°C were found to express extremely low levels of C-C chemokine receptors on their cell surface while rapidly purified (<1 hr) T cells expressed a 2–4-fold greater number of cell surface C-C chemokine receptors as well as a greater chemokine responsiveness (Taub, 1993a; Taub, 1995d; Xu, 1995). T cell migration in response to these C-C chemokines also required the presence of extracellular matrix proteins to facilitate cell adhesion and migration (Taub, 1993a). Fibronectin was found to provide the best adhesive substrate by which T cell migrate in response to chemokines (Gilat, 1995; Taub, 1993a; Taub, 1993b; Taub, 1995b; Taub, 1995c; Taub, 1995d). These C-C chemokine findings have been subsequently supported by several laboratories as described in Table 2. In many of these assays, different T cell subsets were found to

Table 2 *In vitro* effects of chemokines on human lymphocyte migration and adhesion.

Reference	Chemokine (s)	Effects
Larsen, 1989	IL-8	CD4+ and CD8+ T cell migration
Schall, 1990	RANTES	CD4+ and CD45RO+ T cell migration
Tanaka, 1993b	MIP-1β	CD8+ T cell migration and adhesion to VCAM-1 and fibronectin
Taub, 1993a; Taub, 1996a	MIP-1α	Preferential CD8+ (both CD45RA+ and CD29+) T cell migration/adhesion toIL-1α-activated HUVEC and matrix proteins
	MIP-1β	Preferential CD4+ (both CD45RA+ and CD29+) T cell migration/adhesion to IL-1α-activated HUVEC and matrix proteins
	RANTES	CD4+ and CD8+ (both CD45RA+ and CD29+) T cell migration and adhesion to IL-1α-activated HUVEC and matrix proteins
Taub 1993b	IP-10	Ptimed CD4+ and CD8+ (both CD45RA+ and CD29+) T cell migration and adhesion to IL-1α-activated HUVEC and matrix proteins
Bacon, 1993; Bacon, 1995a	IL-8	CD4+, CD8+, and CD45RA+ T cell migration
Schall, 1993	MIP-1α	Preferential CD8+ T cell migration at high concentrations and CD4+ T cell migration at low concentrations
	MIP-1α	Preferential CD4+ T cell migration
Loetscher, 1994;	MCP-1, -2, -3	CD4+ and CD8+ T cell migration and adhesion to matrix proteins
Carr, 1994; Taub, 1995b; Roth, 1995		
Xu, 1995	IL-8	CD4+ and CD8+ T cell migration
Milat, 1995	MIP-1β, RANTES	CD4+ T cell adhesion to fibronectin
Del Pozo, 1995	MIP-1β	CD8+ T cell uropod formation and ICAM-3 redistribution
	MIP-1α, RANTES	CD4+ T cell uropod formation and ICAM-3 redistribution
Murphy, 1996	MIP-1β, RANTES	CD4+ and CD8+ T cell activation and migration *in vitro* and *in vivo* as well as mediating human T cell engraftment in huPBL-SCID mice
Taub, 1996d	IL-8	CD4+ and CD8+ human T cell migration *in vivo* and *in vitro*
Taub, 1996e	IP-10	Primed CD4+ and CD8+ human T cell migration *in vivo* and *in vitro*

selectively migrate in response to C-C chemokines. Several laboratories have shown that MIP-1α preferentially but not selectively induced human CD8+ T cell migration *in vitro* while MIP-1β preferentially induced CD4+ T cell migration (Schall, 1993; Taub, 1993a; Taub, 1993b). Other laboratories have observed distinct CD4+ or CD8+ as well as CD45RA+ or CD45RO+ T cell subset migration in response to MIP-1α and MIP-1β depending on the chemokine concentration used (Schall, 1993; Tanaka, 1993a; Taub, 1993a). Subsequent studies by Carr (1994), Loetscher (1994), Taub (1995b), and Roth (1995) have also found that MCP-1, MCP-2, and MCP-3 are also chemoattractants for both CD4+ and CD8+ T cells. IL-8 preferentially attracts CD45RA+ T cells when compared with CD45RO+ control population but does not distinguish between $CD4^+$ and $CD8^+$ T cells (Bacon, 1993; Bacon, 1995a; Xu, 1995). Thus, the differences in the migratory responses of various T cell subsets in response to C-C chemokines may be dependent on a number of factors including the T cell activation state, memory-naive status, the presence of adhesive ligands, and the T cell donors. As the chemokines appear to attract different subsets of lymphocytes, it is believed that these chemoattractants facilitate the sorting out of lymphocytes into different lymphoid compartments (Butcher, 1996; Springer, 1995; Tanaka, 1993c; Taub, 1994b). The migration observed in response to all of the C-C chemokines was found to be highly dose-dependent with optimal chemotactic activity being observed between 0.1–1 nM. However, higher chemokine doses ($\geq$ 10nM), which failed to induce T cell migration in microchemotaxis assays (Schall, 1993; Tanaka, 1993a; Taub, 1993a; Taub, 1993b; Taub, 1995d), were needed to facilitate *in vitro* endothelial transmigration of human T cells across an endothelial barrier (Roth, 1995). As with neutrophils and monocytes, this decreased ability of T cells to migrate in response to high concentrations of chemokine is believed to be due to chemokine receptor desensitization as well as disruption of the soluble chemokine gradient *in vitro.*

As described above, the chemokine-mediated T cell migration observed by several laboratories required the presence of T cell adhesive molecules within the chemotaxis assay (Taub, 1993a; Taub, 1993b; Taub, 1995d). Differential T cell migratory responses were observed in response to the C-C chemokines depending on the adhesive ligand utilized (ie. fibronectin vs. collagen) within the assay suggesting a relationship between cell adhesion and cell migration (Taub, 1995d). More directly, Shaw and coworkers (Tanaka, 1993a; Tanaka, 1993c) demonstrated that MIP-1β promotes the rapid (>15 minutes) adhesion of CD8+ but not CD4+ T cells to both immobilized fibronectin and VCAM-1-Ig in the presence of heparin-BSA or the heparin proteoglycans. As observed with the effects on migration, this chemokine-mediated adhesion was dose-and time-dependent. As proteoglycan CD44 is expressed on the endothelial cell surface, is upregulated upon activation, and appears to bind chemokines through its heparan sulfate side chains, it seems that this molecule is the most likely physiological candidate for the *in vivo* presentation of chemokines on the endothelial cell surface (Haynes, 1989). A role for CD44 was demonstrated by the finding that when purified CD44 co-immobilized with VCAM-1, it can capture MIP-1β and activate T cell β_1 integrins (Tanaka, 1993a). At the same time, Lloyd and Taub (Lloyd, 1996; Taub, 1993a; Taub, 1993b) demonstrated that MIP-1α and MIP-1β preferentially promote human T cell adhesion to IL-1α-activated human umbilical vein endothelial cell monolayers within 30 minutes after the initiation of the assay. However, this C-C chemokine induction to cytokine-activated endothelial cells was found to be dependent on both the donor of the T cells and the endothelial cells. Additional studies by these same investigators have demonstrated that MIP-1α, MIP-1β, and RANTES also induce the adhesion of rapidly purified quiescent T cells and T cell clones to the purified adhesive ligands, ICAM-1 and VCAM-1, as well as the extracellular matrix proteins, fibronectin, collagen type IV, and laminin (Lloyd, 1996). As with the Shaw study, this chemokine-induced increase in T cell adhesion to purified ligands was also rapid (<15 minutes) and required the presence of either β_2 or β_1 integrins depending on the substrate utilized. These studies are in agreement with the previous observations that T cell homing to lymphoid organs as well as trafficking to inflammatory tissues in vivo utilizes both β_2- and β_1-dependent pathways (Tanaka, 1993a). This increase in lymphocyte adhesion induced by C-C chemokines did not require the synthesis of new β_2 or β_1 integrins or CAMs as no increased expression was observed after 30 minutes or 4 hours of C-C chemokine treatment *in vitro*. Bacon and coworkers (1995b) have demonstrated that stimulation of a human CD4+ T cell clone with 1 μM of RANTES results in T cell activation/proliferation in the absence of a TCR signal. In addition, RANTES stimulation also results in increased cell surface expression of the β_2 (CD18) and β_1 (CD29) integrin molecules. However, this concentration of RANTES is quite high and may not be achieved physiologically even within inflammatory tissues. In contrast, Tanaka (1993a) and Lloyd (1993a) have failed to observe any modulation in the expression of β_1 or β_2 integrins on primary human T lymphocytes in response to α or β chemokines at concentrations of 0.1–10 nM. Recent studies by this laboratory have revealed that several C-C chemokines are capable of inducing the cell surface expression of β_1 and β_2 integrins on several antigen-specific human Th1, Th2, and Th0 cell clones but not primary T cells at chemokine concentrations of 1–100 nM. In most of these studies, like neutrophils and monocytes, the concentrations of chemokine required to trigger rapid and firm lymphocyte adhesion to CAMs failed to induce optimal T cell migration while doses of chemokines optimal for cell migration only modestly promoted lymphocyte adhesion suggesting that the concentration of a chemokine agonist may conversely regulate the observed biological effect, in this case, lymphocyte migration versus adhesion.

Gilat and coworkers (1994) have also demonstrated that extracellular matrix-bound MIP-1β and RANTES prolonged the time course of β_1-mediated interactions between CD4+ T cells and extracellular matrix molecules. The proadhesive effects of MIP-1β and RANTES required the presence of heparan sulfate-bound glycosaminoglycans on the extracellular matrix molecules as heparinase-treated matrix proteins failed to bind chemokine resulting in an abrogation in chemokine-mediated T cell adhesion. Similarly, T cell adhesion was also abrogated by the pretreatment of chemokines with soluble heparin or heparan sulfate. Thus, the adhesive interactions mediated by chemokines, which are regulated by heparin-like molecules, recruit T lymphocytes into extravascular compartments and haptotactically direct the migration of the matrix-bound leukocytes toward adjacent inflammatory tissue.

Several morphological changes occur during lymphocyte activation and migration (Butcher, 1996; del Pozo, 1995). Spherical resting T cells become polarized and develop cytoplasmic projections termed *uropods*. Del Pozo and coworkers (1995) have recently shown that RANTES, MCP-1 and to a lesser extent MIP-1α, MIP-1β, and IL-8 induce uropod formation and ICAM-3 redistribution in T lymphoblasts adhered to ICAM-1 or VCAM-1 as well as fragments of fibronectin. In addition to ICAM-3, ICAM-1, CD43, and CD44 are also redistributed to the lymphocyte uropod during RANTES stimulation. As with the above studies, chemokine-induced uropod formation was also T cell subset-specific with MIP-1β targeting CD8+ T cells and RANTES and MIP-1α targeting CD4+ T lymphocytes. The effect of each chemokine appears to involve G proteins and a cAMP-dependent signal transduction pathway. Overall, these findings suggest that chemokines may promote the redistribution of adhesion molecules to the leading edge of a lymphocyte uropod upon adhesion on integrin substrates possibly facilitating integrin:CAM interactions with endothelial cells as well as other cell populations.

Unfortunately, many of the above studies examine lymphocytes under *in vitro* static conditions which may not be physiologically relevant. In addition, the apparent contradictions in lymphocyte subset preferences and binding kinetics may be due to a number of variables including differences in levels and signaling through both promiscuous and specific chemokine receptors on the lymphocyte cell surface and variations in the responsiveness of individual T cells and their subsets. These difficulties can be overcome by using lymphoid cells lines transfected with classical chemoattractant receptors as recently described by Butcher and colleagues (Laudanna, 1996). Stimulation of lymphoid cells transfected with fMLP or IL-8 receptors results in the nucleotide exchange on the small G protein RhoA in seconds and promotes cell adhesion to various adhesive ligands. Using a murine B cell lymphoid cell line transfected with the human fMLP receptor or with the human IL-8 receptor type A, a rapid and strong VLA-4-dependent adhesion to purified VCAM-1 was observed upon stimulation with IL-8 or fMLP *in vitro*. This adhesion induction was found to be independent of diacylglycerol- and calcium-dependent isoforms of protein kinase C as the inhibitor calphostin C failed to blocked fMLP and IL-8 induced adhesion to VCAM-1. However, inactivation of Rho by C3 transferase exoenzyme, which inhibits Rho by ADP ribosylation, blocked fMLP and IL-8-induced VLA-4-mediated lymphocyte adhesion to VCAM-1 as well as neutrophil CD18 integrin adhesion to fibrinogen. These findings are the first to directly demonstrate a specific role for chemokine receptor-associated G proteins, in particular Rho, in the rapid triggering of leukocyte adhesion. This alteration in integrin-mediated adhesion is most likely due to avidity alterations in the cell surface integrin molecules via a G protein-mediated activation event. However, avidity determinations are required to directly prove this hypothesis.

CHEMOKINE EFFECTS IN VIVO

Many of the proadhesive and promigratory effects of chemokines described thus far were identified and characterized using *in vitro* assays with purified cell populations. However, the only true measure of the ability of a given chemokine to mediate leukocyte or leukocyte subset recruitment is through the localized injections of chemokines *in vivo*. Many chemokines have also been shown to induce the selective recruitment of various leukocytic subpopulations to sites of injection (Baggiolini, 1994; Matsushima, 1992; Oppenheim, 1991; Schall, 1991; Taub, 1993c; Taub, 1994b). *In vivo* injection of both C-X-C and C-C chemokines into rodent and rabbit models have been shown to induce a number of proinflammatory effects (Table 3) (Baggiolini, 1994; Matsushima, 1992; Oppenheim, 1991; Schall, 1991; Taub, 1993c; Taub, 1994b). Subcutaneous and intradermal injections of IL-8, GROα, GCP-2, and NAP-2 into humans, rabbits, rats and mice have been reported to induce predominantly neutrophil infiltration,

Table 3 *In vivo* effects of C-X-C and C-C chemokine family members (Baggiolini, 1995; Taub, 1994b).

Chemokine	Effects
IL-8/NAP-1	Neutrophil infiltration in human, primates, mouse, rat, rabbit, and dog
GCP-1	T cell infiltration in rats and dogs
	Human T cell infitration in SCID mouse
	Plasma leakage in rabbits
	Neutrophilia
	Angiogensis
	Synovial inflammation
GRO-α/MGSA	Neutrophil infiltration in rabbits
GCP-2	Granulocyte infiltration in rabbits
PF-4	Fibrosis
	Neutrophil and mononuclear cell infiltration in mice
	Inhibits angiogensis
MIP-2	Neutrophil infiltration in mice
IP-10	Protection against tumor growth
	Human T cell infiltration in SCID mice
MCP-1	Monocyte infiltration in mice, rats, and dogs
	Human T cell infiltration in SCID mice
	Protection against tumor growth
MCP-2	Monocyte infiltration in rabbits
MCP-3	Monocyte infiltration in rabbits
MIP-1α	Neutrophil infiltration in mice and rabbits
	Monocyte infiltration in rabbits and dogs
	Non-prostaglandin mediated fever induction
	Human T cell infiltration in SCID mice
MIP-1β	Neutrophil infiltration in mice and rabbits
	Monocyte infiltration in rabbits
	Non-prostaglandin mediated fever induction
	Human T cell infiltration in SCID mice
RANTES	Eosinophil and monocyte infiltration in mice, dogs, and humans
	Human T cell infiltration in SCID mice
Eotaxin	Eosinophils and monocytes in guinea pigs and mice

beginning as soon as 3 hr post injection (Colditz, 1989; Colditz, 1990; Colditz, 1992; Collins, 1992; Foster, 1989; Harada, 1993; Hechtman, 1990; Larsen, 1989; Leonard, 1991; Proost, 1993; Rampart, 1989; Van Osselaer, 1991; Swensson, 1991). Very few monocytes or lymphocytes have been observed in these studies; however, only short term histological analyses (<24hr) were performed by these investigators. In addition, MIP-2 appears to be a more potent mediator of neutrophil recruitment in mouse skin and brain tissues than IL-8, GROα, NAP-2, or the C-C chemokines (Bell, 1996a; Bell, 1996b). However, IL-8 has also been shown to induce lymphocyte infiltration in rats by 48 hr after subcutaneous and intradermal injection (Larsen, 1989). Interestingly, these same studies demonstrating that injection or expression of MIP-2 in mouse hypothalamus induces neutrophil infiltration also demonstrate the ability of MIP-2 to permeabilize the blood-brain barrier thus permitting the entry of leukocytes across the parenchyma (Bell, 1996a). Intracutaneous injections of low doses of IL-8 in humans induced a rapid local neutrophilic infiltration without any significant edema or erythema possibly through the avidity alterations and the induction of β_2 integrins (Leonard, 1991; Swensson, 1991). In contrast, intravenous administration of IL-8 induces a profound neutrophilia most probably due to the demargination of tissue and bone marrow granulocytes (Hechtman, 1990; Hechtman, 1991; Jagels, 1992; Ley, 1994; Van Zee, 1992). In contrast to the findings with subcutaneous and intradermal injections with IL-8, intravenous administration of IL-8 also specifically reduces the local inflammatory response to subcutaneously injected IL-8, but also to the neutrophil chemotactic activators, fMLP or C5a (Hechtman, 1990; Hechtman, 1991; Van Zee, 1992). As discussed earlier, this inhibitory effect is most likely due to the heterologous desensitization of neutrophils, the profound neutrophilia induced upon injection, and the lack of an IL-8 concentration gradient. Injections of MIP-1α, MIP-1β, RANTES, MCP-1, and I-309/TCA-3 have been reported to cause edema and a mild early neutrophil accumulation followed by a more prominent monocytic infiltration (Jose, 1991; Luo, 1994; Matsushima, 1989; Meurer, 1993; Miller, 1989; Sherry, 1988; Taub, 1994b; Wolpe, 1988; Wolpe, 1989; Yoshimura, 1987). In contrast, subcutaneous injection of human MCP-1 into the ears of rats induces a monocytic but not neutrophilic infiltrate beginning as early as 3 hr and peaking by 18–24 hr (Matsushima, 1989; Yoshimura, 1987). Purified preparations of mixtures of MIP-1α/β have also been reported to induce local acute neutrophil accumulation at injection sites, whereas intracisternal injections of such mixtures of MIP-1α/β cause an influx of neutrophils into the cerebrospinal fluid by 4–6 hr, followed by a later monocytic infiltrate (Wolpe, 1988; Wolpe, 1989). Intradermal injection of MCP-2 and MCP-3 in rabbits resulted in selective monocyte recruitment *in vivo* (Yoshimura, 1987) while the intradermal injection of eotaxin into guinea pigs induces eosinophil accumulation in the skin (Jose, 1991; Jose, 1994). Interestingly, the human C-C chemokines, RANTES, MIP-1α, and MCP-1, were found to be ineffective in the guinea pig (Jose, 1994). While the most extensive and detailed studies have been performed with IL-8, it is believed that many C-X-C chemokines with neutrophil activating properties will have similar activities *in vivo*. In contrast to IL-8, IP-10 has been shown to induce a murine mononuclear cell infiltrate (mostly monocytes) within 4 hr of injection into mice with little to no neutrophil recruitment, supporting numerous *in vitro* chemotaxis studies (Taub, 1996d).

More recent studies in mice with severe combined immunodeficiency (SCID) engrafted with human peripheral blood lymphocytes (huPBL-SCID) have demonstrated that human CD3$^+$ T cells migrate by 72 hr into injection sites in response to recombinant preparations of human IL-8 and IP-10 (Taub, 1996c; Taub, 1996d). The injection of human IL-8 into these chimeric SCID mice also led to an early (4 hr) infiltration by murine neutrophils. Interestingly, 72 hr after IL-8 injection, a high number of human T lymphocytes infiltrated the injection site. These IL-8/huPBL-SCID studies suggest that neutrophil infiltration may be an important prerequisite for lymphocyte entry into inflammatory sites. Systemic depletion of neutrophils within the huPBL-SCID prior to chemokine administration blocked human T cell recruitment to the IL-8 injection site supporting the hypothesis that the chemotactic effects of IL-8 on human T cells were indirect (Chertov, 1996; Taub, 1996c). Similarly, subcutaneous injections of rhIP-10 into these huPBL-SCID mice induced significant murine leukocyte infiltration, particularly monocytes, as well as human CD3+ T cells infiltration within 24–48 hr of IP-10 injection (Taub, 1996d). In addition, the C-C chemokines, MIP-1α, MIP-1β, MCP-1, and RANTES, have been shown to induce significant infiltration of murine monocytes within 4 hr of injection in huPBL-SCID mice (Murphy, 1994; Murphy, 1996; Taub, 1996c). Significant human CD3+ T cell recruitment can also be observed within these injection sites between 4–48 hr of chemokine administration. In these same animals, MIP-1β and RANTES treatment also promoted human T cell trafficking to the thymus and various peripheral lymphoid tissues of these mice (Murphy, 1996). Similar results have been observed using human growth hormone (rhGH) in this model (Taub, 1994a). This human lymphocyte trafficking appears to require the presence of active integrin molecules as antibodies to both β_2 (CD18) and β_1 (CD29) block chemoattractant-mediated T cell engraftment (Murphy, 1996; Taub, 1994a). Together, these results suggest that chemokine injection or localized chemokine production contributes to the trafficking of cells to peripheral tissue sites as well as lymphoid tissues.

CHEMOKINES AND PATHOLOGY

Chemokines have been detected in numerous inflammatory and non-inflammatory disease states (Table 4). In humans, IL-8 and MCP-1 production has been associated with inflammation in multiple diseases, including sepsis, atherosclerosis, cystic fibrosis, idiopathic pulmonary fibrosis, as well as many other diseases and syndromes (Baggiolini, 1994; Brown, 1996; Cook, 1996; Edgington, 1993; Gomez-Chiarri, 1993; Kunkel, 1996; Oppenheim, 1996; Schall, 1994; Standiford, 1995; Standiford, 1996; Strieter, 1996a; Strieter, 1996b; Taub, 1993c; Taub, 1994b). In many of the inflamed lesions, the presence of C-X-C and/or C-C chemokines is highly correlated with the level of neutrophil and/or monocyte infiltration within the diseased tissue. Neutralization of chemokines with specific antibodies within an animal asthma or autoimmunity model typically results in decreased cell infiltration as well as a decrease in disease severity while neutralization of specific chemokines within a sepsis or microbial inflammation model usually results in disease progression and mortality. Thus, under some conditions, the chemokines rather than having self-destructive effects facilitate leukocyte adhesion and transmigration into compromised tissues resulting in the phagocytosis and killing of microbes or allogeneic cells. During these circumstances, chemokines promote host defense and assist in disease regression. In other states, aberrant chemokine production, due to an inappropriate or a genetically predisposed inflammatory disease state such as autoimmunity or allergy, appears to contribute to the induction and progression of disease pathology. In these diseases, chemokines induce the infiltration and activation of leukocytes into the susceptible tissues promoting tissue destruction. The generation of humanized antibodies to the various chemokines and/or chemokine-specific antagonists may aid in the treatment of these aberrant disease states.

CHEMOKINES AND LYMPHOCYTE RECIRCULATION/HOMING

Despite the fact that chemokines induce the adhesion and migration of different lymphocyte subsets, their contribution to the process of lymphocyte homing into

Table 4 Role of chemokines in disease pathology (Oppenheim, 1996; Taub, 1994b).

Pathological State	Chemokines
Pulmonary Fibrosis	IL-8
Bronchial carcinoma	IL-8
Sarcoidosis	IL-8, MCP-1
Cystic fibrosis lavage fluids/tissues	IL-8, ENA-78, MCP-1
Pulmonary reperfusion injury	IL-8, MCP-1
Rejection of lung transplant	IL-8, MCP-1
Endotoxemia and sepsis	IL-8, MIP-1α, RANTES, MIP-2
Rheumatoid synovial fluids	IL-8, ENA-78, MCP-1, MIP-1α, RANTES
Osteoarthritic synovial fluids	MIP-1β
Psoriatic scale extracts	IL-8, GROα, MCP-1, IP-10
Gingivitis	IL-8
Inflammatory bowel syndrome	IL-8
Gastritis	IL-8
Alcoholic hepatitis	IL-8
Post-major surgery	Elevated plasma IL-8
Peritonitis	Elevated plasma IL-8
Acute glumerulonephritis	IL-8
Uveoretinitis	IL-8, IP-10, MIP-1α, MIP-1β, RANTES, MCP-1
S. mansoni-induced granulomas	MIP-1α
Meningococcal infections	IL-8
Lepromatous/tubercuolid leprosy lesions	IP-10
Mycobacterial infections	IL-8
Crypotococcal infections	MIP-1α, MCP-1, MIP-2
Angiogenesis	IL-8, GROα, GROβ, ENA-78, GCP-2, NAP-2
Angiostatic	PF-4, IP-10, mig
Many cancer cells	Produce C-X-C and C-C chemokines as growth factors and angiogenic factors
Atherosclerosis	MIP-1α, MIP-1β, MCP-1, RANTES, IL-8
Cardiac allograft	MIP-1α, MIP-1β, RANTES
Wound healing	MCP-1

various lymphoid organs remains very circumstantial (Butcher, 1996; Picker, 1994). However, recent studies demonstrating the expression of chemokines in some normal healthy tissues suggests that these attractants may play a role in leukocyte homing *in vivo* (Butcher, 1996; Rot, 1996; Tanaka, 1993c). Constitutively induced chemokine production by secretory epithelial cells has been described in skin, gastrointestinal mucosa, and mammary glands (Rot, 1992a; Rot, 1993; Rot, 1996). Under normal physiological conditions, lymphocytes are the infiltrating cell type within these tissues in response to these attractants. In addition, MIP-1β, a potent T cell chemoattractant, was observed on the cell surface of HEV in normal mouse and human lymph nodes and may participate in populating the lymph nodes with lymphocytes (Tanaka, 1993c). Both normal and HIV-infected lymph nodes were also found to express constitutive levels of C-C chemokines (Tedla, 1996). The presence of IP-10 in normal murine thymic tissue and its ability to induce lymphocyte migration in vivo and in vitro suggests a possible role in populating the thymus (Gattass, 1994; Taub, 1993b; Taub, 1996c). Similarly, the C chemokine, Ltn, which was isolated and cloned from murine thymocytes and appears to be constitutively expressed within this organ (Kelner, 1994). In addition, a role for chemokines in thymocyte homing is also greatly supported by studies demonstrating that the ability of murine thymocytes to gain access to peripheral compartments is PT-sensitive (Chaffin, 1991; Morse, 1970). In these studies, mature lineage T cells are able to accumulate in the thymuses of transgenic lck-PT mice, which express the catalytic subunit of PT in all thymocytes, but not in the peripheral lymphoid organs. These mature T cells are biologically active and resemble normal peripheral blood lymphocytes but perform poorly in *in vivo* homing assays. However, this migratory defect does not result in deficient expression of common homing receptors or alterations in intracellular cAMP suggesting a G protein signaling pathway(s) requirement for thymocyte emigration. Furthermore, in situ video microscopic analyses of lymphocyte interactions with HEV in exteriorized Peyer's patches reveal that PT has no effect on the selectin-mediated rolling displayed by many lymphocytes on HEV but inhibits the proadhesive integrin signaling event required for lymphocyte arrest (Bargatze, 1993). This PT effect is dependent on G-proteins modifying ADP-ribosyltransferase activity and can be reversed by phorbol esters (which can bypass cell surface receptor triggering of integrin activation). These studies suggest that lymphocyte adhesion to HEV involves a G-protein signaling pathway, most probably chemokines, which facilitate firm lymphocyte attachment.

Studies by Murphy and coworkers (1996) have also shown that human lymphocytes treated with MIP-1β or RANTES in vitro, but not IL-8, facilitated human CD3+ T cell engraftment in the thymus of huPBL-SCID mice. This thymic trafficking was found to be chemokine-specific and integrin-dependent. These findings, as well as the ability of C-C chemokines to mediate lymphocyte adhesion to purified VCAM-1, ICAM-1, various extracellular matrix proteins (Taub, 1996b), and IL-1-activated endothelial cells (Taub, 1993a), strongly suggest that lymphocyte trafficking into normal lymphoid organs could be guided by chemokines in a manner similar those utilized in inflammatory sites. Together, these studies strongly support a role for chemokines in lymphocyte trafficking and homing to the various lymphoid organs *in vivo.*

CONCLUSIONS

The role of chemokines in leukocyte extravasation is best explained by the multistep model of leukocyte trafficking into sites of inflammation (Butcher, 1991; Butcher, 1996; Springer, 1994; Springer, 1995; Tanaka, 1993a; Tanaka, 1993c). The key factors of these molecules are the sequential interactions of selectin:carbohydrate ligand, chemokine:proteoglycan:receptor, and integrin family members. Within a compromised tissue, damaged peripheral tissue cells and activated leukocytes produce and release a number of inflammatory mediators including IL-1, TNFα, chemokines, and eicosinoids into the local tissue microenviroment. These agents activate the local endothelial cell barrier upregulating cell surface selectin and CAM expression. Initial selectin-mediated tethering and rolling of circulating leukocytes along this "inflamed" endothelial lining results in enhanced exposure of leukocytes to chemokines located on the endothelial cell surface via heparan proteoglycan interactions (Rot, 1992a; Rot, 1993; Rot, 1996; Tanaka, 1993c). Exposure of leukocytes to chemokines not only provides the chemotactic signal to the cells but also facilitates their release from selectin binding and promotes firm adhesion to endothelial cells through the leukocyte β_1 and β_2 integrin interactions with the CAMs on the endothelial cell surface. In some instances, this leukocyte:endothelial cell interaction leads to the further release chemokines into the basilar space of the endothelial layer which may further impart directionality to the leukocyte emigration. Due to the leukocyte-subset selectivity of the chemokines, this adhesion and the subsequent transmigration events may involve multiple adhesion pathways depending on the leukocyte population. Furthermore, the ability of chemokines to promote metalloprotease expression and secretion may assist in transversing of the endothelial barrier as well as the various layers of the basement membrane. Once across the vascular barrier, leukocytes directionally adhere and migrate along the various extracellular matrix proteins and CAMs within the tissue toward the sites of chemokine production. The chemotactic gradients are believed to be maintained via chemokine binding to extracellular matrix proteins (many of which possess GAG-binding sites) and heparin proteoglycans found within the tissues. As leukocytes infiltrate sites of increased chemokine concentrations, chemokine receptor desensitization occurs and inhibits further migration in response to a given chemokine gradient. However, many additional chemokine functions require increased chemokine levels and thus may facilitate other biological activities including adhesion, degranulation, cytotoxicity, and phagocytosis. At present, it appears that each of the chemokines have both unique and overlapping activities. Chemokines are implicated as major participants in acute and chronic inflammatory reactions and angiogenesis (Strieter, 1996). There is recent evidence from this laboratory demonstrating that chemokines also play a role in the induction of immune reactions and promote the effector limb of immunity (Murphy, 1996; Taub, 1996a; Taub, 1996b). The likely possibility that chemokines may contribute to the normal homing and distribution of leukocytes is also currently under evaluation. It seems quite likely that chemokines may facilitate T cell homing to lymph nodes and possibly the thymus during T cell development, maturation, and recirculation. Recent immunohistological studies demonstrating the presence of chemokines on the surface of HEV (Rot, 1996; Tanaka, 1993c) and within thymic tissues (Gattass, 1994) would suggest a possible role in lymphocyte recirculation. Thus, chemokines are attractive targets for the development of new anti-inflammatory agents as their inhibition may be an effective anti-inflammatory strategy. Conversely, promoting chemokine activity might enhance wound healing and tissue repair as well as the trafficking and engraftment of leukocytes to various lymphoid organs.

REFERENCES

Abbassi, O., Kishimoto, T.K., McIntire, L.V., Anderson, D.C. and Smith, C.W. (1993) E-selectin supports neutrophil rolling in vitro under conditions of flow. *J. Clin. Invest.*, **92**, 2719.

Alam, R., P.A. Forsythe, S. Stafford, M.A.Lett-Brown and Grant, J.A. (1992) Macrophage inflammatory protein-1α activates basophils and mast cells. *J. Exp. Med.*, **176**, 781.

Alam, R., Stafford, S., Forsythe, P., Harrison, R., Faubion, D., Lett-Brown, M.A. and Grant, J.A. (1993) RANTES is a chemotactic and activating factor for human eosinophils. *J. Immunol.*, **150**, 3442.

Allavena, P., Paganin, C., Zhou, D., Bianchi, G., Sozzani, S., Mantovani, A. (1994) Interleukin-12 is chemotactic for natural killer cells and stimulated their interaction with vascular endothelial cells. *Blood*, **84**, 2261.

Amatruda, T.T., Gerard, N.P., Gerard, C. and Simon, M.I. (1993) Specific interactions of chemoattractant factor receptors with G-proteins. *J. Biol. Chem.*, **268**, 10139.

Angiolillo, A.L., Sgadari, C., Taub, D.D., Farber, J.M., Maheshwari, S., Kleinman, H.K. and Tosato, G. (1995) Human interferon-inducible protein-10 is a potent inhibitor of angiogenesis in vivo. *J. Exp. Med.*, **182**, 155.

Arfors, K.E., Lundberg, C., Lindbom, L., Lundberg, K., Beatty, P.G. and Harlan, J.M. (1987) A monoclonal antibody to the membrane glycoprotein complex CD18 inhibits polymorphonuclear leukocyte accumulation and plasma leakage in vivo. *Blood*, **69**, 338.

Bacon, K.B., Quinn, D.G., Aubry, J.-P. and Camp, R.D.R. (1993) Interleukin-8–stimulated polyphosphoinositide hydrolysis in human peripheral blood lymphocytes. *Blood*, **81**, 430.

Bacon, K.B., Flores-Romo, L., Aubry, J.-P., Wells, T.N.C. and Power, C.A. (1994) Interleukin-8 and RANTES induce the adhesion of the human basophilic line KU-812 to human endothelial cell monolayers. *Immunol.*, **82**, 473.

Bacon, K.B., Flores-Romo, L., Life, P.F., Taub, D.D., Premack, B.A., Arkinstall, S.J., Wells, S.J. and Schall, T.J. (1995a) IL-8–induced signal transduction in T lymphocytes involves receptor-mediated activation of phospholipase C and D. *J. Immunol.*, **154**, 3654.

Bacon, K.B., Premack, B.A., Gardner, P. and Schall, T.J. (1995b) Activation of dual T cell signaling pathways by the chemokine RANTES. *Science*, **269**, 1727.

Baggiolini, M. (1991) Chemotactic and inflammatory cytokines – CXC and CC proteins. In: Westwick, J., Kunkel, S. and I.J.D. Lindley (eds). *Chemotactic Cytokines: Biology of the inflammatory peptide gene superfamily.* Londom:Plenum Press, pg. 1–11.

Baggiolini, M., Dewald, B., Moser, B. (1994) Interleukin-8 and related chemotactic cytokines - CXC and CC Chemokines. *Adv Immunol.*, **55**, 97.

Bainton, D.F., Miller, L.J., Kishimoto, T.K. and Springer, T.A. (1987) Leukocyte adhesion receptors are stored in peroxidase-negative granules in human neutrophils. *J. Exp. Med.*, **166**, 1641.

Bargatze, R.F. and Butcher, E.C. (1993) Rapid G protein-regulated activation event involved in lymphocyte binding to high endothelial venules. *J. Exp. Med.*, **178**, 367.

Bell, M., Taub, D.D. and Perry, V.H. (1996a) Overriding the brain's intrinsic resistance to leukocyte recruitment with intraperitoneal injections of recombinant chemokines. *Submitted for publication.*

Bell, M.,Taub, D.D., Kunkel, S.L., Strieter, R.M., Foley, R., Goldie, J. and Perry, V.H. (1996b) Recombinant human adenovirus with rat MIP-2 gene insertion causes prolonged PMN recruitment to the murine brain. *Submitted for publication.*

Berg, E.L., Magnani, J., Warnock, R.A., Robinson, M.K. and Butcher, E.C. (1992) Comparison of L-selectin and E-selectin ligand specificities: The L-selectin can bind the E-selectin ligands sialyl Le^x and sialyl Le^a. *Biochem. Biophys. Res. Commun.*, **184**, 1048.

Berger, M., O'Shea, J., Cross, A.S., Folks, T.M., Chused, T.M., Brown, E.J. and Frank, M.M. (1984) Human neutrophils increase expression of C3bi as well as C3b receptors upon activation. *J. Clin. Invest.*, **74**, 1566.

Berger, M.S., Taub, D.D., Orlofsky, A., Kleyman, T.R., Coupaye-Gerard, B., Eisner, D., Cohen, S.A. The chemokine C10: Immunological analysis of the sequence encoded by the novel second exon. *Cytokine* (in press).

Bevilacqua, M.P., Pober, J.S., Wheeler, E., Cotran, R.S. and Gimbrone, M.A. (1985) Interleukin-1 acts on cultureed human vascular endothelium to increase the adhesion of polymorphonuclear leukocytes, monocytes and related leukocyte cell lines. *J. Clin. Invest.*, **76**, 2003.

Bischoff, S.C., M. Krieger, T. Brunner, A. Rot, V.v. Tscharner, M. Baggiolini and Dahinden, C.A. (1993) RANTES and related chemokines activate human basophil granulocytes through different G protein-coupled receptors. *Eur. J. Immunol.*, **23**, 761.

Bokoch, G.M. (1995) Chemoattractant signaling and leukocyte activation. *Blood*, **86**, 1649.

Brett, J., Gerlach, H., Nawroth, P., Steinberg, S., Godman, G. and Stern, D. (1989) Tumor necrosis factor/cachectin increases permeability of endothelial cell monolayers by mechanisms involving regulatory G proteins. *J. Exp. Med.*, **169**, 1977.

Brevario, F., Bertocchio, F., Dejana, E. and Bussolino, F. (1988) IL-1–induced adhesion of polymorphonuclear leukocytes to cultured endothelial cells, role of platelet activating factor. *J. Immunol.*, **141**, 3391.

Brown, Z., Gerritsen, M.E., Carley, W.W., Strieter, R.M., Kunkel, S.L. and Westwick, J. (1994) Chemokine gene expression and secretion by cytokine-activated human microvascular endothelial cells: Differential regulation of monocyte chemoattractant protein-1 and interleukin-8 in response to interferon-γ. *Am. J. Pathol.*, **145**, 913.

Brown, Z., Robson, R.L. and Westwick, J. (1996) Regulation and expression of chemokines: potential role in glomerulonephritis. *J. Leuk. Biol.*, **59**, 75.

Bussolino, F., Breviario, F., Tetta, C., Anglietta, M., Mantovani, A. and Dejana, E. (1986) Interleukin-1 stimulates platelet activating factor production in cultured human endothelial cells. *J. Clin. Invest.*, 77, 2027.

Butcher, E.C. (1991) Leukocyte-endothelial cell recognition: Three (or more) steps to specificity and diversity. *Cell*, **67**, 1033.

Butcher, E.C. and Picker, L.J. (1996) Lymphocyte homing and homeostasis. *Science*, **272**, 60.

Camussi, G., Tetta, C. and Baggiolini, C. (1990) The role of platelet-activating factor in inflammation. *Curr. Immun. Immunopathol.*, **57**, 331.

Camussi, G., Montrucchio, G., Lupia, E., De Martino, A., Perona, L., Arese, M., Vercellone, A., Toniolo, A. and Bussolino, F. (1995) Platelet-activating factor directly stimulates in vitro migration of endothelial cells and promotes in vivo angiogensis by a heparan-dependent mechanism. *J. Immunol.*, **154**, 6492.

Carlos, T.M. (1994) Leukocyte-endothelial adhesion molecules. *Blood*, **84**, 2068.

Carr, M.W., Roth, S.J., Luther, E., Rose, S.S., Springer, T.A. (1994) Monocyte chemoattractant protein 1 acts as a T-lymphocyte chemoattractant. *Proc. Natl. Acad. Sci. USA*, **91**, 3652.

Carveth, H.J., Bohnsack, J.F., McIntyre, T.M., Baggiolini, M., Prescott, S.M. and Zimmermann, G.A. (1989) Neutrophil activating factor (NAF) induces polymorphonuclear leukocyte adherence to endothelial cells and to subendothelial matrix proteins. *Biochem. Biophys. Res. Commun.*, **162**, 387.

Cavender, D.E., Haskard, D.O., Joseph, B. and Ziff, M. (1986) Interleukin-1 increases the binding of human B and T lymphocytes to endothelial cell monolayers. *J. Immunol.*, **136**, 203.

Cavender, D.E., Edelbaum, D. and Welkovich, L. (1991) Effects of inflammtory cytokines and phorbol esters on the adhesion of U937 cells, a human monocyte-like cell line, to endothelial cell monolayers and extracellulae matrix proteins. *J. Leuk. Biol.*, **49**, 566.

Chaffin, K.E. and Perlmutter, R.M. (1991) A pertussis toxin-sensitive process controls thymocyte emigration. *Eur. J. Immunol.*, **21**, 2565.

Charo, I.F., Yuen, C. and Goldstein, I.M. (1985) Adherence of human polymorphonuclear leukocytes to endothelial monolayers: effects of temperature, divalent cations and chemotactic factors on the strength of adherence measured with a new centrifugation assay. *Blood*, **65**, 473.

Charo, I.F., Yuen, C., Perez, H.D. and Goldstein, I.M. (1986) Chemotactic peptides modulate adherence of human polymorphonuclear leukocytes to monolayers of cultured endothelial cells. *J. Immunol.*, **136**, 3412.

Chertov, O., Michiel, D.F., Xu, L., Wang, J.M., Pereira, A.H., Taub, D.D. and Oppenheim, J.J. (1996) Identification of defensin-1, defensin-2 and azurocidin as T cell chemoattractant proteins released from IL-8–stimulated neutrophils. *J. Biol. Chem.*, **271**, 2935.

Chuluyan, H.E., Schall, T.J., Yoshimura, T. and Issekutz, A.C. (1995) IL-1 activation of endothelium supports VLA-4 (CD49d/CD29)-mediated monocyte transendothelial migration to C5a, MIP-1α, RANTES and PAF but inhibits migration to MCP-1: a regulatory role for endothelium-derived MCP-1. *J. Leuk. Biol.*, **58**, 71.

Colditz, I.G. (1985) Margination and emigration of leukocytes. *Surv. Synth. Pathol. Res.*, **4**, 44.

Colditz, I., Zwahlen, R., Dewald, B. and Baggiolini, M. (1989) In vivo inflammatory activity of neutrophil-activating factor, a novel chemotactic peptide derived from human monocytes. *Am. J. Pathol.*, **134**, 755.

Colditz, I.G., Zwahlen, R.D. and Baggiolini, M. (1990) Neutrophil accumulation and plasma leakage induced in vivo by neutrophil-activating peptide-1. *J. Leuk. Biol.*, **48**, 129.

Colditz, I.G. and Watson, D.L. (1992) The effect of cytokines and chemotactic agonists on the migration of T lymphocytes into skin. *Immunology*, **76**, 272.

Collins, P.D., Jose, P.J. and Williams, T.J. The sequential generation of neutrophil chemotattractant proteins in acute infalmmation in the rabbit in vivo: relationship between C5a and a protein with the charcateristics of IL-8. *J. Immunol.*, **146**, 677.

Cook, D.N. (1996) The role of MIP-1α in inflammation and hematopoiesis. *J. Leuk. Biol.*, **59**, 61.

Cramer, E.B., Migliorisi, G., Pologe, L., Abrahams, E., Pawlowski, N., Cohn, Z. and Scott, W.A. (1984) Effect of leukotrienes on endothelium and the transendothelial migration of neutrophils. *J. Allergy Clin. Immunol.*, **74**, 386.

Cybulsky, M.I., McComb, D.J. and Movat, H.Z. (1988) Neutrophil leukocyte emigration induced by endotoxin: mediator roles of interleukin 1 and tumor necrosis factor α. *J. Immunol.*, **140**, 3144.

Darbonne, W.C., Rice, G.C., Mohler, M.A., Apple, T., Herbert, C.A., Valente, A.J. and Baker, J.B. (1991) Red blood cells are a sink for interleukin 8, a leukocyte chemotaxin. *J. Clin. Invest.*, **88** (4), 1362.

del Pozo, M.A., Sanchez-Mateos, P., Nieto, M. and Sanchez-Madrid, F. (1995) Chemokines regulate cellular polarization and adhesion receptor redistribution during lymphocyte interaction with endothelium and extracelluar matrix. Involvement of cAMP signaling pathway. *J. Cell Biol.*, **131**, 495.

Detmers, P.A., Lo, S.K., Olsen-Egbert, E., Walz, A., Baggiolini, M. and Cohn, Z.A. (1990) Neutrophil-activating protein 1/interleukin-8 stimulates the binding activity of leukocyte adhesion receptor CD11b/CD18 on human neutrophils. *J. Exp. Med.*, **171**, 1155.

Devi, S., Laning, J., Luo, Y. and Dorf, M.E. (1995) Biologic activities of the β-chemokine TCA-3 on neutrophils and macrophages. *J. Immunol.*, **154**, 5376.

Ebisawa, M., Yamada, T., Bickel, C., Klunk, D. and Schleimer, R.P. (1994) Eosinophil transendothelial migration induced by cytokines. III. Effect of the chemokine RANTES. *J. Immunol.*, **153**, 2153.

Edgington, S.M. (1993) Chemokines in cardiovascular disease. *Biotechnology*, **11**, 676.

El-Naggar, A.K., Van Epps, D.E., Williams, Jr., R.C. (1981) Effect of culturing on the human lymphocyte locomotion response to casein, C5a and fMet-Leu-Phe. *Cell Immunol.*, **60**, 43.

Fernandez, H.N., Henson, P.M., Otani, A. and Hugli, T. (1978) Chemotactic response to human C3a and C5a anaphylatoxins. I. Evaluation of C3a and C5a leukotaxis in vitro and under stimulated in vivo conditions. *J. Immunol.*, **120**, 109.

Foster, S.J., Aked, D.M., Schroder, J.M. and Christophers, E. (1989) Acute inflammatory effects of a monocyte-derived neutrophil-activating peptide in rabbit skin. *Immunol.*, **67**, 181.

Furie, M.B. and McHugh, D.D. (1989) Migration of neutrophils across endothelial monolayers is stimulated by treatment of the monolayers with interleukin-1 or tumor necrosis factor-alpha. *J. Immunol.*, **143**, 3309.

Furie, M.B., Tancinco, C.A. and Smith, C.W. (1991) Monoclonal antibodies to leukocyte integrins CD11a/CD18 and CD11b/CD18 or intercellular adhesion molecule-1 inhibits chemoattractant-stimulated neutrophil transnedothelial cell migration in vitro. *Blood*, **78**, 2089.

Gamble, J.R., Harlan, J.M., Klebanoff, S.J. and Vadas, M.A. (1985) Stimulation of the adherence of neutrophils to umbilical vein endothelium by human recombinant tumor necrosis factor. *Proc. Natl. Acad. Sci. USA*, **82**, 8667.

Gao, J.-L., Kuhns, D.B., Tiffany, H.L., McDermott, Li, X., Francke, U. and Murphy, P.M. (1993) Structure and functional expression of the human macrophage inflammatory protein 1α/RANTES receptor. *J. Exp. Med.*, **177**, 1421.

Gattass, C.R., King, L.B., Luster, A.D., Ashwell, J.D. (1994) Constitutive expression of interferon-γ-inducible protein 10 in lymphoid organs and inducible expression in T cells and thymocytes. *J. Exp. Med.*, **179**, 1373.

Gilat, D., Hershoviz, R., Mekori, Y.A., Vlodavsky, I., Lider, O. (1994) Regulation of adhesion of CD4+ T lymphocytes to intact or heparinase-treated subendothelial extracellular matrix by diffusible or anchored RANTES and MIP-1β. *J. Immunol.*, **153**, 4899.

Gimbrone, M.A., Jr., Brock, A.F. and Schafer, A.I. (1984) Leukotriene B_4 stimulates polymorphonuclear leukocyte adhesion to cultured vascular endothelial cells. *J. Clin. Invest.*, **74**, 1552.

Gimbrone, M.A.J., Obin, M.S., Brock, A.F., Luis, E.A., Hass, P.E., Hebert, C.A., Yip, Y.K., Leung, D.W., Lowe, D.G., Kohr, W.J., Darbonne, W.C., Bechtol, K.B. and Baker, J.B. (1989) Endothelial interleukin-8: a novel inhibitor of leukocyte-endothelial interactions. *Science*, **246**, 1601.

Goetzl. E.J. and Pickett, W.C. (1981) Novel structural determinants of the human neutrophil chemotactic activity of leukotriene B_4. *J. Exp. Med.*, **153**, 482.

Gomez-Chiarri, M., Ortiz, A., Seron, D., Gonzalez, E. and Egido, J. (1993) The intercrine superfamily and renal disease. *Kidney Int.*, **43**, 881.

Hadley, T.J., Lu, Z., Wasniowska, K., Martin, A.W., Peiper, S.C., Hesselgesser, J. and Horuk, R. (1994) Postcapillary venule endothelial cells in kidney express a multispecific chemokine receptor that is structurally and functionally identical to the erythroid isoform, which is the Duffy blood group antigen. *J. Clin. Invest.*, **94**, 985.

Hakkert, B.C., Kuijpers, T.W., Leeuwenberg, J.F.M., van Mourik, J.A. and Roos, D. (1991) Neutrophil and monocyte adherence to and migration across monolayers of cytokine-activated endothelial cell: the contribution of CD18, ELAM-1 and VLA-4. *Blood*, **78**, 2721.

Harada, A., Sekido, N., Kuno, K., Akiyama, M., Kasahara, T., Nakanishi, I., Mukaida, N. and Matsushima, K. (1993) Expression of recombinant rabbit IL-8 in Escherichia coli and establishment of the essential involvement of IL-8 in recruiting neutrophils into lipopolysaccharide-induced inflammatory site of rabbit skin. *Inter. Immunol.*, **5**, 681.

Haynes, B.F., Thelen, M.J., Hale, L.P. and Denning, S.M. (1989) CD44, a molecule involved in leukocyte adherence and T cell activation. *Immunol. Today*, **10**, 423.

Heagy, W., Laurance, M., Cohen, E., Finberg, R. (1990) Neurohormones regulate T cell function. *J. Exp. Med.*, **171**, 1625.

Hechtman, D.H., Cybalsky, M.I., Baker, J.B. and Gimbrone, Jr, J.A. (1990) Intravenous IL-8 reduces neutophil accumulation at intradermal sites of inflammation. *FASEB J.*, **4**, 890.

Hechtman, D.H., Cybulsky, M.I., Fuchs, H.J., Baker, J.B. and Gimbrone Jr, M.A. (1991) Intravascular IL-8. Inhibitor of polymorphonuclear leukocyte accumulation at sites of acute inflammation. *J. Immunol.*, **147**, 883.

Henning, B., Goldblum, S.E. and McClain, C.J. (1987) Interleukin-1 (IL-1) and tumor necrosis factor/cachectin (TNF) increase endothelial permeability in vitro. *J. Leuk. Biol.*, **42**, 551.

Hoover, R.L., Briggs, R.T. and Karnovsky, M.J. (1978a) The adhesive interaction between polymorphonuclear leukocytes and endothelial cells in vitro. *Cell*, **14**, 423.

Hoover, R.L., Folger, R., Haering, W.A., Ware, B.R. and Karnovsky, M.J. (1978b) Adhesion of leukocytes to endothelium: roles of divalent cations, surface charge, chemotactic agents and substrate. *J. Cell. Sci.*, **45**, 73.

Hoover, R.L., Karnovsky, M.J., Austen, K.F., Corey, E.J. and Lewis, R.A. (1984) Leukotriene B_4 action on endothelium mediates augmented neutrophil/endothlium adhesion. *Proc. Natl. Acad. Sci. USA*, **81**, 2191.

Horuk, R. (1994) Molecular properties of the chemokine receptor family. *Trends Pharm Sci.*, **15**, 159.

Horuk, R., Martin, A., Hesselgesser, J., Hadley, T., Lu, Z., Wang, Z. and Peiper, S.C. (1996) The Duffy antigen receptor for chemokines: structural analysis and expression in the brain. *J. Leuk. Biol.*, **59**, 29.

Huber, A.R., Kunkel, S.L., Todd, R.F. III and Weiss, S.J. (1991) Regulation of transendothelial neutrophil migration by endogenous interleukin-8. *Science*, **254**, 99.

Hynes, R.O. (1992) Integrins: versatility, modulation and signaling in cell adhesion. *Cell*, **69**, 11.

Issekutz, T.B. (1995) In vivo blood monocyte migration to acute inflammatory reactions, IL-1α, TNF-α, IFN-γ and C5a utilizes LFA-1, Mac-1 and VLA-4. *J. Immunol.*, **154**, 6533.

Jagels, M.A. and Hugli, T.E. (1992) Neutrophil chemotactic factors promote leukocytosis: A common mechanism for cellular recruitment from bone marrow. *J. Immunol.*, **148**, 1119.

Jagels, M.A. and Hugli, T.E. (1994) C5a and TNFα induce alterations in the expression of β_2 integrins and L-selectin on human neutrophils through a common calcium/calmodulin-dependent pathway. *J. Cell Biochem. (Suppl.)*, **18A**, E509.

Jiang, Y., Beller, D.I., Frendl, G. and Graves, D.T. (1992) Monocyte chemoattractant protein-1 regulates adhesion molecule expression and cytokine production in human monocytes. *J. Immunol.*, **148**, 2423.

Johnston, J.A., Taub, D.D., Lloyd, A.R., Conlon, K., Oppenheim, J.J. and Kelvin, D.J. (1994) Human T lymphocyte chemotaxis and adhesion induced by vasoactive intestinal peptide. *J. Immunol.*, **153**, 1762.

Jose, P.J., Collins, P.D., Perkins, J.A., Beaubein, N.F., Totty, N.F., Waterfield, M.D., Hsuan, J. and Williams, T.J. (1991) Identification of a second neutrophil chemoattractant cytokine generated during an inflammatory reaction in the rabbit peritoneal cavity in vivo: purification, partial amino acid sequence and structural relationship to melanoma growth stimulating activity. *Biochem. J.*, **278**, 493.

Jose, P.J., Griffiths-Johnson, D.A., Collins, P.D., Walsh, D.T., Moqbel, R, Totty, N.F., Truong, O., Hsuan, J.J. and Williams, T.J. (1994) Eotaxin: A potent eosinophil chemoatrractant cytokine detected in a guinea pig model of allergic airways inflammation. *J. Exp. Med.*, **179**, 881.

Kameyoshi, Y., Dorschner, A., Mallet, A.I., Christophers, E. and Schroder, J.M. (1992) Cytokine RANTES released by thrombin-stimulated platelets is a potent attractant for human eosinophils. *J. Exp. Med.*, **176**, 587.

Kaplanski, G., Porat, R., Aiura, K., Erban, J.K., Gelfand, J.A. and Dinarello, C.A. (1993) Activated platelets induce endothelial secretion of interleukin-8 in vitro via an interleukin-1–mediated event. *Blood*, **81**, 2492.

Keller, H.U., Hess, M.W. and Cottier, H. (1977) The chemokinetic effect of serum albumin. *Experientia*, **33**, 1386.

Keller, H.U., Wissler, J.H. and Damerau, B. (1981) Diverging effects of chemotactic serum peptides and synthetic f-met-leu-phe on neutrophil locomotion and adhesion. *Immunology*, **42**, 379.

Kelner, G.S., Kennedy, J., Bacon, K., Kleyensteuber, S., Largaespada, D.A., Jenkins, N.A., Copeland, N.G., Bazan, J.F., Moore, K.W., Schall, T.J. and Zlotnik, A. (1994) Lymphotactin: A cytokine that represents a new class of chemokines. *Science*, **266**, 1395.

Kelvin, D.J., Michiel, D.F., Johnston, J.A., Lloyd, A.R., Sprenger, H., Oppenheim, J.J. and Wang, J.M. (1993) Chemokines and serpentines: The molecular biology of chemokine receptors. *J. Leuk. Biol.*, **54**, 604.

Kishimoto, T.K., Jutila, M.A., Berg, E.L. and Butcher, E.C. (1989) Neutrophil Mac-1 and MEL-14 adhesion proteins inversely regulated by chemotactic factors. *Science*, **245**, 1238.

Koch, A.E., Polverini, P.J., Kunkel, S.L., Harlow, L.A., DiPietro, L.A., Elner, V.M., Elner, S.G. and Strieter, R.M. (1992) Interleukin-8 as a macrophage-dependent mediator of angiogenesis. *Science*, **258**, 1798.

Kuijpers, T.W., Hakkert, B.C., Hoogerwerf, M., Leeuwenberg, J.F.M., Roos, D. (1991) Role of endothelial leukocyte adhesion molecule-1 and platelet-activating factor in neutrophil adherence to IL-1–prestimulated endothelial cells. Endothelial leukocyte adhesion molecule-1–mediated CD18 activation. *J. Immunol.*, **147**, 1369.

Kuijper, T.W., Hakkert, B.C., Hart, M.H.L. and Roos, D. (1992) Neutrophil migration across monolayers of cytokine-prestimulated endothelial cells: A role for platelet-activating factor and IL-8. *J. Cell Biol.*, **117**, 565.

Kuna, P., S.R. Reddigari, D. Rucinski, J.J. Oppenheim and Kaplan, A.P. (1992a) Monocyte chemotactic and activating factor is a potent histamine-releasing factor for basophils. *J. Exp. Med.*, **175**, 489.

Kuna, P., S.R. Reddigari, T.J. Schall, D. Rucinski, M.Y. Viksman and Kaplan, A.P. (1992b) RANTES, a monocyte and T lymphocyte chemotactic cytokine releases histamine from human basophils. *J. Immunol.*, **149**, 636.

Kuna, P., Reddigari, S.R., Schall, T.J., Rucinski, D., Sadick, M. and Kaplan, A.P. (1993) Characterization of the human basophil response to cytokines, growth factors and histamine releasing factors of the intercrine/chemokine family. *J. Immunol.*, **150**, 1932.

Kunkel, S.L., Lukacs, N., Kasama, T. and Strieter, R.M. (1996) The role of chemokines in inflammatory joint disease. *J. Leuk. Biol.*, **59**, 6.

Lacasse, C., Rola-Pleszczynski, M. (1991) Immune regulation by platelet-activating factor: II. Mediation of suppression by cytokine-stimulated endothelial cells in vitro. *J. Leuk. Biol.*, **49**, 245.

Larsen, C. G., Anderson, A.O., Appella, E., Oppenheim, J.J., Matsushima, K. (1989) Neutrophil activating factor (NAP-1) is also chemotactic for T lymphocytes. *Science*, **243**, 1464.

Laudanna, C., Campbell. J.J., Butcher, E.C. (1996) Role of Rho in chemoattractant-activated leukocyte adhesion through integrins. *Science*, **271**, 981.

Lawrence, M.B. and Springer, T.A. (1991) Leukocytes roll on a selectin at physiologic flow rates: Distinction from and prerequisite for adhesion through integrins. *Cell*, **65**, 859.

Leonard, E.J., Yoshimura, T., Tanaka, S. and Raffeld, M. (1991) Neutrophil recruitment by intradermally injected neutrophil attractant/activation protein-1. *J. Invest. Dermat.*, **96** (**5**), 690.

Lewis, R.A., Austen, K.F. (1984) The biologically active leukotrienes. *J. Clin. Invest.*, **73**, 889.

Ley, K., Baker, J.B., Cybulsky, M.I., Gimbrone Jr., M.A. and Luscinskas, F.W. (1994) Intravenous interleukin-8 inhibits granulocyte emigration from rabbit mesenteric venules without altering L-selectin expression of leukocyte rolling. *J. Immunol.*, **151**, 6347.

Linbom, L., Hedqvist, P., Dahlen, S.E., Lingren, J.A. and Arfors, K.E. (1982) Leukotriene B_4 induces extravasation and migration of polymorphonuclear leukocytes in vivo. *Acta Physiol. Scand.*, **116**, 105.

Lo, S.k., Detmers, P.A., Levin, S.M. and Wright, S.D. (1989) Transient adhesion of neutrophils to endothelium. *J. Exp. Med.*, **169**, 1779.

Loetscher, P., Seitz, M., Clark-Lewis, I., Baggiolini, M., Moser, B. (1994) Monocyte chemotactic proteins MCP-1, MCP-2 and MCP-3 are major attractants for human CD4+ and CD8+ T lymphocytes. *FASEB J.*, **8**, 1055.

Lorant, D.E., Worthen, G.S., Henson, P.M. and Hyde, D.M. (1995) Activation of polymorphonuclear leukocytes reduces their adhesion to P-selectin and causes redistribution of ligands for P-selectin on their surface. *J. Clin. Invest.*, **96**, 171.

Lukacs, N.W., Strieter, R.M., Elner, V.M., Evanoff, H.L., Burdick, M., Kunkel, S.L. (1994) Intercellular adhesion molecule-1 mediates the expression of monocyte-derived MIP-1α during monocyte-endothelial cell interactions. *Blood*, **83**, 1174.

Luo, Y., Laning, J., Devi, S., Mak, J., Schall, T.J. and Dorf, M.E. (1994) Biological activities of the murine β-chemokine, TCA-3. *J. Immunol.*, **153**, 4625.

Luscinskas, F.W., Brock, A.F., Arnout, M.A. and Gimbrone, M.A. (1989) Endothelial-leukocyte adhesion molecule-1–dependent and leukcoyte CD11/18–dependent mechanisms contribute to polymorphonuclear leukocyte adhesion to cytokine-activated human vascular endothelium. *J. Immunol.*, **142**, 2257.

Luster, A.D., Greenberg, S.M., Leder, P. (1995) The IP-10 chemokine binds to a specific cell surface heparan sulfate site shared with platelet factor 4 and inhibits endothelial cell proliferation. *J. Exp. Med.*, **182**, 219.

Maghazachi, A.A., Al-Aoukaty, A., Schall, T.J. (1994) C-C chemokines induce chemotaxis of NK and IL-2-activated NK cells. Role for G proteins. *J. Immunol.*, **153**, 4969.

Maione, T.E., Gray, G.S., Petro, J., Hunt, A.F., Donner, A.L., Bauer, S.I., Carson, H.F. and Sharpe, R.J. (1990) Inhibition of angiogenesis by recombinant human platelet factor-4 and related peptides. *Science*, **247**, 77.

Mantovani, A., Bussolino, F. and Dejana, E. (1992) Cytokine regulation of endothelial cell function. *FASEB J.*, **6**, 2591.

Marasco, W.A. Phan, S.H., Krutch, H., Showell, H.J., Feltner, D.E., Nairn, R., Beckner, E.L. and Ward, P.A. (1984) Purification and identification of formyl-methionyl-leucyl-phenylalanine as the major peptide neutrophil chemotactic factor produced by *Escherichia coli*. *J. Biol. Chem.*, **259**, 5430.

Marder, S.R., Chenoweth, D.E., Goldstein, I.M. and Perez, H.D. (1985) Chemotactic responses of human peripheral blood monocytes to the complement-derived peptides C5a and C5a des Arg. *J. Immunol.*, **134**, 3325.

Matsushima, K., Baldwin, E.T., Mukaida, N. (1992) Interleukin-8 and MCAF: Novel leukocyte recruitment and activating cytokines. *Chem. Immunol.*, **51**, 236.

Matsushima, K., Morishita, K., Yoshimura, T., Lavu, S., Kobayashi, Y., Lew, W., Appella, E., Kung, H.F., Leonard, E.J. and Oppenheim, J.J. (1988) Molecular cloning of cDNA for a human monocyte derived neutrophil chemotactic factor (MDNCF) and the induction of MDNCF mRNA by interleukin-1 and tumor necrosis factor. *J. Exp. Med.*, **167**, 1883.

Matsushima, K., Larsen, C.G., DuBois, G.C. and Oppenheim, J.J. (1989) Purification and characterization of a novel monocyte chemotactic and activating factor produced by a human myelomonocytic cell line. *J. Exp. Med.*, **169**, 1485.

May, M. and Ager, A. (1992) ICAM-1–independent lymphocyte transmigration across high endothelium: differential up-regulation by interferon-γ, tumor necrosis factor-α and interleukin- 1β. *Eur. J. Immunol.*, **22**, 219.

Meerschaert, J. and Furie, M.B. (1994) Monocytes use either CD11/CD18 or VLA-4 to migrate across human endothelium in vitro. *J. Immunol.*, **152**, 1915.

Meurer, R., Van Riper, G., Feeney, W., Cunningham, P., Hora Jr., D., Springer, M.S., MacIntyre, D.E. and Rosen, H. (1993) Formation of eosinophilic and monocytic intradermal inflammatory sites in the dog by injection of human RANTES but not monocyte chemoattractant protein-1, human macrophage inflammatory protein -1 alpha, or human interleukin-8. *J. Exp. Med.*, **178**, 1913.

Migliorisi, G., Pawlowski, N., Scott, W.A., Fulop, A. and Cramer, E. (1984) Effect of leukotriene (LTB_4) on the transendothelial migration of human monocytes and neutrophils (PMNs). *Fed. Proc.*, **43**, 845.

Migliorisi, G., Folkes, E., Pawlowski, N. and Cramer, E.B. (1987) In vitro studies of human monocyte migration across endothelium in response to leukotriene B_4 and f-met-leu-phe. *Am. J. Pathol.*, **127**, 157.

Miller, M.D., Krangel, M.S. (1992) Biology and biochemistry of the chemokines: A family of chemotactic and inflammatory cytokines. *Crit. Rev. Immunol.*, **12**, 17.

Morse, S.I. and Barron, B.A. (1970) Studies on the leukocytosis and lymphocytosis induced by Bordetella pertussis. III. The distribution of transfused lymphocytes in pertussis-treated and normal mice. *J. Exp. Med.*, **155**, 663.

Moser, R., Schleiffenbaum, B., Groscurth, P and Fehr, J. (1989) Interleukin-1 and tumor necrosis factor stimulate human vascular endothelial cells to promote transendothelial neutrophil passage. *J. Clin. Invest.*, **83**, 444.

Munro, J.M., Pober, J.S. and Cotran, R.S. (1989) Tumor necrosis factor and interferon-gamma induce distnct patterns of endothelial activation and associated leukocyte accumulation in skin of *Papio anubis*. *Am. J. Pathol.*, **135**, 121.

Murphy, P.M. and Tiffany, H.L. (1991) Cloning of complementary DNA encoding a functional interleukin-8 receptor. *Science*, **253**, 1280.

Murphy, P.M. (1994) The molecular biology of leukocyte chemoattractant receptors. *Annu. Rev. Immunol.*, **12**, 593.

Murphy, W.J., Taub, D.D., Anver, M., Conlon, K., Oppenheim, J.J., Kelvin, D.J. and Longo, D.L. (1994) Human RANTES induces the migration of human T lymphocytes into the peripheral tissues of mice with severe combined immune deficiency. *Eur. J. Immunol.*, **24(8)**, 1823.

Murphy, W.J., Tian, Z.-G., Asai, O., Funakoshi, S., Rotter, P., Henry, M., Strieter, R.M., Kunkel, S.L., Longo, D.L. and Taub, D.D. (1996) Chemokines and T lymphocyte activation. II. Facilitation of human T cell trafficking in severe combined immunodeficiency mice. *J. Immunol.*, **156**, 2104.

Neote, K., DiGregorio, D., Mak, J.Y., Horuk, R. and Schall, T.J. (1993a) Molecular cloning, functional expression and signaling characteristics of a C-C chemokine receptor. *Cell*, 72, 415.

Neote, K., Darbonne, W., Ogez, J., Horuk, R. and Schall, T.J. (1993b) Identification of a promiscuous inflammatory peptide receptor on the surface of red blood cells. *J. Biol. Chem.*, **268** (17), 12247.

Oppenheim, J.J., Zachariae, C.O.C., Mukaida, N. and Matsushima, K. (1991) Properties of the novel proinflammatory supergene "intercrine" cytokine family. *Ann. Rev. Immunol.*, **9**, 617.

Oppenheim, J.J., Wang, J.M., Chertov, O., Taub, D.D. and Ben-Baruch, A. (1996) The role of chemokines in transplantation. In: *Transplantation Biology: Cellular and Molecular Aspects*; eds, Tilney, N.L., Strom, T.B., *et al.*, Lippincott-Raven Publishers, Philadelphia, PA., pg. 21.1.

Osborne, L. (1990) Leukocyte adhesion to endothelium in inflammation. *Cell*, **62**, 3.

Oseas, R., Yang, H.H., Baehner,,R.L. and Boxer, L.A. (1981) Lactoferrin: a promoter of polymorphonuclear adhesiveness. *Blood*, 57, 939.

Palmblad, J., Malmsted, C., Uden, A-M., Radmark, O. and Samuelsson, B. (1981) Leukotriene B_4 is a potent and stereospecific stimulator of neutrophil chemotaxis and adherence. *Blood*, **58**, 658.

Picker, L.J. (1994) Control of lymphocyte homing. *Curr. Opin. Immunol.*, **6**, 394.

Pohlman, T.H., Stanness, K.A., Beatty, P.G., Ochs, H.D. and Harlan, J.M. (1986) An endothelial cell surface factor(s) induced in vitro by lipopolysaccharide, interleukin 1 and tumor necrosis factor-α increases neutrophil adherence by a Cdw18-dependent mechanism. *J. Immunol.*, **136**, 4548.

Proost, P., Wolf-Peeters, C.D., Conings, R., Opdenakker, G., Billiau, A. and Damme, J.V. (1993) Identification of a novel granulocyte chemotactic protein (GCP-2) from human tumor cells: In vitro and in vivo comparison with natural forms of GRO, IP-10 and IL-8. *J. Immunol.*, **150**, 1000.

Rampart, M., Van Damme, J., Zonnkeyn, L. and Herman, A.C. (1989) Granulocyte chemotactic protein/interleukin-8 induces plasma leakage and neutrophil accumulation in rabbit skin. *Am. J. Pathol.*, **135**, 21.

Rollins, B.J., Yoshimura, T., Leonard, E.J. and Pober, J.S. (1990) Cytokine-activated human endothelial cells synthesize and secrete a monocyte chemoattractant, MCP-1/JE. *Am. J. Pathol.*, **136**, 1229.

Rollins, B.J. and Pober, J.S. (1991) Interleukin-4 induces the synthesis and secretion of MCP-1/JE by human endothelial cells. *Am J. Pathol.*, **138**, 1315.

Rot, A. (1992a) Endothelial binding of NAP-1/IL-8: role in neutrophil emigration. *Immunol. Today*, **13**, 291.

Rot, A., M. Krieger, T. Brunner, S.C. Bischoff, T.J. Schall and Dahinden, C.A. (1992b) RANTES and macrophage inflammatory protein 1α induce the migration and activation of normal human eosinophil granulocytes. *J. Exp. Med.*, **176**, 1489.

Rot, A. Chemokines link the two steps of leukocyte adhesion to endothelium. *Immunologist* 1993; 1(5):145.

Rot, A., Hub, E., Middleton, J., Pons, F., Rabeck, C., Thierer, K., Wintle, J., Wolff, B., Zsak, M. and Dukor, P. (1996) Some aspects of IL-8 pathophysiology III. Chemokine interaction with endothelial cells. *J. Leuk. Biol.*, **59**, 39.

Roth, S.J., Woldemar-Carr, M. and Springer, T.A. (1995) C-C chemokines, but not the C-X-C chemokines interleukin-8 and interferon-γ inducible protein-10, stimulate transendothelial chemotaxis of T lymphocytes. *Eur. J. Immunol.*, **25**, 3482.

Sarris, A.H., Esgleyes-Ribot, T., Crow, M., Broxmeyer, H.E., Karasavvas, N., Pugh, W., Grossman, D., Deisseroth, A. and Duvic, M. (1995) Cytokine loops involving interferon-γ and IP-10, a cytokine chemotactic for CD4+ lymphocytes: An explanation for the epidermotropism of cutaneous T-cell lymphoma? *Blood*, **86**, 651.

Schall, T.J., Bacon, K., Toy, K.I., Goedell, D.V. (1990) Selective attraction of monocytes and T lymphocytes of the memory phenotype by the cytokine RANTES. *Nature*, **347**, 669.

Schall, T.J. (1991) Biology of the rantes/sis cytokine family. *Cytokine*, **3**, 165.

Schall, T.J., Bacon, K., Camp, R.D.R., Kaspari, J.W., Goeddel, D.V. (1993) Human macrophage inflammatory protein-1β (MIP-1α) and MIP-1β chemokines attract distinct populations of lymphocytes. *J. Exp. Med.*, **177**, 1821.

Schall, T.J. and Bacon, K.B. (1994) Chemokines, leukocyte trafficking and inflammation. *Curr. Opin. Immunol.*, **6**, 865.

Schiffmann, E., Corcoran, B.A. and Wahl, S.M. (1975) Formylmethionyl peptides as chemoattractants for leukocytes. *Proc. Natl. Acad. Sci. USA*, **72**, 1059.

Schroder, J.M., Noso, N., Sticherling, M. and Christophers, E. (1996) Role of eosinophil-chemotactic C-C chemokines in cutaneous inflammtion. *J. Leuk. Biol.*, **59**, 1.

Senior, R.M., Griffin, G.L., Huang, J.S., Walz, D.A., Deuel, T.F. (1983) Chemotactic activity of platelet alpha granule proteins for fibroblasts. *J. Cell Biol.*, **96**, 283.

Sherry, B., Tekamp-Olson, P., Gallegos, C., Bauer, D., Davatelis, G., Wolpe, S.D., Masiarz, F., Cort, D. and Cerami, A. (1988) Resolution of the two components of macrophage inflammatory protein 1 and cloning and characterization of one of those components, macrophage inflammatory protein 1β. *J. Exp. Med.*, **168**, 2251.

Shimizu, Y., Newman, W., Tanaka, Y. and Shaw, S. (1992) Lymphocyte interactions with endothelial cells. *Immunol. Today*, **13**, 106.

Shukula, S.D. (1992) Platelet-activating factor: receptors and signal transduction. *Biochem. J.*, **292**, 617.

Shyy, Y.-J., Wickham, L.L., Hagan, J.P., Hsieh, H.-J., Hu, Y.-L., Telian, S.H., Valente, A.J., Paul Sung, K.-L. and Chien, S. (1993) Human monocyte colony-stimulating factor stimulates the gene expression of monocyte chemotactic protein-1 and increases the adhesion of monocytes to endothelial monolayers. *J. Clin. Invest.*, **92**, 1745.

Sica, A., Matsushima, K., Van Damme, J., Wang, J.M., Polentarutti, N., Dejana, E., Colotta, F. and Montovani, A. (1990a) IL-1 transcriptionally activates the neutrophil chemotactic factor/IL-8 gene in endothelial cells. *Immunology*, **69**, 548.

Sica, A., Wang, J.M., Colotta, F., Dejana, E., Mantovani, A., Oppenheim, J.J., Larsen, C.G., Zachariae, C.O.C. and Matsushima, K. (1990b) Monocyte chemotactic and activating factor gene expression induced in endothelial cells by IL-1 and TNF. *J. Immunol.*, **144**, 3034.

Smith, C.W., Hollers, J.C., Patrick, R.A. and Hasselt, C. (1979) Motility and adhesiveness in human neutrophils. Effects of chemotactic factors. *J. Clin. Invest.*, **63**, 221.

Smith, C.W., Marlin, S.D., Rothlein, R., Toman, C. and Anderson, D.C. (1989) Cooperative interactions of LFA-1 and Mac-1 with intercellular adhesion molecule-1 in facilitating adherence and transendothelial migration in vitro. *Blood*, **78**, 2089.

Smith, M.J.H., Ford-Hutchinson, A.W. and Bray, M.A. (1980) Leukotriene B_4: A potential mediator of inflammation. *J. Pharm. Pharmacol.*, **32**, 517.

Smith, R.P.C., Lackie, J.M. and Wilkinson, P.C. (1979) The effects of chemotactic factors on the adhesiveness of rabbit neutrophil granulocytes. *Exp. Cell Res.*, **122**, 169.

Smith, W.B., Gamble, J.R., Clark-Lewis, I. and Vadas, M.A. (1991) Interleukin-8 induces neutrophil transendothelial migration. *Immunology*, **72**, 65.

Sozzani, S., Luini, W., Molino, M., Jilek, P., Bottazzi, B., Cerletti, C., Matsushima, K., Montovani, A. (1991) The signal transduction pathway involved in the migration induced by a monocyte chemotactic cytokine. *J. Immunol.*, **147**, 2215.

Spagnuolo, P.J., Ellner, J.J., Hassid, A. and Dunn, M.J. (1980) Thromboxane A_2 mediates augmented polymorphonuclear adhesiveness. *J. Clin. Invest.*, **66**, 406.

Spangrude, G.J., Sacchi, F. Hill, H.R., van Epps, D.E. and Daynes, R.A. (1985) Inhibition of lymphocyte and neutrophil chemotaxis by pertussis toxin. *J. Immunol.*, **135**, 4135.

Springer, T.A. (1990) Adhesion receptors of the immune system. *Nature*, **346**, 425.

Springer, T.A. (1994) Traffic signals for lymphocyte recirculation and leukocyte emigration: the multistep paradigm. *Cell*, **76**, 301.

Springer, T.A. (1995) Traffic signals on endothelium for lymphocyte recirculation and leukocyte emigration. *Annu. Rev. Physiol.*, **57**, 827.

Standiford, T.J., Kunkel, S.L., Lukacs, N.W., Greenberger, M.J., Danforth, J.M., Kunkel, R.G. and Strieter, R.M. (1995) Macrophage inflammatory protein-1α mediates lung leukocyte recruitment, lung capillary leak and early mortality in murine endotoxemia. *J. Immunol.*, **55**, 1515.

Standiford, T.J., Kunkel, S.L., Greenberger, M.J., Laichalk, L.L. and Strieter, R.M. (1996) Expression and regulation of chemokines in bacterial pneumonia. *J. Leuk. Biol.*, **59**, 24.

Strieter, R.M. Kunkel. S.L., Showell, H.J., Remick, D.G., Phan, S.H., Ward, P.A. and Marks, R.M. (1989) Endothelial cell gene expression of neutrophil chemotactic factor by TNF-α, LPS and IL-1β. *Science*, **243**, 1467.

Streiter, R.M. and Kunkel, S.L. (1993) The immunopathology of chemotactic cytokines. In: Westwick, J., Lindley, J., Kunkel, S., eds., *The Chemokines*. New York:Plenum Press, pg.19.

Strieter, R.M., Standiford, T.J., Huffnagle, G.B., Colletti, L.M., Lukacs, N.W. and Kunkel, S.L. (1996a) The good, the bad and the ugly. The role of chemokines in models of human disease. *J. Immunol.*, **156**, 3583.

Strieter, R.M., Polverini, P.J., Arenberg, D.A., Walz, A., Openakker, G., Van Damme, J. and Kunkel, S.L. (1996b) The role of C-X-C chemokines as regulators of angiogenesis in lung cancer. *J. Leuk. Biol.* (in press).

Swensson, O., Schubert, C., Christophers, E. and Schroder, J.M. (1991) Inflammatory properties of neutrophil-activating protein-1/interleukin-8 (NAP-1/IL-8) in human skin: A light- and electronmicroscopic study. *J. Invest. Dermatol.*, **96**, 682.

Tanaka, Y., Adams, D.H. and Shaw, S. (1993a) Regulation of leukocyte recruitment by proadhesive cytokines immobilized on endothelial proteoglycan. *Curr. Top. Microbiol. Immunol.*, **184**, 99.

Tanaka, Y., Adams, D.H., Hubscher, S., Hirano, H., Siebenlist, U., Shaw, S. (1993b) T-cell adhesion induced by proteoglycan-immobilized cytokine MIP-1β. *Nature*, **361**, 81.

Tanaka, Y., Adams, D.H. and Shaw, S. (1993c) Proteoglycans on endothelial cells present adhesion-inducing cytokines to leukocytes. *Immunol. Today*, **14**, 111.

Taub, D.D., Conlon, K., Lloyd, A.R., Oppenheim, J.J., Kelvin, D.J. (1993a) Preferential migration of activated $CD4^+$ and $CD4^+$ T cells in response to MIP-1α and MIP-1β. *Science*, **260**, 355.

Taub, D.D., Lloyd, A.R., Conlon, K., Wang, J.M., Ortaldo, J.R., Harada, A., Matsushima, K., Kelvin, D.J., Oppenheim, J.J. (1993b) Recombinant human interferon-inducible protein 10 is a chemoattractant for human monocytes and T lymphocytes and promotes T cell adhesion to endothelial cells. *J. Exp. Med.*, **177**, 1809.

Taub, D.D., Oppenheim, J.J. (1993c) Review of the chemokine meeting: The third international symposium of chemotactic cytokines. *Cytokine*, **5** (**3**), 175.

Taub, D.D., Tsarfaty, G., Lloyd, A.R., Durum, S.K., Longo, D.L. and Murphy, W.J. (1994a) Growth hormone promotes human T cell adhesion and migration to both human and murine matrix proteins in vitro and directly promotes xenogeneic engraftment. *J. Clin. Invest.*, **94**, 293.

Taub, D.D. and Oppenheim, J.J. (1994b) Chemokines, inflammation and the immune system. *Ther. Immunol.*, **1**, 229.

Taub, D.D., Dastych, J., Inamura, N., Upton, J., Kelvin, D., Metcalfe, D. and Oppenheim, J.J. (1995a) Bone marrow-derived murine mast cells migrate, but do not degranulate, in response to chemokines. *J. Immunol.*, **154**, 2393.

Taub, D.D., Proost, P., Murphy, W.J., Anver, M., Longo, D.L., Van Damme, J. and Oppenheim, J.J. (1995b) Monocyte chemotactic protein-1 (MCP-1), -2 and -3 are chemotactic for human T lymphocytes. *J. Clin. Invest.*, **95**, 1370.

Taub, D. D., Sayers, T., Carter, C., Ortaldo, J. (1995c) α and β Chemokines induce NK cell migration and enhance NK cell cytolytic activity via cellular degranulation. *J. Immunol.*, **155**, 3877.

Taub, D.D., Clark, D. and Turcovski-Corrales, S.M. (1995d) Chemotaxis of T lymphocytes on extracellular matrix protein: Analysis of the in vitro method to quantitate chemotaxis of human T cells. *J. Immunol. Meth.*, **184**, 187.

Taub, D.D., Lloyd, A.R., Wang, J.M. and Oppenheim, J.J. (1995e) The α and β chemokine superfamily. In *Current Protocols in Immunology*, Unit 6.12, eds. Coligan, J.E., Kruisbeek, A., Margulies, D., Shevach, E. and Strober, W., Greene Publishing Associates and Wiley Interscience, New York.

Taub, D.D., Ortaldo, J.R., Turcovski-Corrales, S.M., Key, M.L., Longo, D.L. and Murphy, W.J. (1996a) β chemokines costimulate lymphocyte cytolysis, proliferation and lymphokine production. *J. Leuk. Biol.*, **59**, 81.

Taub, D.D., Turcovski-Corrales, S.M., Key, M.L., Longo, D.L. and Murphy, W.J. (1996b) I. Chemokines and T lymphocytes: α and β chemokines costimulate human T lymphocyte proliferation and lymphokine production in vitro. *J. Immunol.*, **156**, 2095.

Taub, D.D., Anver, M., Oppenheim, J.J., Longo, D.L. and Murphy, W.J. (1996c) T cell recruitment by interleukin-8 (IL-8): IL-8–induced degranulation of neutrophils releases potent chemoattractants for human T lymphocytes both in vivo and in vitro. *J. Clin. Invest.*, **97**, 1931.

Taub, D.D., Longo, D.L. and Murphy, W.J. (1996d) Human IP-10 induces mononuclear cell infiltration in mice and promotes the migration of human T lymphocytes into the peripheral tissues of huPBL-SCID mice. *Blood*, **87**, 1423.

Tedla, N., Palladinetti, P., Kelly, M., DiGirolamo, N., Chattophadhay, U., Cooke, B., Truskett, P., Dwyer, J., Wakefield, D. and Lloyd, A. (1996) Chemokines and T lymphocyte recruitment to lymph nodes in HIV infection. *Submitted for publication.*

Thureson-Klein, A., Hedqvist, P. and Lindbom, L. (1984) Ultrastructure of polymorphonuclear leukocytes in post capillary venules after exposure to leukotriene B_4 in vivo. *Acta. Physiol. Scand.*, **122**, 221.

Tonnesen, M.G., Smedly, L.A. and Henson, P.M. (1984) Neutrophil-endothelial cell interactions. Modulation of neutrophil adhesiveness induced by complement fragments C5a and C5a des Arg and formyl-methionyl-leucyl-phenylalanaine. *J. Clin. Invest.*, **74**, 1581.

Turner, L., Ward, S.G. and Westwick, J. (1995) RANTES-activated human T lymphocytes: A role for phosphoinositol 3–kinase. *J. Immunol.* 1995; **155**, 2437–2444.

Ulich, T.R., Del Castillo, J., Keys, M., Granger, G.A. and Ni, R.X. (1987) Kinetics and mechanisms of recombinant human interleukin-1 and tumor necrosis factor-α-induced changes in circulating numbers of neutrophils and lymphocytes. *J. Immunol.*, **139**, 3406.

Vaddi, K. and Newton, R.C. (1994) Regulation of monocyte integrin expression by β-family chemokines. *J. Immunol.*, **153**, 4721.

Van Damme, J., Proost, P., Lenaerts, J.-P. and Opdenakker, G. (1992) Structural and functional identification of two human, tumor-derived monocyte chemotactic proteins (MCP-2 and MCP-3) belonging to the chemokine family. *J. Exp. Med.*, **176**, 59.

Van Osselaer, N., Van Damme, J., Rampart, M. and Herman, A.C. (1991) Increased microvascular permeability in vivo in response to intradermal injection of neutrophil-activating protein (NAP-2) in rabbit skin. *Am. J. Pathol.*, **138**, 23.

Van Zee, K.J., Fischer, E., Hawes, A.S., Hebert, C.A., Terrell, T.G., Baker, J.B., Lowry, S.F. and Moldawer, L.L. (1992) Effects of intravenous IL-8 administration in nonhuman primates. *J. Immunol.*, **148**, 1746.

Vedder, N.B. and Harlan, J.M. (1988) Increased surface expression of CD11b/Cd18 (Mac-1) is not required for stimulated neutrophil adherence to cultured endothelium. *J. Clin. Invest.*, **81**, 676.

Wang, J.M., Xu, L., Murphy, W.J., Taub, D.D. and Chertov, O. (1996) IL-8–induced lymphocyte migration: Direct as well as indirect mechanisms. In *Methods in Enzymology*, Academic Press, San Diego, CA.

Ward, P. (1974) Leukotaxis and leukotactic disorders. *Am. J. Pathol.*, 77, 520.

Webb, L.M.C., Ehrengruber, M.U., Clark-Lewis, I., Baggiolini, M. and Rot, A. (1993) Binding of heparin sulfate or heparin enhances neutrophil responses to interleukin-8. *Proc. Natl. Acad. Sci. USA*, **90**, 7158.

Wells, T.N.C., Power, C.A., Lusti-Narasimhan, M., Hoogewerf, A.J., Cooke, R.M., Chung, C., Peitsch, M.C. and Proudfoot, A.E.I. (1996) Selectivity and antagonism of chemokine receptors. *J. Leuk. Biol.*, **59**, 53.

Wolpe, S.D., Davatelis, G., Sherry, B., Beutler, B., Hesse, D.G., Nguyen, H.T., Moldawer, L.L., Nathan, C.F., Lowry, S.F. and Cerami, A. (1988) Macrophages secrete a novel heparin-binding protein with inflammatory and neutrophil chemokinetic properties. *J. Exp. Med.*, **167**, 570.

Wolpe, S.D. and Cerami, A. (1989) Macrophage inflammatory proteins 1 and 2: members of a novel superfamily of cytokines. *FASEB J.*, **3**, 2565.

Wu, D., LaRosa, G.J., Simon, M.I. (1993) G-protein coupled signal transduction pathways for interleukin-8. *Science*, **261**, 101.

Xu, L., Kelvin, D., Ye, G.Q., Ben-Baruch, A., Taub, D.D., Oppenheim, J.J. and Wang, J.M. (1995) Effect of receptor expression on T lymphocyte chemotactic response to IL-8. *J. Leuk. Biol.*, **57**, 35.

Yoshimura, T., Matsushima, K., Tanaka, S., Robinson, E.A., Appella, E., Oppenheim, J.J. and Leonard, E.J. (1987) Purification of a human monocyte-derived neutrophil chemotactic factor that has peptide sequence similarity to other host defense cytokines. *J. Exp. Med.*, **169**, 1449.

Yu, C-L., Haskard, D.O., Cavender, D., Johnson, A.R. and Ziff, M. (1985) Human gamma interferon increases the binding of T lymphocytes to endothelial cells. *Clin. Exp. Immunol.*, **62**, 554.

Yue, T.L., Rabinovici, R. and Feuerstein, G. (1991) Platelet-activating factor (PAF), a putative mediator in inflammatory tissue injury. *Adv. Exp. Med. Biol.*, **314**, 223.

Zimmermann, G.A., Prescott, S.M. and McIntyre, T.M. (1992) Endothelial cell interactions with granulocytes: tethering and signaling molecules. *Immunol. Today*, **13**, 93.

6. THE ONTOGENY OF LYMPHOCYTES AND HOMING TO THE THYMUS

BEAT A. IMHOF[1], PHILIPPE NAQUET[2], OLLI VAINIO[3], and DOMINIQUE DUNON[4]

[1] *Department of Pathology, Centre Médical Universitaire, Geneva, Switzerland*
[2] *Centre d'Immunologie INSERM-CNRS, Marseille, France*
[3] *Medical Microbiology, Turku University, Turku, Finland*
[4] *Université Pierre et Marie Curie, CNRS URA, Paris, France*

SUMMARY

Hemopoietic stem cells circulate between embryonic compartments shortly after the time of emergence. During early embryogenesis lymphoid progenitors have to reach the appropriate microenvironment of hemopoietic organs which is necessary for their maturation into B or T cells (Imhof and Dunon, 1995; Kisielow and von Boehmer, 1995; Savagner *et al.*, 1988; Shortman and Wu, 1996; van Ewijk, 1991; Zuniga-Pflücker and Lenardo, 1996). During later embryogenesis or in the adult, "naive" differentiated lymphocytes subsequently colonize peripheral organs where they can be prepared for the defence against viral or bacterial infection (see other contributers) (Butcher and Picker, 1996; Imhof and Dunon, 1995; Springer, 1994). In this chapter we shall summarize the actual knowledge of the embryonic origin of hemopoietic progenitors and we will describe the adhesion molecules which are involved in the migration of pro-T lymphocytes to the thymus.

EMERGENCE OF HEMOPOIETIC CELLS DURING EARLY EMBRYOGENESIS

The cell migration process during gastrulation creates the three embryonic germ layers that form endodermal, mesodermal and ectodermal organs (Gilbert, 1994; Psychoyos and Stern, 1996; Stern and Canning, 1990; Stern *et al.*, 1992). It is hypothesized that the hemopoietic compartment originates from the primary mesenchyme which is generated by gastrulation at the primitive streak stage. Hemopoiesis occurs in lymphoid organs as the embryo develops. In mouse embryos, the yolk sac appears at day 7 of gestation, the liver primordium at day 8–9, the thymic anlage at day 9 and the spleen and bone marrow at day 15–16 of gestation (Figure 1). It has been shown that the yolk sac, derived from mesoderm and endoderm, is the first organ where erythroid and myeloid cells develop (Dieterlen-Lièvre, 1984; Godin *et al.*, 1995). As early as day 8 it also generates lymphoid progenitors of B and T lineages (Palacios and Imhof, 1993). No lymphoid precursors were found in the embryo proper at this stage

Correspondence: Beat A. Imhof, Department of Pathology, Centre Médical Universitaire, rue Michel-Servet 1, CH-12M Geneva, Switzerland. Tel: (41)-22-702 5745; Fax: (41)-22-702 5746.

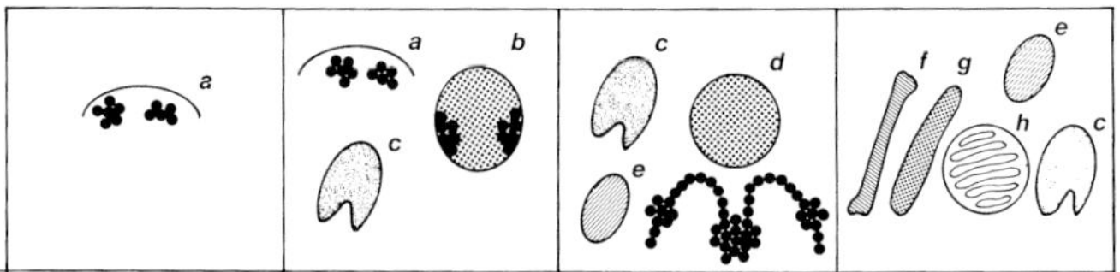

chicken	day of development	1	3-4	6-8	9-11
	location	yolk sac[a]	yolk sac basolateral intra-aorta[b]	dorsal mesentery, ventral to aorta[d] thymus[e]	bone marrow[f] spleen[g] bursa of fabricius[h] thymus
mouse	day of development	8	10.5 - 11.5	12.5 - 13.5	14.5 - 16
	location	yolk sac	fetal liver[c] para-aortic splanchnopleura	fetal liver thymus	bone marrow spleen fetal liver thymus
human	day of development	20	34-38	45-52	56-62
	carnegie stage	9	14 - 16	18 - 20	22 - 24
	somites	1 - 3	35 - 45	50 - 60	

Figure 1 The location of emerging hematopoietic stem cells (HSC) in embryos at different periods of development. Equivalent stages for chicken, mouse and human embryos are indicated, and the hemopoietic organs are given as symbols in the top row.

of mouse development (3–6 pairs of somites). This was based on functional in vitro and in vivo assays, flowcytometry and the determination of T-cell receptor and Ig gene rearrangement. The mammalian yolk sac cells which can differentiate into lymphocytes, are most likely multipotent hemopoietic stem cells (HSC) rather than cell-lineage determined T- and /or B- cell precursors. Since blood circulation to the yolk sac has not started at day 8, the possibility can be ruled out that the hemopoietic cells detected in the yolk sac were circulating cells from the embryo proper. It seems that the yolk sac itself derives from a region of the early embryonic hypoblast, and cells with the potential for hemopoietic differentiation may have immigrated during these early embryonic stages (Stern and Canning, 1990). In fact, at presomite stages, cells which can undergo hemopoietic differentiation are found in the embryo proper.

TRAFFIC OF LYMPHOID PROGENITORS IN THE VASCULARIZED EMBRYO

The earliest hemopoietic stem cells of the embryo proper emerge in intraembryonic foci in the para-aortic splanchnopleura (Godin *et al.*, 1995; Godin *et al.*, 1993). These data were obtained both from grafting experiments in vivo and by cloning techniques in vitro which support differentiation of B cells, T cells and myeloid cells with enumeration of stem cells at different embryonic stages. Recently, intra-aortic stem cells have also been localized in human embryos (Tavian *et al.*, 1996). Hence the ventral wall of the aorta appears to have an important role in the emergence of hemopoietic stem cells during embryogenesis.

The first mouse stem cells originating from the aortic foci colonize the thymus and seed thc fetal liver (Figure 1), which becomes then highly hemopoietic during the rest of embryogenesis (Ikuta *et al.*, 1992; van Ewijk, 1991). In later staged embryos the vascular system is well established. Since fetal liver cells seed the thymus by the vascular system , it was tempting to search for HSC in the embryonic blood (Rodewald, 1995). Human umbilical cord contained CD34^{+} cells which give rise to myeloid and erythroid cells in colony forming asays (Terstappen *et al.*, 1992), and from embryonic mouse blood it was possible to separate pro-T cells (Thy-1^{+}, c-kit^{lo}, CD3^{-}) from the pluripotent hemopoietic progenitors (Thy-1^{-}, c-kit^{+}) (Rodewald *et al.*, 1994).

In later mouse embryos the bone marrow is colonized by HSC at around day 15 (Figure 1), and becomes the only hemopoietic organ that raises HSC during adult life. HSC have been enriched from adult bone marrow in many laboratories and their phenotype was characterized (Ikuta *et al.*, 1992). It seems that a bone marrow HSC expresses low levels of Thy-1 , high levels of Sca-1, but is negative for the lineage markers TER-119 (erythroid), B220 (B cell), Mac-1 (monocytes), Gr-1 (granulocytes), CD4 (T-cells), and CD8 (T-cells). Such cells allowed the long-term thymus reconstitution of letally irradiated mice (Smith *et al.*, 1991).

MODEL OF THYMUS COLONIZATION

Hemopoietic stem cells are transported from their site of emergence to the thymus via the blood circulation (Dunon *et al.*, 1993; Ikuta *et al.*, 1992; Imhof and Dunon, 1995; Picker and Butcher, 1992). Early progenitors enter the nonvascularized thymic rudiment by the capsule. After vascularization, two migration pathways through blood vessels are used. They enter the organ at the corticomedullary junction or they enter between thymus lobules and migrate into the lobule through the capsule (Dunon et al; in preparation) (Kyewski, 1987; Savagner *et al.*, 1986; van Ewijk, 1991). After extravasation, the T cell precursors find themselves in a perivascular space rich in mesenchymal cells embedded in extracellular matrix. Finally, after invasive migration through the epithelial basal membrane, T cell precursors interact with the thymic epithelium and other cells that constitute this complex environment and consequently undergo differentiation along the T cell pathway (Anderson *et al.*, 1996; Shortman and Wu, 1996). At the molecular level, the attachment of pro-T cells to thymic endothelium and extravasation requires the participation of several adhesion molecules and probably chemokines or cytokines (Imhof and Dunon, 1995). Cell migration in the perivascular space needs a concentration gradient of chemotactic molecules and the adhesion of pro-T cells to extracellular matrix. Before the entry of pro-T cells to the thymus, another set of adhesion molecules is probably required mediating the pro-T cells to interact with the thymic epithelium and its basement membrane.

α6 Integrins

The study of the interaction between T cell progenitors and endothelium became productive with the establishment of progenitor cell lines which were able to populate a thymus in vivo (Dunon *et al.*, 1993). The pro-T cell line FTF1, a nontransformed T cell progenitor clone isolated from fetal thymus was shown to bind to frozen sections of thymus and liver from newborn mice and to an embryonic endothelial cell line. This property of pro-T cells appeared to be restricted to vessels in hemopoietic tissues. A monoclonal antibody raised against an embryonic mouse endothelial cell line was found to block the binding of pro-T cells to thymus-derived endothelium, but it did not affect the adhesion of mature T lymphocytes and myeloid cells. This antibody was subsequently found to recognize $\alpha6$ integrins (Imhof *et al.*, 1991; Ruiz *et al.*, 1993). Although $\alpha6\beta1$ and $\alpha6\beta4$ integrins are laminin receptors, our antibody does not inhibit binding of $\alpha6\beta1$ cells to laminin fragments (Ruiz *et al.*, 1993). This observation suggests that another ligand for $\alpha6$ integrins might exist on pro-T cells .

Recently, the involment of $\alpha6$ integrins in pro-T cell homing to the thymus was confirmed by *in vivo* experiments (Ruiz *et al.*, 1995). Injection of anti-$\alpha6$ antibodies reduced the number of thymocytes up to 80% in a congenic mouse model for thymus colonization by pro-T cells. The fact that anti-$\alpha6$ antibodies did not influence T cell differentiation and development in organotypic cultures demonstrated that $\alpha6$ integrins participate specifically to the thymus homing step. It is interesting to note that $\alpha6$ integrin protein and mRNA are detected by day 10 of embryogenesis in the thymus whereas pro-T cell progenitors start thymus colonization via blood vessels (Ruiz *et al.*, 1995).

$\alpha4\beta1$ Integrins

Recently, analysis of chimeric mice containing the integrin $\alpha4^{-}/^{-}$ or $\beta1^{-}/^{-}$ cells demonstrated that $\alpha4\beta1$ integrin participates in pro-T cell homing to the thymus (Hirsch *et al.*, 1996)(R. Hynes, presented at a Keystone meeting). One of the studies showed that $\alpha4\beta1$ integrin expression by thymus homing cells was essential in the adult animal but not during embryogenesis, where the animal can compensate with other adhesion mechanisms. Whether VCAM-1, the major cell-cell adhesion receptor for $\alpha4\beta1$, is important for thymus homing of HSCs is still a matter of debate, VCAM-1 expression is below detection level on endothelium but it is prominent on thymic epithelium (Reza and Ritter, 1994). It is interesting to note that thymus homing of mature CD4^{+} T cells involved also the $\alpha4\beta1$-VCAM-1 pathway (Bell *et al.*, 1995). Recently it was demonstrated that $\alpha4$ integrins can also interact in a homophilic way, this is interesting since endothelial cells can express this integrin (Altevogt *et al.*, 1995). Thus, thymus homing is governed by two different pathways, an embryonic $\alpha4$-integrin independent and an adult pathway in which the $\alpha4\beta1$ integrin plays a major role. The $\alpha6$ integrins may be involved in both pathways.

CD44

CD44 is a family of hyaluronic acid binding proteins, some of which are present on pro-T cells (Lesley *et al.*, 1993; Mackay *et al.*, 1993; Shortman and Wu, 1996). A role in thymus colonization has been ascribed to the hemopoietic form of CD44. Bone marrow cells which express CD44 are able to repopulate the thymus of irradiated mice at long term, and anti-CD44 can block the homing of these cells (Dunon *et al.*, 1993). This observation awaits further investigation to identify the CD44 ligand on vascular endothelium.

HEMCAM

HEMCAM is a molecule expressed by hemopoietic progenitor cells in bone marow of the chicken and the mouse (O. Vainio *et al.*, 1996). HEMCAM$^+$ bone marrow cells coexpressing the receptor tyrosine kinase c-kit could differentiate into myeloid and eythroid CFU's, which suggested that this population contained precursors for multiple lineages. Indeed, intrathymic injection of only three HEMCAM$^+$ c-kit$^+$ cells in congenic irradiated animals was sufficient to give rise to T cells. The cloning of HEMCAM cDNA revealed that this protein belongs to the immunoglobulin superfamily and consists of the V-V-C2-C2-C2 Ig domains. There are three mRNA splice variants. One has a short cytoplasmic tail; another has a long tail; while the third lacked transmembrane and cytoplasmic regions. HEMCAM is almost identical to gicerin, a molecule involved in neurite outgrowth and Wilm's kidney tumor progression in the chicken and it is significantly homologous with MUC18 a molecule involved in melanoma progression and metastasis in human beings. HEMCAM is an adhesion molecule because it promotes cell-cell adhesion of transfected cells. Recently we found that HEMCAM participates in thymus homing of T-cell progenitors since thymus regeneration of irradiated animals was inhibited by the injection of anti-HEMCAM antibody in chicken as well as in mouse. In the latter species crosslinking of HEMCAM leads to cell spreading of T lymphocyte progenitors adhering to vascular adhesion molecules PECAM-1 and VCAM-1. This last observation suggests that HEMCAM transduces a signal leading to the activation of integrins. In the multistep model of leukocyte interaction to endothelium, HEMCAM would be involved in the triggering step following rolling of the leukocyte on the endothelium. However, these data are preliminary and the attribution of a precise role for HEMCAM in thymus homing awaits further investigation.

Chemotactic Molecules

The role of chemotactic factors in thymus homing seemed to be restricted to the migration of T cell progenitors from the vascular endothelium through the perithymic mesenchyma to the thymic epithelium. Chemotactic peptides for pro-T cells have been partially purified from medium conditioned by thymic epithelial cells taken from avian embryos (Champion *et al.*, 1986). Several chemotactic molecules secreted by thymic stromal cells have also been characterized in mammals: TMFs and TCFs (Deugnier *et al.*, 1989; Imhof *et al.*, 1988). To date the best characterized chemotactic molecules involved in thymus homing of mammals and birds is β2-microglobulin (Dargemont *et al.*, 1989). After migration against a concentration gradient of β2-microglobulin in Boyden chambers, the bone marrow cells were enriched for pro-T cell progenitors (Dunon *et al.*, 1990; Dunon *et al.*, 1993). However, the thymus colonization in adult β2-microglobulin deficient mice is normal. This suggests that other molecular entities with chemotactic properties must exist. Notably, the recent multistep model of leukocyte-endothelium interaction involves numerous chemokines and suggests that some of these chemokines could be involved in the arrest of pro-T cells on the thymic endothelium (Imhof and Dunon, 1995).

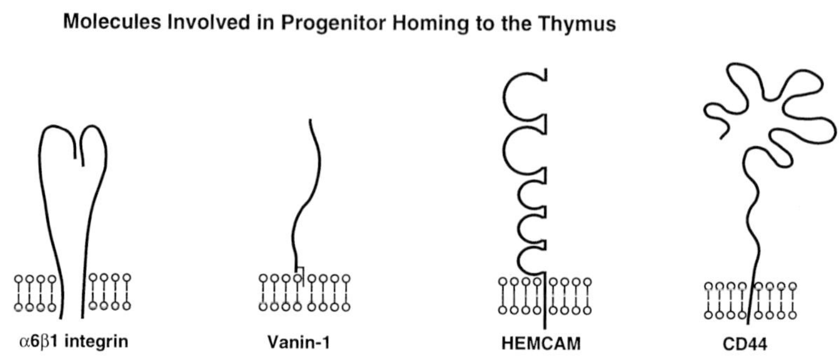

Expressed by:	α6β1 integrin	Vanin-1	HEMCAM	CD44
Progenitor Cells	+	-	+	+
Endothelium	+	?	+	+
Perivascular	+	+	-	+

Figure 2 Four molecules which are involved in thymus homing of HSCs are schematically illustrated. The table indicates the expression patterns of these homing molecules.

Vanin-1

Recently, the migration of hematopoietic precursor cells to the thymus has been shown to depend upon a novel molecule, Vanin-1 (M. Aurrand-Lions *et al.*, 1996). An anti-Vanin-1 antibody blocked specifically the entry of hemopoietic precursor cells into irradiated thymus *in vivo.* This antibody had no effect on T cell differentiation. Transfection of thymic stromal cells with the vanin cDNA increases their ability to interact with thymocytes in *in vitro* adhesion assays. The Vanin-1 protein belongs to a novel class of adhesion molecules, since this protein is a GPI anchored 70 kD protein which does not present clear homologies with known molecules. Vanin-1 is expressed by perivascular cells, including thymic epithelial cells but is absent of lymphoid cells. This location suggests that Vanin-1 is involved in the last step of our adhesion model which concerns the interaction of pro-T cells with thymic epithelial cells.

CONCLUSION

The existence of the multistep model which was proposed to describe the leukocyte homing to inflammatory tissues has been definitely confirmed during the last two years. It becomes now clear that the homing of HSCs and pro-T cells to the thymus is also a multistep adhesion event. However, the composition of molecules in thymus homing seems to be different from the described sets being involved in homing of leukocytes to inflammatory tissues. At present four adhesion molecules seem to play a role in thymus homing (Figure 2).

ACKNOWLEDGEMENTS

Part of this work was supported by the association pour la Recherche contre le Cancer (ARC), institutional funds from CNRS and INSERM, and the Academy of Finland. This work was supported by grant 3100-49241.96 from the Swiss National Science Foundation.

REFERENCES

Altevogt, P., Hubbe, M., Ruppert, M., Lohr, J., von Hoegen, P., Sammar, M., Andrew, D.P., McEvoy, L., Humphries, M.J. and Butcher, E.C. (1995). The $\alpha 4$ integrin chain is a ligand for $\alpha 4\beta 7$ and $\alpha 4\beta 1$. *J. Exp. Med.*, **182**, 345–355.

Anderson, G., Moore, N.C., Owen, J.J.T. and Jenkinson, E.J. (1996). Cellular interactions in thymocyte development. *Ann. Rev. Immunol.*, **14**, 73–99.

Aurrand-Lions, M., Galland, F., Bazin, H., Zakharyev, M., Imhof, B.A. and Naguet, P. (1996). Vanin-1, a novel GPI-linked perivascular molecule involved in thymus homing. *Immunity*, **5**, 391–405.

Bell, E.B., Sparshott, S.M. and Ager, A. (1995). Migration pathways of CD4 T cell subsets in vivo : the CD45RC- subset enters the thymus via $\alpha 4$ integrin-VCAM-1 interaction. *Int. Immunol.*, 7, 1861–1871.

Butcher, E.C. and Picker, L.J. (1996). Lymphocyte homing and homeostasis. *Science*, **272**, 60–66.

Champion, S., Imhof, B.A., Savagner, P. and Thiery, J.P. (1986). The embryonic thymus produces chemotactic peptides involved in the homing of hemopoietic precursors. *Cell*, **44**, 781–790.

Dargemont, C., Dunon, D., Deugnier, M.A., Denoyelle, M., Girault, J.M., Lederer, F., Le, K.H.D., Godeau, F., Thiery, J.P. and Imhof, B.A. (1989). Thymotaxin, a chemotactic protein, is identical to $\beta 2$-microglobulin. *Science*, **246**, 803–806.

Deugnier, M.A., Imhof, B.A., Bauvois, B., Dunon, D., Denoyelle, M. and Thiery, J.P. (1989). Characterization of rat T cell precursors sorted by chemotactic migration toward thymotaxin. *Cell*, **56**, 1073–1083.

Dieterlen-Liévre, F. (1984). Blood in chimeras. In Chimeras in developmental biology, N.M. Le Douarin and A. McLaren, eds. (London, New York: Academic Press), pp. 133–163.

Dunon, D., Kaufman, J., Salomonsen, J., Skjoedt, K., Vainio, O., Thiery, J.P. and Imhof, B.A. (1990). T cell precursor migration towards beta 2–microglobulin is involved in thymus colonization of chicken embryos. *Embo J.*, **9**, 3315–22.

Dunon, D., Ruiz, P. and Imhof, B.A. (1993). Pro-T cell homing to the thymus. *Curr. Top. Microbiol. Immunol.*, **184**, 139–150.

Gilbert, S.F. (1994). Developmental Biology, 4th Edition (Sunderland, MA: Sinauer Associates, Inc.).

Godin, I.E., Dieterlen-Lièvre, F. and Cumano, A. (1995). Emergence of multipotent hemopoietic cells in the yolk sac and paraaortic splanchnopleura in mouse emryos, beginning at 8.5 days postcoitus. *Proc. Natl. Acad. Sci. USA*, **92**, 773–777.

Godin, I.E., Garcia-Porrero, J.A., Coutinho, A., Dieterlen-Lièvre, F. and Marcos, M.A.R. (1993). Para-aortic splanchnopleura from early mouse embryos contains B 1α cell progenitors. *Nature*, **364**, 67–69.

Hirsch, E., Iglesias, A., Potocnik, A.J., Hartmann, U. and Fässler, R. (1996). Impaired migration but not differentiation of haematopoietic stem cells in the absence of b1 integrins. *Nature*, **380**, 171–175.

Ikuta, K., Uchida, N., Friedman, J. and Weissman, I.L. (1992). Lymphocyte development from stem cells. *Annu. Rev. Immunol.*, **10**, 759–783.

Imhof, B.A., Deugnier, M.A., Girault, J.M., Champion, S., Damais, C., Itoh, T. and Thiery, J.P. (1988). Thymotaxin: a thymic epithelial peptide chemotactic for T-cell precursors. *Proc. Natl. Acad. Sci. USA*, **85**, 7699–7703.

Imhof, B.A. and Dunon, D. (1995). Leukocyte migration and adhesion. *Adv. Immunol.*, **58**, 345–416.

Imhof, B.A., Ruiz, P., Hesse, B., Palacios, R. and Dunon, D. (1991). EA-1, a novel adhesion molecule involved in the homing of progenitor T lymphocytes to the thymus. *J. Cell Biol.*, **114**, 1069–1078.

Kisielow, P. and von Boehmer, H. (1995). Development and selection of T cells: facts and puzzles. *Adv. Immunol.*, **58**, 87–209.

Kyewski, B.A. (1987). Seeding of thymic microenvironments defined by distinct thymocyte-stromal interactions is developmentally controlled. *J. Exp. Med.*, **166**, 520–538.

Lesley, J., Hyman, R. and Kincade, P.W. (1993). CD44 and its interaction with extracellular matrix. *Adv. Immunol.*, **54**, 271–335.

Mackay, C.R., Terpe, A., Stauder, R., Marston, W.L., Stark, H. and Günthert, U. (1993). CD44 isoforms are preferentially expressed on proliferative epithelia and can be induced by TNFα. *J. Exp. Med.*, Submitted.

Palacios, R. and Imhof, B.A. (1993). At day 8–8.5 of mouse development the yolk sac, not the embryo proper, has lymphoid precursor potential in vivo and *in vitro*. *Proc. Natl. Acad. Sci. USA*, **90**, 6581–6585.

Picker, L.J. and Butcher, E.C. (1992). Physiological and molecular mechanisms of lymphocyte homing. *Annu. Rev. Immunol.*, **10**, 561–591.

Psychoyos, D. and Stern, C.D. (1996). Fates and migratoryroutes of primitive streak cells in the chick embryo. *Development*, **122**, 1523–1534.

Reza, J.N. and Ritter, M.A. (1994). Differential expression of adhesion molecules within the human thymus. *Dev. Immunol.*, **4**, 55–64.

Rodewald, H.R. (1995). Pathways from hemopoietic stem cells to thymocytes. *Curr. Opin. Immunol.*, 7, 176–187.

Rodewald, H.R., Kretzschmar, K., Takeda, S., Hohl, C. and Dessing, M. (1994). Identification of pro-thymocytes in murine fetal blood: T lineage commitment can precede thymus colonization. *EMBO J.*, **13**, 4229–4240.

Ruiz, P., Dunon, D., Sonnenberg, A. and Imhof, B.A. (1993). Suppression of mouse melanoma metastasis by EA-1, a monoclonal antibody specific for alpha 6 integrins. *Cell Adh. Comm.*, **1**, 67–81.

Ruiz, P., Wiles, M.V. and Imhof, B.A. (1995). $\alpha 6$ integrins participate in pro-T cell homing to the thymus. *Eur. J. Immunol.*, **25**, 2034–2041.

Savagner, P., Bauvois, B., Deugnier, M.A., Imhof, B.A. and Thiery, J.P. (1988). Aspects of haemopoietic cell dynamics: ontogeny and targeted migration. *Ann. Inst. Pasteur/Immunol.*, **139**, 409–431.

Savagner, P., Imhof, B.A., Yamada, K.M. and Thiery, J.P. (1986). Homing of hemopoietic precursor cells to the embryonic thymus: Characterization of an invasive mechanism induced by chemotactic peptides. *J. Cell Biol.*, **103**, 2715–2727.

Shortman, K. and Wu, L. (1996). Early T lymphocyte progenitors. *Annu. Rev. Immunol.*, **14**, 29–47.

Smith, L.G., Weissman, I.L. and Meimfeld, S. (1991). Clonal analysis of hemopoietic stem-cell differentiation *in vivo*. *Proc. Natl. Acad. Sci. USA*, **88**, 2788–2792.

Springer, T.A. (1994). Traffic signals for lymphocyte recirculation and leukocyte emigration: The multistep paradigm. *Cell*, **76**, 301–314.

Stern, C.D. and Canning, D.R. (1990). Origin of cells giving rise to mesoderm and endoderm in chick embryo. *Nature*, **343**, 273–5.

Stern, C.D., Hatada, Y., Selleck, M.A. and Storey, K.G. (1992). Relationships between mesoderm induction and the embryonic axes in chick and frog embryos. *Dev. Suppl.*, 151–6.

Tavian, M., Coulombel, L., Luton, D., San Clemente, H., Dieterlen-Lièvre, F. and Péault, B. (1996). Aorta-associated CD34+ hematopoietic cells in the early human embryo. *Blood*, **87**, 67–72.

Terstappen, L.W., Huang, S. and Picker, L.J. (1992). Flow cytometric assessment of human T-cell differentiation in thymus and bone marrow. *Blood*, **79**, 666–677.

van Ewijk, W. (1991). T-cell differentiation is influenced by thymic microenvironments. *Annu. Rev. Immunol.*, **9**, 591–615.

Vainio, O., Dunon, D., Aïssi, F., Dangy, J.-P., McNaghy, K.M. and Imhof, B.A. (1996). HEMCAM, an adhesion molecule expressed by c-kit^{+} hemopoietic progenitors. *J. Cell. Biol.*, **135**, 1655–1668.

Zuniga-Pflücker, J.C. and Lenardo, M.J. (1996). Regulation of thymocyte development from immature progenitors. *Curr. Opin. Immunol.*, **8**, 215–224.

7. ROLES OF $\alpha_4\beta_7$ AND $\alpha_E\beta_7$ INTEGRINS IN THE INTERACTIONS OF LYMPHOCYTES WITH THE INTESTINAL MUCOSA

ALEXANDRE BENMERAH, DELPHINE GUY-GRAND and NADINE CERF-BENSUSSAN*

Développement Normal et Pathologique du Système Immunitaire INSERM U429, Hôpital Necker-Enfants Malades, Paris, France

INTRODUCTION

The intestinal mucosa, constantly exposed to the highly antigenic intraluminal content, is a physiological inflammatory site. Its contains a large number of competent immune effector cells which contribute to its protection against pathogens. Numerous studies have demonstrated the unique properties of mucosal effector cells and have delineated their pathways of migration. These studies have been already thoroughly reviewed and will therefore be briefly summarized herein. More recent work has adressed the molecular mechanisms involved in the interactions of mucosal effector immune cells with the intestinal mucosa. In this chapter, we will emphasize on the role of two membrane receptors, the $\alpha_4\beta_7$ and $\alpha_E\beta_7$ integrins which enable the interactions of intestinal lymphocytes with the intestinal endothelium and epithelium respectively.

LYMPHOCYTE MIGRATION INTO THE INTESTINAL MUCOSA

The Intestinal Mucosa Contains a Large Number of Mature Effector Cells

The intestinal mucosa is made of connective tissue, the lamina propria, covered by a cylindrical epithelium which arises from crypt cells and forms villi in the small intestine. Mucosal immune effector cells are disseminated throughout the intestinal mucosa of the small and large intestine and consist of plasma cells and T cells.

Intestinal plasma cells are exclusively localized in the lamina propria. They represent approximately 75% of all plasma cells and produce a considerable amount of polymeric IgA which are transported into the intestinal lumen where they can complex dietary antigens and pathogens (reviewed in: Brantzaeg *et al*, 1989).

Mature effector T cells are distributed in lamina propria and in epithelium. In both sites, they bear markers of activated memory T cells but possess distinct phenotypical and functional properties (reviewed in: Cerf-Bensussan *et al.*, 1993; Guy Grand and Vassalli, 1993). Lymphocytes in the epithelium (intraepithelial lymphocytes: IEL) are mainly mature CD8+ cytotoxic T cells. They have been thoroughly characterized in mice. In this species, IEL are made of two ontogenically distinct subsets. Approximately

*Corresponding author.

50% of IEL in normal mice are conventional thymus-processed T cells that express an $\alpha\beta$ T cell receptor. About 10% of these bear CD4 whereas 90% bear classical heterodimeric CD8α/β molecules. The others are thymoindependent T cells which are found in the intestine of athymic animals. The latter cells can be easily recognized in normal mice as they bear an unusual homodimeric CD8α/α molecule, express a TcR$\alpha\beta$ or $\gamma\delta$, use a distinct CD3 transducing module and coexpress NK receptors that they can use to kill NK targets (Guy-Grand and Vassalli, 1993; Guy-Grand *et al.*, 1996). Both populations of thymodependent and thymoindependent TcR$\alpha\beta$+ IEL are oligoclonal but possess a distinct repertoire (Regnault *et al.* 1994). Lamina propria T lymphocytes have been less well characterized than IEL. Most of them are conventional thymus-processed T cells bearing an $\alpha\beta$ T cell receptor. Only a minority are CD8$^+$, the majority being mature CD4$^+$ lymphocytes producing large amounts of cytokines (Mega *et al.*, 1991; Zeitz *et al.*, 1991).

In mice, the majority of thymodependent intestinal T cells and a large number of intestinal IgA plasma cells derive from blastic precursors sensitized to intraluminal antigens in Peyer's patches which circulate through an hemolymphatic cycle and specifically home back into the intestinal mucosa (Figure 1).

Migration studies in rodents have clearly established the existence of the hemolymphatic cycle of lymphoblasts of intestinal origin. In these experiments, lymphoblasts derived from various lymphoid organs (approximately 1% of total lymphocytes) were labeled with H^3-thymidine and their migration in different tissues followed by autoradiography coupled to phenotype studies. Blastic precursors derived from mesenteric lymph nodes (MLN) and thoracic duct (TD) migrated massively into the intestinal mucosa within 24 hours. In contrast, blastic cells derived from extraintestinal sites or obtained by *in vitro* stimulation did not migrate into the intestinal mucosa (Gowans and Knight, 1964; Griscelli *et al.*, 1969; Hall, 1970; Guy-Grand *et al.*, 1974; Parrott and Ferguson, 1974; McWilliams *et al.*, 1975). Labeled cells in gut-associated lymphoid organs and in the intestinal mucosa were either IgA-containing B cells (Guy-Grand *et al.*, 1974; McWilliams *et al.*, 1975) or T cells bearing Thy-1, an $\alpha\beta$TcR, CD4 or CD8α/β (Guy-Grand *et al.*, 1978; 1991).

A preferential migration of small lymphocytes derived from the intestinal lymph or MLN either back into the intestinal lymph (Cahill *et al.*, 1977; Scollay *et al.*, 1976) or into the intestinal mucosa (Guy-Grand *et al.*, 1978) has also been observed. However, the differences with the migration of small lymphocytes of peripheral origin are much less significant. Finally, it has been shown that a small proportion of IgA-containing B cells isolated from lamina propria (Tseng 1984) and of T lymphocytes isolated from the epithelium (Guy-Grand *et al.*, 1978) can return to the intestinal mucosa after intravenous injection. These cells may either represent recently arrived mesenteric lymphoblasts or alternatively represent a population of cells able to recirculate.

Experiments based on the transfer of PP cells (Craig and Cebra 1971; Rudzik *et al.*, 1975) and on the selective labeling or irradiation of PP (Guy-Grand *et al.*, 1978) have shown that mesenteric and thoracic duct B and T lymphoblasts which migrate into the intestinal mucosa derive from blastic precursors stimulated by intraluminal antigens in PP. However, PP-derived cells do not migrate rapidly into the intestinal mucosa. After injection of PP cells to an irradiated rabbit, IgA plasma cells take 6 to 8 days

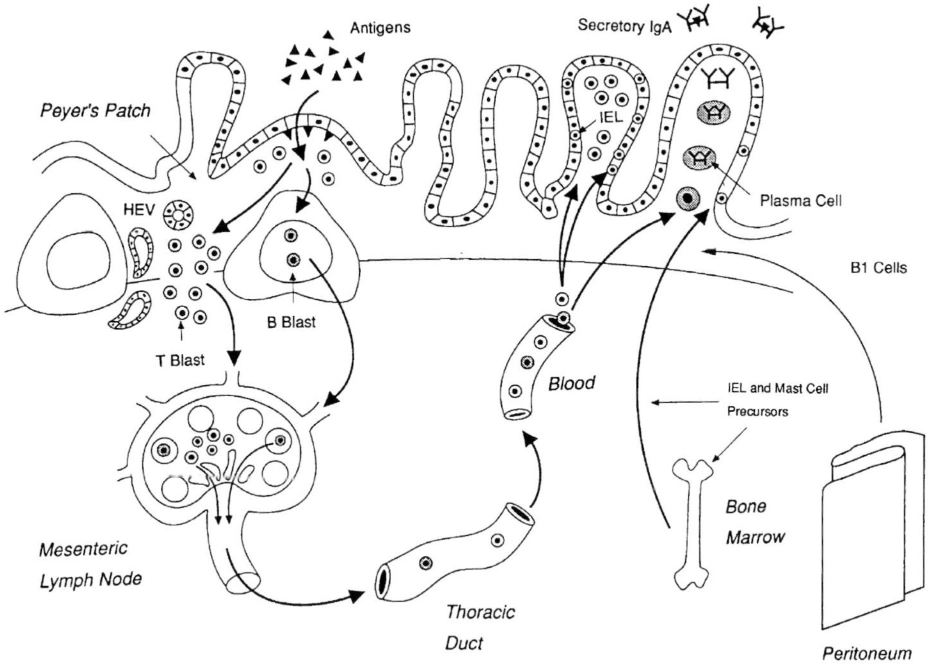

Figure 1 Traffic of intestinal lymphocytes in mice. A large number of mouse IgA plasma cells and of thymodependent intestinal T cells derive from blastic precursors sensitized to intraluminal antigens in Peyer's patches. After circulation in mesenteric lymph nodes, thoracic duct lymph and blood, these lymphoblasts home back into the intestinal mucosa where they become mature effector cells. In addition, two other pathways of lymphocyte migration in the intestine have been described. First, a large population of thymoindependent T lymphocytes derive from bone marrow precursors which migrate directly into the intestinal mucosa where they differentiate into mature T cells and distribute mainly in the epithelium. Second, some IgA plasma cells derive from peritoneal B1 cells which migrate into the intestinal wall via a pathway which is not yet elucidated.

to appear in the intestinal mucosa (Rudzik *et al.*, 1975) and no radiolabelled cells are observed in the intestinal mucosa 24 hours after transfer of radiolabeled PP-derived lymphoblasts. Double-transfer experiments indicate that PP-derived lymphocytes first migrate either in MLN (Roux *et al.*, 1981) or in spleen (Tseng, 1981). After one to five days, cells recovered from these organs become able to migrate into the intestinal mucosa. Therefore, a maturation step may be required for PP-derived lymphoblasts to home into the intestinal mucosa. Accordingly, a progressive differentiation of the intestinal precursors is observed during their hemolymphatic cycle. B blasts progressively acquire intracytoplasmic IgA (Guy-Grand *et al.*, 1974) and T blasts secrete increasing amounts of cytokines (Guy-Grand *et al.*, 1984).

Intraluminal antigens are necessary to trigger the proliferation of sensitized blasts in PP and initiate the hemolymphatic cycle. Thus, studies using isolated loops have shown that PP are required to induce an antigen-specific IgA response in the intestinal mucosa (Husband and Gowans, 1978; Robertson and Cebra, 1976). Furthermore, in newborn and axenic mice, PP do not contain dividing precursors and there are no T cells nor plasma cells in the intestinal mucosa. One week after exposure of axenic mice to a conventional environment, PP contains numerous dividing cells and one week later, the intestinal mucosa contains a normal number of IgA plasma cells and T cells (Guy-Grand *et al.*, 1988) . Altogether, these experiences suggest that PP provide a microenvironment which favor the committment of naive lymphocytes into antigen-sensitized lymphocytes endowed with intestinal homing properties. The circulation of lymphoblasts sensitized to intraluminal antigens in PP within the hemolymphatic cycle allows to redistribute the immune response generated in one site of the intestine throughout the intestinal mucosa. In addition, transfer experiments as well as studies of antibodies synthesized in reponse to intestinal pathogens have shown that the IgA immune response initiated in the intestine can be redistributed to other mucosae and particularly to the lactating mammary gland (reviewed in: Phillips-Quagliata *et al.*, 1994).

Several mechanisms able to drive the migration of PP-derived blasts into the intestinal mucosa have been considered. The role of intraluminal antigens has been intially investigated. The study of fetal grafts able to grow subcutaneously protected from antigenic stimulation clearly showed that antigens do not drive the migration of IgA-containing B cell precursors and of T cell precursors into the intestinal mucosa. Thus, three weeks after their implantation in adult rodents, these grafts contained IgA plasma cells and T cells (Gowans and Knight, 1964). In addition, in transfer experiments, labeled blasts from MLN were recovered in the fetal grafts as IgA plasma cells and T cells (Guy-Grand *et al.*, 1974). However, the local presence of the antigen may increase the number of specifically-sensitized cells by favoring their local proliferation (Husband and Gowans, 1978). In addition, strong inflammatory reactions such as those induced by gut parasitic stimulation (Rose *et al.*, 1976), graft-versus-host reaction (Guy-Grand *et al.*, 1988) or in humans, inflammatory bowel diseases (Burgio *et al.*, 1995; Salmi *et al.*, 1994) may favor the local migration of lymphocytes of peripheral origin which do not migrate into the intestinal mucosa in normal conditions. Thus, intraluminal antigens are necessary to trigger the proliferation of sensitized blasts in PP and to initiate the hemolymphatic cycle but do not play a primary role in the final step of migration.

The hemolymphatic cycle of PP-derived lymphoblasts is well demonstrated in rodents. Its existence in humans is infered but not demonstrated.

In mice, recent studies have provided evidence of other pathways of lymphocyte migration into the intestinal mucosa (Figure 1).

Thymoindependent intraepithelial T lymphocytes (identified by the absence of CD4 or of CD8β chains) do not circulate in the intestinal lymph. Transfer of bone marrow cells into thymectomized mice indicate that they derive from bone marrow precursors which migrate directly via the blood into the intestinal mucosa (Guy-Grand *et al.*, 1991; Rocha *et al.*, 1994). Several observations suggest that they differentiate locally in the intestinal microenvironnement into mature effector T cells. Particularly, transcripts for RAG-1, an enzyme required for T cell rearrangements, were detected by in situ hybridization in a small subset of immature mouse CD3$^-$IEL (Guy-Grand *et al.*, 1992). Interestingly, RAG-1 has also been detected by PCR in human IEL suggesting that a comparable pathway of differentiation may also exist in humans (Lundquist *et al.*, 1995).

In addition, studies of chimeric mice constructed by reconstituting lethally irradiated mice with a mixture of syngeneic bone marrow cells and peritoneal cells from Ig heavy chain allotype congenic donors have suggested that a large number of mouse intestinal IgA plasma cells may derive from a population of unconventional B cells (B_1 lymphocytes) that predominate in peritoneal and pleural cavities and descend from precursors in the fetal liver and omentum (Kroese *et al.*, 1989). These B lymphocytes which produce natural antibodies against autoantigens and common microbial epitopes, may represent together with thymoindependent IEL a first line of defense.

The mechanisms which control the migration of PP-derived lymphoblasts, and of the precursors of unconventional intestinal T and B lymphocytes across lamina propria endothelium as well as the mechanisms which drive the migration of T lymphocytes, mainly CD8+, into the epithelial layer are not entirely elucidated. However, the recent characterization of two integrins sharing the same $\beta 7$ chain, $\alpha_4\beta_7$ and $\alpha_E\beta_7$ has brought some insight into the mechanisms of interactions between lymphocytes and the intestinal mucosa.

PROPERTIES OF $\alpha_4\beta_7$ AND $\alpha_E\beta_7$ INTEGRINS

Integrins form a large family of membrane glycoproteins participating in cell adhesion, morphogenesis, migration and differentiation. They are made of two non covalently-bound subunits α and β which interact in their extracellular N-terminal domains to form a binding site for their ligands, extracellular matrix proteins or membrane receptors. Their intracytoplasmic domains interact direcly or indirectly with intracellular proteins which can modulate their conformation or transduce activating signals transduction (Hynes 1992; Clark and Brugge 1995). Eight β chains and 16 α chains have been identified. One β chain can associate with different α chains, defining subfamilies sharing the same β chain. The β_7 chain associated with α_4 or α_E defines a recently individualized subfamily made of two members which participate in the interactions of lymphocytes with the intestinal mucosa.

Structural Properties of $\alpha_4\beta_7$ and $\alpha_E\beta_7$ Integrins

The study of the primary structure of $\alpha_4\beta_7$ and $\alpha_E\beta_7$ integrins indicates that some of the functional domains identified in other integrins are conserved in these two proteins.

The β_7 chain contains approximately 800 amino acids and shows 32 to 46% homology with other integrin β chains, the strongest homology being with the β_2 chain (CD18) shared by leukocyte integrins (Erle *et al.*, 1991; Hu *et al.*, 1992; Jiang *et al.*, 1992; Krissansen *et al.*, 1992). The short intracytoplasmic domain of the human β_7 contains 52 amino acids including 20 identical and 7 homologous to those observed in other human β chains. The study of the $\beta_{1,-2,-3}$ integrins has shown that the intracytoplasmic domain of the β chain has an important role in the regulation of the adhesive properties of integrins. Via its binding to actin-based cytoskeleton proteins, it favors the redistribution of integrin molecules in areas of cell to cell contact or of cell to extracellular matrix contact. One of the two binding sites for α-actinin identified in the intracytoplasmic domain of β_1 (Otey *et al.*, 1993; Williams *et al.*, 1994) is almost entirely conserved in β_7 whereas the other binding site for α-actinin and the binding site for talin are less conserved. In addition, the juxtamembrane region of β_7 is very comparable to that of β_1 and β_3 chains, and peptides derived from this region bind the tyrosine kinase pp125^{FAK}, a kinase which participates in signal transduction via integrins and perhaps in the regulation of their interactions with the cytoskeleton (Schaller and Parsons, 1994; Schaller *et al.*, 1995).

The α_4 and α_E chains contain approximately 1000–1150 aminoacids. Their sequences share several common features with other integrin α chains (Shaw *et al.*, 1994; Smith *et al.*, 1994; Takada *et al.*, 1989). In the short intracytoplasmic domain, the juxtamembrane sequence GFFKR is conserved. This sequence contributes to the regulation of the conformational changes of integrins. Thus, its mutation in the α_{IIb} chain expressed on platelets in association with β_3 blocks the integrin in a high affinity conformation. Recent work suggests that this region binds to an intracellular protein, calreticulin, which, upon binding, may stabilize the high affinity state (Coppolino *et al.*, 1995). In the long extracellular domains of α_4 and α_E chains, most cysteine residues important for the tridimensional structure are conserved as well as the seven repeated sequences including the three last ones which contain an EF-hand motif (DXXXDXXXD-) binding divalent cations. The study of integrin α chains, including α_4, has shown that the concentration and the nature of divalent cations in the medium can modulate the conformation of these chains and thus their affinity for their ligands because the EF-hand motifs are present within the binding site or very close to it (Masumoto *et al.*, 1993). In addition, the concentration in calcium may favor the aggregation of integrin molecules on the cell surface and thus increases their avidity for their ligands (Lub *et al.*, 1995). Accordingly, the avidity of $\alpha_4\beta_7$ for its ligand MadCAM-1 can be markedly increased in the presence of Mg^{++} and Mn^{++} (Berlin *et al.*, 1993) and the binding of $\alpha_E\beta_7$ to its epithelial ligand E-cadherin requires the presence of Ca^{++} and Mg^{++} (Cepek *et al.*, 1993). As some other integrin α chains, α_4 et α_E contain a site of cleavage. The cleavage site of α_4, located in the middle of the extracellular domain, produces an inconstant and uncomplete cleavage of α_4 in two fragments of 70 and 80 kD (Takada *et al.*, 1989). The site of cleavage of α_E, present in its N-terminus, releases a peptide of approximately 20 kD, without homology in the sequence of other a chains and which remains bound to the main peptide by a disulfure bridge (Shaw *et al.*, 1994; Smith *et al.*, 1994). The functions of this unusual peptide and of these cleavages are not yet known. In addition, the N-terminus of α_E contains an I domain homologous to the I domain present in several integrin α chains. The role of the I domain of α_E is not known but may be important for binding of this integrin to its ligand. Indeed, the I domain present in LFA-1 ($\alpha_L\beta_2$) participates in the binding of this integrin to its ligands ICAM-1 et ICAM-3 (Landis *et al.*, 1994).

Distribution of $\alpha_4\beta_7$ and $\alpha_E\beta_7$ Integrins (Table 1)

The distribution of $\alpha_4\beta_7$ has been studied in humans with the ACT-1 monoclonal antibody which recognizes the $\alpha_4\beta_7$ heterodimer. Initially considered as an activation marker (Lazarovits *et al.*, 1984), $\alpha_4\beta_7$ is in fact detected on the surface of many mature immune cells although with an heterogeneous expression (Erle *et al.*,1994). In the peripheral blood of adults, approximately 50% of B lymphocytes and of CD4$^+$ and

Table 1 Main properties of the $\alpha_4\beta_7$ and $\alpha_e\beta_7$ integrins.

	Distribution in normal tissues	Distribution in pathological tissues	Ligand(s)	Function(s)
$\alpha_4\beta_7$	peripheral T and B lymphocytes – adults: approximately 50% – newborns: approximately 90%	Intestinal malignant lymphomatous polyposis	MadCAM-1	mucosal homing
			VCAM-1	?
	NK lymphocytes, eosinophils		fibronectin	?
			$\alpha 4$	homophilic adhesion
$\alpha_E\beta_7$	IEL (gut, bronchi) > 90% mouse intraepidermic T cells	IEL in inflammatory sites	E-cadherin	– adhesion to epithelial cells – activation of IEL?
	lamina propria lymphocytes: 20–40%	Tumoral IEL: – enteropathy-associated-T cell lymphomas – Mycosis fungoides	$\alpha_E\beta_7$?	homophilic adhesion?
	activated peripheral lymphocytes ($CD8^+ \gg CD4^+$) some mouse dendritic cells and mast cells	Hairy B cell leukemia		

CD8+ T lymphocytes express $\alpha_4\beta_7$ with a variable intensity depending on cells and individuals. On most lymphocytes, expression of $\alpha_4\beta_7$ is inversely proportional to that of the β_1 integrin and on CD4+ memory cells its expression is exclusive with that of the CLA antigen expressed on lymphocytes endowed with skin homing properties (Erle *et al.*, 1994; Schweighofffer *et al.*, 1993). A stronger and more homogenous expression of $\alpha_4\beta_7$ is observed on T and B lymphocytes of newborns, on NK cells and eosinophils. In contrast, neutrophils and monocytes do not express this integrin (Erle *et al.*, 1994). In tissues, $\alpha_4\beta_7$ is expressed by numerous lymphocytes in T areas and in germinal centers in lymph nodes and spleen. A few positive cells are detected in the thymic medulla (Lazarovits *et al.*, 1984). In intestinal tissues, $\alpha_4\beta_7$ is expressed by many PP lymphocytes, is strongly expressed by lamina propria lymphocytes and more weakly expressed by intraepithelial lymphocytes (Parker *et al.*, 1992; Salmi *et al.*, 1995 and unpublished data). In pathological circumstances in humans, $\alpha_4\beta_7$ is strongly expressed by cells of malignant lymphomatous polyposis characterized by the presence of multiple lymphomatous polypes throughout the intestine (Pals *et al.*, 1994). It is also expressed by histiocytic cells in the rare cases of histiocytosis X associated with an intestinal dissemination (Geissman *et al*, submitted).

The expression of the $\alpha_E\beta_7$ integrin (recently clustered as CD103) is much more restricted than that of $\alpha_4\beta_7$ and is mainly detected on lymphocytes associated with epitheliums. Thus in normal humans, $\alpha_E\beta_7$ is expressed by over 90% IEL in the intestinal epithelium and in the bronchial epithelium. In the human intestine, it is expressed by all subsets of IEL either bearing a TcR$\alpha\beta$ and CD4+ (10%) or CD8+ (70%), or bearing a TcR$\gamma\delta$ and CD8+ or CD8−CD4−(15%) or lacking the CD3-TcR complex but CD7+ (5%). In the intestinal lamina propria, approximately 20–40% of the T lymphocytes express $\alpha_E\beta_7$, mainly those expressing the CD8 molecule. In contrast, in lymphoid organs, in blood and thoracic duct lymph, $\alpha_E\beta_7$ is expressed on less than 5–7% of lymphocytes and in sections of thymus, it is detected on rare small cells in the medulla (Cerf-Bensussan *et al.*, 1987; Jarry *et al.*, 1990; Kruschwitz *et al.*, 1991; Micklem *et al.*, 1991). In animals, $\alpha_E\beta_7$ has been identified in mice (Kilshaw and Murant 1990; 1991; Yuan *et al.*, 1991, Smith *et al.*, 1994) and a molecule comparable to $\alpha_E\beta_7$ has been described in rats (Cerf-Bensussan *et al.*, 1986) and chickens (Haury *et al.*, 1993). In mice, the distribution of $\alpha_E\beta_7$, as defined by the M290 antibody, is very comparable to that observed in humans with the HML-1 antibody (Kilshaw and Murant 1990). A somewhat broader distribution on peripheral CD8+ T cells has been observed using another anti-mouse $\alpha_E\beta_7$ antibody, 2E7, which also showed $\alpha_E\beta_7$ expression on resident intraepidermic T lymphocytes (Lefrançois *et al.*, 1994). In pathological conditions, $\alpha_E\beta_7$ is expressed by normal T lymphocytes infiltrating inflammatory epitheliums (epidermis, mammary glands, salivary glands) (Cerf-Bensussan *et al.*, 1988; Kruschwitz *et al.*, 1991) or tumoral epitheliums (Jarry *et al.*, 1988) as well as on T lymphomas with epitheliotropism, enteropathy-associated T cell lymphomas (Spencer *et al.*, 1988; Stein *et al.*, 1988) and Mycosis fungoïdes (Sperling *et al.* 1989, Kruschwitz *et al.*, 1991; Simonitsch *et al.*, 1994). Notably, the expression of $\alpha_E\beta_7$ decreases when these lymphomas become invasive and loose their epitheliotropism, this decrease being correlated with a poor prognosis (Simonitsch *et al.* 1994). However, expression of $\alpha_E\beta_7$ is not entirely restricted to IEL. In humans and in mice, its expression can be induced on peripheral lymphocytes, mainly on CD8+ lymphocytes, after in vitro activation by various mitogens (Schiefferdecker *et al.*, 1990; Kruschwitz *et al.* 1991). In addition, $\alpha_E\beta_7$ is expressed by Hairy B cell leukemia malignant cells (Kruschwitz *et al.*, 1991; Micklem *et al.*, 1991). Finally, expression of $\alpha_E\beta_7$ has been detected on mouse mucosal mast cells (Smith *et al.*, 1994) and on mouse dendritic cells in lymph nodes but not in other sites (Kilshaw *et al.*, 1993). The expression of $\alpha_E\beta_7$ on the latter cells in humans remains a matter of debate (Kruschwitz *et al.*, 1991).The selective distribution of $\alpha_E\beta_7$ might be explained at least partially by its regulation by TGFβ, a cytokine synthetized by epithelial cells and by activated lymphocytes and macrophages. Thus, it was shown that TGFβ inhibits α_4 expression and induces α_E expression (Kilshaw and Murrant 1990;1991; Parker *et al.*, 1992). The large amount of TGFβ synthesized by enterocytes (Barnard *et al.*, 1993) may downregulate the expression of $\alpha_4\beta_7$, high on most lamina propria lymphocytes, but 5 to 10 times weaker on IEL and up-regulate the expression of $\alpha_E\beta_7$, present on only 20 to 50% of lamina propria lymphocytes but strongly expressed by more than 90% IEL (personal unpublished data).

Role of $\alpha_4\beta_7$ in Intestinal Lymphocyte Homing

The role of the $\alpha_4\beta_7$ in the intestinal lymphocyte homing has been delineated by Weissman, Butcher, and their coworkers. These authors had initially observed that a murine lymphoma, TK-1, selectively bound *in vitro* to high endothelial venules (HEV) of Peyer's patches but not of peripheral lymph node and that it colonized *in vivo* the intestinal tissues but not the peripheral lymphoid organs (Butcher *et al.*, 1980). To identify the putative homing receptor expressed on TK-1, they selected monoclonal antibodies blocking the interaction between TK-1 and intestinal venules. The molecular characterization of the receptor defined by these antibodies allowed its identification with the $\alpha_4\beta_7$ integrin (Holzmann *et al.*,1989; Hu *et al.*, 1992).

Following a comparable approach, Butcher and his group next identified the ligand of this integrin, MadCAM-1, on the murine intestinal endothelium and, more recently, on macaque and human intestinal endothelium (Streeter *et al.*, 1988; Nakache *et al.*, 1989; Briskin *et al.*, 1993; Shyjan, 1996). Structural studies have shown that MadCAM-1 is a member of the immunoglobulin (Ig) superfamily sharing homology with ICAM-1 and VCAM-1. The conservation of its sequence between species is poor except in its first two extracytoplasmic Ig domains. The first N-terminal Ig domain contains the sequence LTDSL, which aligns with a consensus motif for integrin-Ig family member interactions in ICAM-1,-2, and -3 and in VCAM-1, four other endothelial receptors for integrins. Accordingly, point mutations in this sequence resulted in loss of $\alpha_4\beta_7$ binding. The second MadCAM-1 Ig domain, also preserved during evolution, is homologous to the fifth domain of VCAM-1. Mucine-like regions are detected both in murine and human MadCAM-1, although their sequence are quite divergent. In mice, this mucin-like region plays an important function (Shyjan *et al.*, 1996). Thus, in PP and MLN, this region of MadCAM-1 is decorated by carbohydrates which bind to L-selectin, a homing receptor mainly expressed by naive small lymphocytes (Berg *et al.*, 1993). The other extra- and intracellular parts of MAdCAM-1 are less well conserved (Shyjan *et al.*, 1996).

The distribution of MadCAM-1 is compatible with its preferential role in mucosal homing, at least in adults. Thus, in adult mice, MadCAM-1 is detected on HEV in PP and MLN and on the flat endothelium of vessels in intestinal lamina propria and lactating mammary gland (where some lymphocytes of intestinal origin can migrate) but not on HEV of peripheral lymphoid organs (Berlin *et al.*, 1993; Briskin *et al.*, 1993; Streeter *et al.*, 1988). In humans, mRNA encoding MadCAM-1 is selectively detected in small and large intestine (Shyjan *et al.*, 1996). However, MadCAM-1 expression is not entirely restricted to intestine: mRNA was detected in human spleen (Shyjan *et al.*, 1996) and MadCAM-1 expression was observed on mouse cells lining splenic sinuses (Kraal *et al.*, 1995). In addition, in mice, distribution of MadCAM-1 is diffuse in the neonatal period and increases in chronically-inflammed tissues such as pancreas, inflammatory bowel, central nervous system. The latter observation is likely related to the fact that *in vitro*, MadCAM-1 expression can be induced on murine endothelial cells by treatment with proinflammatory cytokines (Sikorski *et al.*, 1993).

In mice, the *in vivo* study of labeled lymphocytes has demonstrated the role of $\alpha_4\beta_7$ and of its ligand MadCAM-1 in intestinal lymphocyte homing. Thus, after preincubation with blocking anti-α_4 or anti-$\alpha_4\beta_7$ antibodies, migration of small lymphocytes and of immunoblasts determined one hour after transfer was decreased by 50% in Peyer's patches and intestine but was not modified in other lymphoid or non lymphoid organs (Hamann *et al.*, 1994). The effect of anti-MadCAM-1 antibody injected 4 hours before the labeled lymphocytes was even more drastic since it almost abolished the migration of small lymphocytes into Peyer's patches, reduced by 37% their migration in mesenteric lymph nodes and also decreased the migration of immunoblasts derived from mesenteric lymph nodes into the intestinal mucosa (Streeter *et al.*, 1988). In contrast, it had no effect on the migration of lymphocytes into other organs.

In vitro and *in vivo* videomicroscopic analyses of lymphocyte interactions with purified MadCAM-1 and with PP and lamina propria endothelium have allowed to delineate the role of $\alpha_4\beta_7$ and MadCAM-1 in intestinal lymphocyte homing in mice (Berlin *et al.*, 1993; Bargatze *et al.*, 1995; Berlin *et al.*, 1995). Their mechanism of action is reviewed in the chapter by Bargatze and Butcher. Briefly, when small naive lymphocytes which express low level of $\alpha_4\beta_7$, enter the PP, they initially roll on endothelium due to the interaction of their homing receptor, L-selectin, with carbohydrates presented by several endothelial receptors including MadCAM-1 and, to a small extent, due to the interaction of $\alpha_4\beta_7$ with MadCAM-1. The next step of arrest and stable lymphocyte adhesion to the wall of the venules is mediated via the interaction of $\alpha_4\beta_7$ with MadCAM-1 and is favored by the interactions of another integrin, LFA-1, with its endothelial ligands, ICAM-2 and probably ICAM-1 expressed on intestinal vessels (Bargatze *et al.*, 1995). This step may be favored by the local presence of chemokines able to induce a high avidity conformation of $\alpha_4\beta_7$ and LFA-1 (Bargatze *et al.*, 1993). These interactions allow the entrance in the PP of many naive small lymphocytes recirculating between secondary lymphoid organs but do not allow their entrance in lamina propria. Indeed, in lamina propria, MadCAM-1 does not bear sugars recognized by L-selectin (Berg *et al.* 1993) and the low expression of $\alpha_4\beta_7$ on naive lymphocytes is not sufficient to allow their efficient interaction with this molecule (Berlin *et al.*, 1993; Bargatze *et al.*, 1995). In contrast, activated lymphocytes such as PP-derived immunoblasts should preferentially enter the lamina propria. On the one hand, activated lymphocytes loose L-selectin which is necessary to home into peripheral lymphoid organs (Kahn *et al.*, 1994). On the other hand, some of them express high level of $\alpha_4\beta_7$ integrin in a high avidity state, such that interactions between $\alpha_4\beta_7$ and MadCAM-1 become sufficient to stop the blasts in lamina propria venules and to initiate subsequent phases of diapedesis (Bargatze *et al.*, 1995; Berlin *et al.*, 1995).

These data partially explain the selective migration of blasts sensitized in PP into the normal intestinal mucosa. Thus, it is likely that lymphocytes stimulated in PP by intraluminal antigens will loose L-selectin and thus will become unable to migrate in peripheral lymphoid organs. Concomittantly, their activation will increase their expression of $\alpha_4\beta_7$ and probably induces a high avidity conformation of this integrin. Because of the selective expression of MadCAM-1 on intestinal endothelium, these immunoblasts of intestinal origin, will selectively home back into lamina propria and in gut-associated lymphoid organs. *In vitro* studies of the binding of lymphocytes extracted from the gut mucosa support this hypothesis. Thus, murine IEL and human lamina propria lymphocytes adhere to mucosal but not to peripheral high endothelium (Schmitz *et al.*, 1988; Jalkanen *et al.*, 1989; Salmi *et al.*, 1995). Furthermore, lamina propria immunoblasts (3–7% of the extracted cells) bound 5 times more efficiently to mucosal high endothelium than small mucosal lymphocytes and 4–8 times better than peripheral blood lymphocytes (Jalkanen *et al.*, 1989; Salmi *et al.*, 1995). Finally, binding of mucosal lymphocytes is almost abolished by pretreatment with anti-β_7 or anti-α_4 antibodies (Salmi *et al.*, 1995). Interestingly, it has been shown that the binding of lymphocytes extracted from an inflammed lamina propria is modified. Thus, lamina propria small lymphocytes and immunoblasts of patients with inflammatory bowel bound as well to peripheral as to mucosal endothelium (Salmi *et al.*, 1994). This finding may reflect the ability ot the inflammed mucosal endothelium to recruit peripheral lymphocytes. As indicated above, migration studies have revealed the increased entrance of small lymphocytes of peripheral origin in animal models of intestinal inflammation. In addition, abnormal recruitment of naive T cells and peripheral monocytes into the intestinal mucosa is observed in patients with Crohn's disease (Burgio *et al.*, 1995). This abnormal recruitment appears to be related to the upregulation of vascular adhesion molecules normally absent or weakly expressed on the lamina propria endothelium during inflammation, such as PNad, a ligand for L-selectin normally exclusively expressed on high endothelium in peripheral lymph nodes, and E-selectin, a ligand for monocytes and a subset of T lymphocytes (Salmi *et al.*, 1994).

Although the studies described above provide a good insight into the mechanisms which regulate the migration of intestinal lymphocytes and pinpoint to the important role of the interactions between $\alpha_4\beta_7$ and MadCAM-1 in the selective migration of intestinal lymphocytes, several points deserve more studies. First, it remains unclear why blasts stimulated in peripheral lymphoid organs do not migrate into the normal intestine. It is possible that expression of $\alpha_4\beta_7$ is less strongly induced on lymphocytes sensitized in peripheral lymphoid organs or that the latter cells acquire other homing receptors which drive their migration toward other tissues. It is also possible that intestinal blasts possess other homing receptors which favor their homing into the intestine. Accordingly, an antibody against an epitope of CD44, an ubiquitous and pleiomorphic molecule which binds hyaluronate and several extracellular matrix, specifically inhibited the binding of human lamina propria lymphocytes to mucosal but not to peripheral high endothelium (Jalkanen *et al.*, 1989). The selective effect of this antibody may be related to the fact that lymphocyte CD44 can bind to MadCAM-1 (Salmi *et al.*, 1991). Finally, the developmental and/or environmental mechanisms responsible for the selective expression of MadCAM-1 on the high-endothelial venules of PP and MLN and on flat intestinal vessels in lamina propria remain to be defined.

Besides its main ligand, MadCAM-1, the $\alpha_4\beta_7$ integrin has three other ligands. Thus, $\alpha_4\beta_7$ binds to fibronectin, an extracellular matrix protein, and to VCAM-1, (Andrew *et al.*, 1994) an endothelial receptor induced by inflammatory cytokines but poorly expressed or absent on lamina propria vessels. The *in vivo* importance of these interactions has not yet been evaluated. In addition, $\alpha_4\beta_7$ can bind to itself, allowing homophilic adhesion between lymphocytes, the function of which *in vivo* is not understood (Altevogt *et al.*, 1995).

Role of the $\alpha_E\beta_7$/CD103 Integrin in the Interactions Between Lymphocytes and Epitheliums

The distribution of the $\alpha_E\beta_7$ integrin, mainly restricted to IEL, strongly suggested the presence of an epithelial ligand for this integrin. Accordingly, several authors showed that antibodies against $\alpha_E\beta_7$ could specifically inhibit adhesion of murine or human IEL to epithelial cell lines of intestinal or extraintestinal origin whereas the same antibodies had no effet on the adhesion of $\alpha_E\beta_7^+$ lymphocytes to endothelial cells or fibroblasts (Cepek *et al.*, 1993; Roberts *et al.*, 1993; Roberts and Kilshaw 1994; Benmerah *et al.*, 1994). Interestingly, interactions between $\alpha_E\beta_7^+$ lymphocytes and epithelial cells occured only when epithelial tight junctions were absent (in undifferentiated epithelial cells) or opened, suggesting that the ligand was on the laterobasal membrane of enterocytes (Cepek *et al.*, 1993; Benmerah *et al.*, 1994). Production of antibodies against epithelial cells capable to block the adhesion of $\alpha_E\beta_7^+$ lymphocytes on epithelial cells has allowed the group of M. Brenner to identify the epithelial ligand of $\alpha_E\beta_7^+$ with E-cadherin (Cepek *et al.*, 1994). On the one hand, these authors showed that blocking antibodies immunoprecipitated E-cadherin. On the other hand, they showed that Cos cells transfected with $\alpha_E\beta_7$ specifically adhered to murine fibroblasts transfected with E-cadherin and that these interactions were blocked by anti-E-cadherin and anti-$\alpha_E\beta_7$ antibodies (Cepek *et al.*, 1994). Interaction between E-cadherin and $\alpha_E\beta_7$ was subsequently demonstrated in mice (Karecla *et al.*, 1995). E-cadherin is an epithelial membrane protein of crucial importance. Homophilic interactions between E-cadherin molecules on adjacent cells play a determinant function in epithelial morphogenesis. In differentiated epitheliums, E-cadherin is a major constituant of tight junctions and thereby controls the cohesion and permeability of epitheliums (Takeichi, 1991; 1995). In addition, E-cadherin may contribute to the regulation of intestinal growth, as suggested by the studies of mice with disrupted expression and function of E-cadherin in their intestine (Herminston and Gordon 1995a; 1995b) and by the demonstration of interconnections between E-cadherin and several signaling proteins, including the product of the APC tumor suppressor gene (Peifer 1993; 1996). Notably, IEL can bind to a form of E-cadherin distinct from that involved in the formation of the adherent junctions. Indeed, to insure adhesion between epithelial cells, E-cadherin must associate with several intracytoplasmic proteins, the α and β catenins which connect the intracytoplasmic domain of E-cadherin with actin-based cytoskeleton (Kemler 1993). In contrast, the association of E-cadherin with α-catenin and cytoskeleton does not seem to be required for lymphoepithelial interactions since adhesion of lymphocytes via CD103 was normal on an α-catenin-deficient epithelial cell line (personal unpublished observations).

The interactions of $\alpha_E\beta_7$/ CD103 on IEL and E-cadherin on epithelial cells allow the adhesion of lymphocytes to epithelium and probably contribute to the retention of lymphocytes in between epithelial cells. In addition, these interactions may promote exchange of signals between lymphocytes and epithelial cells. As discussed above, E-cadherin is involved in the control of several epithelial functions. Thus, its interaction with $\alpha_E\beta_7$ may perhaps modulate its local functions either directly or by interfering with homophilic E-cadherin-E-cadherin binding. Conversely, $\alpha_E\beta_7$ is likely an important accessory molecule for IEL and contribute to their local activation. Indeed, antibodies against $\alpha_E\beta_7$ modulate the cytotoxic activity of IEL toward epithelial cells (Lefrançois *et al.*, 1994; Roberts *et al.*, 1993; Roberts and Kilshaw 1994). As for other integrins, the role of $\alpha_E\beta_7$ may not be restricted to adhesion. Thus, anti-CD103 antibodies strongly enhanced activation of human gut IELs via the CD3-TcR pathway (Sarnacki *et al.*, 1992). Moreover, although most anti-CD103 antibodies had an inhibitory effect on IEL cytotoxicity, one antibody triggered their cytotoxicity (Lefrançois *et al.*, 1994). In contrast, $\alpha_E\beta_7$-E-cadherin interactions do not seem to play a role in lymphocyte homing into the epithelial layer. Antibodies against the rat and chicken homologues of $\alpha_E\beta_7$ did not prevent lymphocyte migration into the intestinal epithelium (Cerf-Bensussan *et al.*, 1988; Haury *et al.*, 1993). In addition, the HML-1 antibody which blocks $\alpha_E\beta_7$-E-cadherin interactions, did not prevent the intraepithelial migration of lymphocytes in human fetal organ cultures (T.T. MacDonald, personal communication). Therefore, the mechanisms which drive the migration of thymus-dependent and independent lymphocytes into the epithelium remain to be deciphered. Recent work suggest the possible role of chemokines synthesized by epithelial cells. Thus, IEL can migrate along a chemotactic gradient of IL-8 and RANTES (Ebert, 1995), and the study of epithelial cell lines suggests that epithelial cells can secrete several chemokines (Eckmann *et al.*, 1993; Jung *et al.*, 1995).

CONCLUSION

Studies in rodents have delineated the origin and homing properties of mucosal lymphocytes. They have demonstrated the role of the Peyer's patches and of the hemolymphatic cycle in the distribution of antigen-specific T cells and IgA plasma cells throughout the intestinal mucosa. The characterization of complementary receptors on the surface of lymphocytes endowed with gut homing properties and on the surface of the intestinal endothelium has clearly demonstrated the prominent role of the interactions between the lymphocyte $\alpha_4\beta_7$ integrin and the mucosal addressin MadCAM-1 in the selective homing of mucosal lymphocytes. These results suggest that these two molecules could be targeted to modulate intestinal lymphocyte migration during the course of deleterious inflammatory intestinal reactions. Accordingly, an antibody against α_4 prevented the induction of an intestinal response against Nippostrongylus brasiliensis in rats (Bell and Issekutz 1993). Furthermore, an antibody against α_4 was able to attenuate the colitis which spontaneously develops in the cotton-top-tamarin, a New World primate (Podolsky *et al.*,1993).

However, more work is needed to fully elucidate the mechanisms which control the specific homing properties of PP-derived mucosal lymphocytes and a number of questions remain unsolved. First, the mechanisms which control the intestinal migration of peritoneum-derived murine B cells and of murine bone marrow precursors able to undergo an extrathymic pathway of T cell differentiation are not known. Second, the identification of the $\alpha_E\beta_7$ integrin has revealed the specific interactions which develop between lymphocytes and epithelial cells, but did not provide a clue to the mechanisms which drive the intraepithelial migration of T cells, mainly CD8+. These mechanisms remain to be deciphered.

ACKNOWLEGMENTS:

A. Benmerah is supported by a fellowship of the Association pour la Recherche contre le Cancer (ARC, Villejuif, France)

REFERENCES

Altevogt, P., Hubbe, M., Ruppert, M., Lohr, J., Von Hoegen, P., Sammar, M., Andrew, DP., McEvoy, L., Humphries, M.J. and Butcher, EC. (1995). The $\alpha4$ integrin chain is a ligand for $\alpha_4\beta_7$ and $\alpha_4\beta_1$. *J. Exp. Med.*, **182**, 345–355.

Andrew, D.P., Berlin, C., Honda, S., Yoshino, T., Hamann, A., Holzman, B., Kilshaw, P.J. and Butcher, E.C. (1994). Distinct but overlapping epitopes are involved in $\alpha_4\beta_7$-mediated adhesion to vascular cell adhesion molecule-1, mucosal addressin-1, fibronectin, and lymphocyte aggregation. *J. Immunol.*, **153**, 3847–3861.

Bargatze, R.F., Jutila, M.A. and Butcher, E.C. (1995). Distinct roles of L-selectin and integrins $\alpha_4\beta_7$ and LFA-1 in lymphocyte homing to Peyer's patch-HEV *in situ*: the multistep model confirmed and refined. *Immunity*, **3**, 99–108.

Bargatze, R.F. and Butcher, E.C. (1993). Rapid G protein-regulated activation event involved in lymphocyte binding to high endothelial venules. *J.Exp. Med.*, **178**, 367–372.

Barnard, J.A., Warwick, G.J. and Gold, L.I. (1993). Localization of transforming growth factor β isoforms in the normal murine small intestine and colon. *Gastroenterology*, **105**, 67–73.

Bell, R.G. and Issekutz, T. (1993). Expression of a protective intestinal immune response can be inhibited at three distinct sites by treatment with anti-α_4 integrin. *J. Immunol.*, **151**, 4790–4802.

Benmerah, A., Badrichani, A., Ngohou, K., Mégarbané, B., Bègue, B. and Cerf-Bensussan, N. (1994). Homotypic aggregation of CD103 $(\alpha_E\beta_7)^+$ lymphocytes by an anti-CD103 antibody, HML-4. *Eur. J. Immunol.*, **24**, 2243–2249.

Berg, E.L., McEvoy, L.M., Berlin, C., Bargatze, R.F. and Butcher, E.C. (1993). L-selectin-mediated lymphocyte rolling on MadCAM-1. *Nature*, **366**, 695–698.

Berlin, C., Bargatze, R.F., Campbell, J.J., von Andrian, U.H., Szabo, M.C., Hasslen, S.R., Nelson, R.D., Berg, E.L., Erlandsen, S.L. and Butcher, E.C. (1995). $\alpha4$ integrins mediate lymphocyte attachment and rolling under physiologic flow. *Cell*, **80**, 413–422.

Berlin, C., Berg, E., Briskin, M.J., Andrew, D.P., Kilshaw, P.J., Holzmann, B. Weissman, I.L., Hamann, A. and Butcher, E.C. (1993). $\alpha_4\beta_7$ integrin mediates lymphocyte-binding to the mucosal vascular addressin MadCAM-1. *Cell*, **74**, 185–195.

Brantzaeg, P., Halstensen, T.S., Kett, K., Krajci, P., Kvale, D., Rognum, T.O., Scott, H. and Sollid, L.M. (1989). Immunobiology and immunopathology of human gut mucosa: humoral immunity and intraepithelial lymphocytes. *Gastroenterology*, **97**, 1562–1584.

Briskin, M.J., McEvoy, L.M. and Butcher, E.C. (1993). MadCAM-1 has homology to immunoglobulin and mucin-like adhesion receptors and to IgA1. *Nature*, **363**, 461–464.

Burgio, V.L., Fais, S., Boirivant, M., Perrone, A. and Pallone, F. (1995). Peripheral monocyte and naive T cell recruitment and activation in Crohn's disease. *Gastroenterology*, **109**, 1029–1037.

Butcher, E.C., Scollay, R.G. and Weissman, I.L. (1980). Organ specificity of lymphocyte migration: mediation by highly selective lymphocyte interaction with organ-specific determinants on high endothelial venules. *Eur. J. Immunol.*, **10**, 556–561.

Cahill, R.N.P., Poskitt, D.C., Frost, H. and Trnka, Z. (1977). Two distinct pools of recirculating T lymphocytes: migratory characteristics of nodal and intestinal T lymphocytes. *J. Exp. Med.*, **145**, 420–430.

Cepek, K.L., Shaw, S.K., Parker, C.M., Russell, G.J., Morrow, J.S., Rimm, D.R. and Brenner, M.B. (1994). Adhesion between epithelial cells and T lymphocytes mediated by E-cadherin and an integrin, $\alpha_E\beta_7$. *Nature*, **372**, 190–193.

Cepek, K.L., Parker, C.M., Madara, J.L. and Brenner, M.B. (1993). Integrin $\alpha_E\beta_7$ mediates adhesion of T lymphocytes to epithelial cells. *J. Immunol.*, **150**, 3459–3470.

Cerf-Bensussan, N., Jarry,A., Gneragbé, T., Brousse, N., Lisowska-Grospierre, B., Griscelli, C. and Guy-Grand, D. (1988). Monoclonal antibodies specific for intestinal lymphocytes. *Monogr. Allergy*, **24**, 167–172.

Cerf-Bensussan, N., Jarry, A., Brousse, N., Lisowska-Grospierre, B., Guy-Grand, D. and Griscelli, C. (1987). A monoclonal antibody (HML-1) defining a novel membrane molecule present on human intestinal lymphocytes. *Eur. J. Immunol.*, **17**, 1279–1285.

Cerf-Bensussan, N., Guy-Grand, D., Lisowska-Grospierre, B., Griscelli, C. and Bhan, A.K. (1986). A monoclonal antibody specific for rat intestinal lymphocytes. *J. Immunol.*, **136**, 76–82.

Cerf-Bensussan, N., Cerf M. and Guy-Grand, D. (1993). Gut intraepithelial lymphocytes and gastrointestinal diseases. *Cur. Op. Gastroenterol.*, **9**, 953–962.

Clark, E.A. and Brugge, J.S. (1995). Integrins and signal transduction pathways: the road taken. *Science*, **268**, 233–238.

Coppolino, M., Leung-Hagesteijn, C., Dehar, S. and Wilkins, J. (1995). Inducible interactions of integrin $\alpha_2\beta_1$ with calreticulin. *J. Biol. Chem.*, **270**, 23132–23138.

Craig, S.W. and Cebra, J.J. (1971). Peyer's patches, an enriched source of precursors for IgA-producing lymphocytes in the rabbit. *J. Exp. Med.*, **134**, 188–200.

Ebert, E.C. (1995). Human intestinal intraepithelial lymphocytes have potent chemotactic activity. *Gastroenterology*, **109**, 1154–1159.

Eckmann, L., Jung, H.C., Schürer-Maly, C., Panja, A., Morzycka-Wroblewska, E. and Kagnoff, M. (1993). Differential cytokine expression by human intestinal epithelial cell lines: regulated expression of interleukin 8. *Gastroenterology*, **105**, 1689–1697.

Erle, D.J., Rüegg, C., Sheppard, D., Pytela, R. (1991). Complete amino acid sequence of an integrin β subunit (β_7) identified in leukocytes. *J. Biol. Chem.*, **206**, 11009–11016.

Erle, D.J., Briskin, M.J., Butcher, E.C., Garcia-Pardo, A.I. and Tidswell, M. (1994). Expression and function of the MadCAM-1 receptor integrin $\alpha_4\beta_7$, on human leukocytes. *J. Immunol.*, **153**, 517–28.

Gowans, J.L. and Knight, E.J. (1964). The route of recirculation of lymphocytes in the rat. *Proc. R. Soc. London B*, **159**, 257–282.

Griscelli, C., Vassalli, P. and McCluskey, R.T. (1969). The distribution of large dividing lymph node cells in syngeneic recipient rats after intravenous injection. *J. Exp. Med.*, **130**, 1427–1451.

Guy-Grand, D., Cuenod-Jabri, B., Malassis-Seris, M., Selz, F. and Vassalli, P. (1996). Complexity of the mouse gut T cell-immune system: identification of two distinct natural killer-T cell intraepithelial lineages. *Eur. J. Immunol.*, **26**, in press.

Guy-Grand, D., Van Den Broecke, C., Briottet, C., Malassis-Seris, M., Selz, F. and Vassalli, P. (1992). Different expression of the recombination activity gene RAG-1 in various populations of thymocytes, peripheral T cells and gut thymoindependent intraepithelial lymphocytes suggest two pathways of T cell receptor rearrangement. *Eur. J. Immunol.*, 22, 505–510.

Guy-Grand, D., and Vassalli, P. (1993). Gut intraepithelial T lymphocytes. *Curr. Opin. Immunol.*, **5**, 247–252.

Guy-Grand, D., Luffau, G. and Vassalli, P. (1984). Gut mucosal mast cells. Origin, traffic and differentiation. *J. Exp. Med.*, **160**, 12–28.

Guy-Grand, D., Griscelli, C. and Vassalli, P. (1974). The gut-associated-lymphoid system: nature and properties of the large dividing cells. *Eur. J. Immunol.*, 4, 435–443.

Guy-Grand, D., Griscelli, C. and Vassalli, P. (1978). The mouse gut T lymphocyte, a novel type of T cell. *J. Exp. Med.*, **148**, 1661–1677.

Guy-Grand, D., Vassalli, P. (1988). Origin and traffic of gut mucosal lymphocytes and mast cells. In: *Migration and homing of lymphoid cells*, A. Husband, Ed. CRC Press, 99–111.

Guy-Grand, D., Cerf-Bensussan, N., Malissen, B., Malassis-Seris, M., Briottet, C., Vassalli, P. (1991). Two gut intraepithelial CD8+ lymphocyte populations with different T cell receptors: a role for the gut epithelium in T cell differentiation. *J. Exp. Med.*, **173**, 471–481.

Hall, J.G. (1970). Homing of lymph-borne immunoblasts to the gut. *Nature*, **226**, 262–264.

Hamann, A., Andrew, D.P., Jablonski-Westrich, D., Holzmann, B. and Butcher, E.C. (1994). Role of α4-integrins in lymphocyte homing to mucosal tissues *in vivo*. *J. Immunol.*, **152**, 3282–3293.

Haury, M., Kasahara, Y., SchalL, S., Bucy, R.P. and Cooper, M.D. (1993). Intestinal lymphocytes in the chicken express an integrin-like antigen. *Eur. J. Immunol.*, **23**, 313–319.

Herminston, M.L. and Gordon, J.I. (1995a). In vivo analysis of cadherin function in the mouse intestinal epithelium: essential roles in adhesion, maintenance of differentiation and regulation of programmed cell death. *J. Cell. Biol.*, **129**, 489–506.

Herminston, M.L. and Gordon, J.I. (1995b). Inflammatory bowel disease and adenomas in mice expressin a dominant negative N-cadherin. *Science*, **270**, 1203–1207.

Holzmann, B., McIntyre, B.W. and Weissman, I.L. (1989). Identification of a murine Peyer's patch-specific lymphocyte homing receptor as an integrin molecule with an alpha chain homologous to human VLA-4α. *Cell*, **56**, 37–46.

Hu, M.C.-T., Crowe, D.T., Weissman, I.L. and Holzman, B. (1992). Cloning and expression of mouse integrin βp(β_7): a functional role in Peyer's patch-specific lymphocyte homing. *Proc. Natl. Acad. Sci. USA*, **89**, 8254–8258.

Husband, A.J. and Gowans, J.L. (1978). The origin and antigen-dependent distribution of IgA-containing cells in the intestine. *J. Exp. Med.*, **148**, 1146–1160.

Hynes, R.O. (1992). Integrins: versatility, modulation, and signaling in cell adhesion. *Cell*, **69**, 11–25.

Jalkanen, S., Nash, De Los Toyos,J., MacDermott, R.P. and Butcher, E.C. (1989). Human lamina propria lymphocytes bear homing receptor and bind selectively to mucosal high endothelium. *Eur. J. Immunol.*, **19**, 63–68.

Jarry, A., Cerf-Bensussan, N., Brousse N., Guy-Grand D., Muzeau F. and Potet F. (1988). Same peculiar subset of HML-1+ lymphocytes present within normal intestinal epithelium is associated with tumoral epithelium of gastrointestinal carcinomas. *Gut*, **29**, 1632–8.

Jarry, A., Cerf-Bensussan, N., Brousse, N., Selz, F. and Guy-Grand, D. (1990). Subsets of $CD3^+$(TCR$\alpha\beta$ or TCR$\gamma\delta$) and $CD3^-$ lymphocytes isolated from normal human gut epithelium differ from their PBL counterparts. *Eur. J. Immunol.*, **20**, 1097–103.

Jiang, W.-M., Jenkins, D., Yuan, Q., Leung, E., Choo, K.H., Watson, J.D. and Krissansen, G.W. (1992). The gene organization of the human β_7 subunit, the common β subunit of the leukocyte integrins HML-1 and LPAM-1. *Int. Immunol.*, 4, 1031–1041.

Jung, H.C., Eckmann, L., Yang, S-K., Panja, A., Fierer, J. and Morzycka-Wroblewska (1995). A distinct array of proinflammatory cytokines is expressed in human colon epithelial cells in response to bacterial invasion. *J. Clin. Invest.*, **95**, 55–65.

Kahn, J., Ingraham, R.H., Shirley, F., Migaki, G.I. and Kishimoto, T.K. (1994). Membrane proximal cleavage of L-selectin: identification of the cleavage site and a 6-kD trans membrane fragment of L-selectin. *J. Cell. Biol.*, **125**, 461–470.

Karecla, P.I., Bowden, S.J., Green, S.J., Kilshaw, P.J. (1995). Recognition of E-cadherin on epithelial cells by the mucosal T cell integrin αM290β7 (αEβ7). *Eur. J. Immunol.*, **25**, 852–856.

Kemler, R. (1993). From cadherins to catenins: cytoplasmic protein interactions and regulation of cell adhesion. *Trends Genetics*, **9**, 317–321.

Kilshaw, P.J. and Murant, S. (1991). Exprcssion and regulation of β_7(βp) integrins on mouse lymphocytes : relevance to the mucosal immune system. *Eur. J. Immunol.*, **21**, 2591–2596.

Kilshaw, P.J. (1993). Expression of the mucosal T cell integrin $\alpha_{M290}\beta_7$ by a major subpopulation of dendritic cells in mice. *Eur. J. Immunol.*, **23**, 3365–3368.

Kilshaw, P.J. and Murant, S.J. (1990). A new surface antigen on intraepithelial lymphocytes in the intestine. *Eur. J. Immunol.*, **20**, 2201–2207.

Kraal, G., Schornagel, K., Streeter, P.R., Holzmann, B. and Butcher, E.C. (1995). Expression of the mucosal vascular addressin, MadCAM-1, on sinus-lining cells in the spleen. *Am. J. Pathol.*, **147**, 763–771.

Krissansen, G.W., Print, C.G., Prestidge, R.L., Hollander, D., Yuan, Q., Jiang, W-M, Jenkins, D.R., Leung,E., Mead, P., Yong, R., Ameratunga, R.V., Cerf-Bensussan, N. and Watson, J.D. (1992). Immunologic and structural relatedness of the integrin β_7 complex and the human intraepithelial lymphocyte antigen HML-1. *FEBS lett*, **296**, 25.

Kroese, F.G.M., Butcher, E.C., Stall, A.M. and Herzenberg, L.A. (1989). A major peritoneal reservoir of precursors for intestinal IgA plasma cells. *Immunological Investigations*, **18**, 47–58.

Kruschwitz, M., Fritzche, G., Schwarting, R., Micklem, K., Mason, D.Y., Falini, B. and Stein, H. (1991). Ber-ACT8: new monoclonal antibody to the mucosa lymphocyte antigen. *J. Clin. Pathol.*, **44**, 636–645.

Landis, R.C., McDowall, A., Holness, C.L.L., Littler, A.J., Simmons, D.L. and Hogg, N. (1994). Involvement of the "I" domain of LFA-1 in selective binding to ligands ICAM-1 and ICAM-3. *J. Cell. Biol.*, **126**, 529–537.

Lazarovits, A.I., Moscicki, R.A., Kurnick, J.T., Camerini, D., Bhan, A.K., Bairid, L.G., Erikson, M. and colvin, R.B. (1984). Lymphocyte activation antigens. I. A monoclonal antibody, anti-ACT1, defines a new late lymphocyte activation antigen. *J. Immunol.*, **133**, 1857–1862.

Lefrançois, L., Barrett, T.A., Havran, W.L., Puddington, L. (1994). Developmental expression of the $\alpha_{IEL}\beta_7$ integrin on T cell receptor $\gamma\delta$ and T cell receptor $\alpha\beta$ Tcells. *Eur. J. Immunol.*, **24**, 635–640.

Lub, M., van Kooyk, Y. and Figdor, C.G. (1995). Ins and out of LFA-1. *Immunol. Today*, **16**, 479–483.

Lundquist, C., Baranov, V., Hammarstrom, S., Athlin, L. and Hammarstrom, M.L. (1995). Evidence for regional specialization and extrathymic T cell maturation in the human gut epithelium. *Int. Immunol.*, 7, 1473–1487.

Masumoto, A. and Hemler, M.E. (1993). Mutation of putative divalent cation sites in the α_4 subunit of the integrin VLA-4: distinct effects on adhesion to CS1/fibronectin, VCAM-1, and invasin. *J. Cell. Biol.*, **123**, 245–253.

McWilliams, M., Philipps-Quagliata, J.M. and Lamm, M.E. (1975). Characteristics of mesenteric lymph node cells homing to gut-associated lymphoid tissue in syngeneic mice. *J. Immunol.*, **115**, 54–118.

Mega, J., Bruce, M.G., Beagley, K.W., McGhee, J.R., Taguchi, T., Pitts, A.M., McGhee, M.L., Bucy, R.P., Eldrige, J.H., Mestecky, J. and Kiyono, H. (1991). Regulation of mucosal responses by CD4$^+$ T lymphocytes: effects of anti-L3T4 treatment on the gastrointestinal immune system. *Int. Immunol.*, **3**, 793–805.

Micklem, K.J., Dong, Y., Willis, A., Pulford, K.A., Visser, L., Durkop, H., Poppema, S., Stein, H. and Mason, D.Y. (1991). HML-1 antigen on mucosa-associated T cells, activated T cells and Hairy leukemic cells is a new integrin containing the β_7 subunit. *Am. J. Pathol.*, **139**, 1297–1300.

Nakache, M., Berg, E.L., Streeter, P.R. and Butcher, E.C. (1989). The mucosal vascular addressin is a tissue-specific endothelial cell adhesion molecule for circulating lymphocytes. *Nature*, **337**, 179–181.

Otey, C.A., Vasquez, G.B., Burridge, K. and Erickson, B.W. (1993). Mapping of the α-actinin binding site within the β_1 integrin cytoplasmic domain. *J. Biol. Chem.*, **268**, 21193–21197.

Pals, S.T., Drillenburg, P., Dragosics, B., Lazarovits, A.I. and Radaskiewicz, T. (1994). Expression of the mucosal homing receptor $\alpha_4\beta_7$ in malignant lymphomatous polyposis of the intestine. *Gastroenterology*, **107**, 1519–1523.

Parker, C.M., Cepek, K.L., Russell, G.J., Shaw, S.K., Posnett, D.N., Schwarting, R. and Brenner, M.B. (1992). A family of β_7 integrins on human mucosal lymphocytes. *Proc Natl Acad Sci, USA*, **89**, 1924–1928.

Parrott, D.M.V. and Ferguson, A. (1974). Selective migration of lymphocytes within the mouse small intestine. *Immunology*, **26**, 571–588.

Peifer, M. (1993). Cancer, catenins, and cuticle pattern: a complex connection. *Science*, **262**, 1667–1668.

Peifer, M. (1996). Regulating cell proliferation: as easy as APC. *Science*, **272**, 974–975.

Phillips-Quagliata, J.M. and Lamm, M.E. (1994). Lymphocyte homing to mucosal sites. *In: Handbook of Mucosal Immunology* Eds. Ogra, P.L., Mestecky, J., Lamm, M.E., Strober, W., McGhee, J.R., Bienenstock, J. Academic Press Inc., pp. 225–239.

Podolsky, D.K., Lobb, R., King, N., Benjamin, C.D., Pepinsky, P., Seghal, P. and deBeaumont, M. (1993). Attenuation of colitis in the Cotton-top tamarin by anti-α_4 integrin monoclonal antibody. *J. Clin. Invest.*, **92**, 372–380.

Regnault, A., Cumano, A., Vassalli, P., Guy-Grand, D. and Kourilsky, P. (1994). Oligoclonal repertoire of the CD8$\alpha\alpha$ and the CD8$\alpha\beta$ TCR-α/β murine intestinal intraepithelial lymphocytes: evidence for the random emergence of T cells. *J. Exp. Med.*, **180**, 1345–1358.

Roberts, A.I., O'Connell, S.M. and Ebert, E.C. (1993). Intestinal intraepithelial lymphocytes bind to colon cancer cells by HML-1 and CD11a. *Cancer Res.*, **53**, 1608–1611.

Roberts, K. and Kilshaw, S.J., (1994). The mucosal T cell integrin $\alpha_{M290}\beta_7$ recognizes a ligand on mucosal epithelial cell lines. *Eur. J. Immunol.*, **23**, 1630–35.

Robertson, S.M. and Cebra, J.J. (1976). A model for local immunity. *La Ricerca Clin. Lab.*, **6** (**SuppL. 3**), 105–119.

Rocha, B., Vassalli, P. and Guy-Grand, D. (1994). Thymic and extrathymic origins of gut intraepithelial lymphocytes populations in mice. *J. Exp. Med.*, **180**, 681–686.

Rose, M.L., Parrott, D.M.V. and Bruce, R.G. (1976). Migration of lymphoblasts to the small intestine. II. Divergent migration of mesenteric and peripheral immunoblasts to sites of inflammation in the mouse. *Cell. Immunol.*, **27**, 36–46.

Roux, M.E., Mc Williams, M, Phillips-Quagliata, J.M. and Lamm, M.E. (1981). Differentiation pathway of Peyer's patches precursors of IgA plasma cells in the secretory immune system. *Cell. Immunol.*, **61**, 141–153.

Rudzik, O., Clancy, R., Perey, D.Y.E., Day, R.P. and Bienenstock, J. (1975). Repopulation with IgA-containing cells of bronchial and intestinal lamina propria after transfer of homologous Peyer's patch and bronchial lymphocytes. *J. Immunol.*, **114**, 1599–1564.

Salmi, M., Granfors, K., MacDermott, R. and Jalkanen, S. (1994). Aberrant binding of lamina propria lymphocytes to vascular endothelium in inflammatory bowel diseases. *Gastroenterology*, **106**, 596–605.

Salmi, M. and Jalkanen, S. (1991). Regulation of lymphocyte traffic to mucosa-associated lymphatic tissues. *Gastroenterol. Clin. N. Am.*, **20**, 495–510.

Salmi, M.A., D.P., Butcher, E.C. and Jalkanen, S. (1995). Dual binding capacity of mucosal immunoblasts to mucosal and synovial endothelium in humans: dissection of the molecular mechanisms. *J. Exp. Med.*, **181**, 137–149.

Sarnacki, S., Bègue, B., Buc, H., Le Deist, F. and Cerf-Bensussan, N. (1992). Enhancement of CD3–induced activation of human intestinal intraepithelial lymphocytes by stimulation of the β_7-containing integrin defined by HML-1 antibody. *Eur. J. Immunol.*, **22**, 2887–2892.

Schaller, M. D. and Parsons, J.T. (1994). Focal adhesion kinase and associated proteins. *Curr. Op. Cell Biol.*, **6**, 705–710.

Schaller, M.D., Otey, C.A., Hildebrand, J.D. and Parsons, T.J. (1995). Focal adhesion kinase and paxillin bind to peptides mimicking b integrin cytoplasmic domains. *J. Cell. Biol.*, **130**, 1181–1187.

Schiefferdecker, H. L., Ullrich, R., Weiss-Breckwoldt, A.N., Schwarting, R., Stein, H., Riecken, E-O. and Zeitz, M. (1990). The HML-1 antigen of intestinal lymphocytes is an activation antigen. *J. Immunol.*, **144**, 2541–2549.

Schmitz, M., Nunez, D. and Butcher, E.C. (1988). Selective recognition of mucosal lymphoid high endothelium by gut intraepithelial lymphocytes. *Gastroenterology*, **94**, 576–581.

Schweighoffer, T., Tanaka, Y., Tidswell, M., Erle, D., Horgan, K.J., Ginther Luce, G.E., Lazarovits, A.I., Buck, D. and Shaw, S. (1993). Selective expression of integrin $\alpha_4\beta_7$ on a subset of human CD4+ memory T cells with hallmarks of gut-trophism. *J. Immunol.*, **151**, 717–729.

Scollay, R., Hopkins, J. and Hall, J. (1976). Possible role of surface Ig in non-random recirculation of small lymphocytes. *Nature*, **260**, 528–530.

Shaw, S.K., Cepek, K.L., Murphy, E.A., Russell, G.J., Brenner, M.B. and Parker, C.M. (1994). Molecular cloning of the human mucosal lymphocyte integrin α_E subunit. *J. Biol. Chem.*, **269**, 6016–6025.

Shyjan, A.M., Bertagnolli, M., Kenney, C.J. and Briskin, M.J. (1996). Human mucosal addressin cell adhesion molecule-1 (MadCAM-1) demonstrates structural and functional similarities to the $\alpha_4\beta_7$-integrin binding domains of murine MadCAM-1 but extreme divergence of mucin-like sequences. *J. Immunol.*, **156**, 2851–2857.

Sikorski, E.E., Hallman, R., Berg, E.L. and Butcher, E.C. (1993). The Peyer's patch high endothelial receptor for lymphocytes, the mucosal addressin, is induced on a murine endothelial cell line by tumor necrosis factor-a and IL-1. *J. Immunol.*, **151**, 5239–44.

Simonitsch, I., Volc-Platzer, B., Mosberger, I. and Radaszkiewciz, T. (1994). Expression of monoclonal antibody HML-1 defined $\alpha_E\beta_7$ integrin in cutaneous T cell lymphoma. *Am. J. Pathol.*, **145**, 1148–58.

Smith, T.J., Ducharme, L.A., Shaw, S.K., Parker, C.M., Brenner, M.B., Kilshaw, P.J. and Weis, J.H. (1994). Murine M290 integrin expression modulated by mast cell activation. *Immunity*, **1**, 393–403.

Spencer, J., Cerf-Bensussan, N., Jarry, A., Brousse N., Guy-Grand, D., Krajewski, A. and Isaacson, P. (1988). Enteropathy-associated T cell lymphoma (malignant histiocytosis of the intestine) is recognized by a monoclonal antibody (HML-1) that defines a membrane molecule on human mucosal lymphocytes. *Am. J. Pathol.*, **132**, 1–5.

Sperling, M., Kaudewitz, P., Braun-Falco, O. and Stein, H. (1989). Reactivity of T cells in Mycosis fugoides exhibiting marked epidermotropism with the monoclonal antibody HML-1 that defines a membrane molecule on human mucosal lymphocytes. *Am. J. Pathol.*, **134**, 1989.

Stein, H., Dienemann, D., Sperling, M., Zeitz, M. and Riecken, E.O. (1988). Identification of a novel T cell lymphoma category derived from intestinal mucosa associated T cells. *Lancet*, **2**, 1053–1054.

Streeter, P.R., Berg, E.L., Rouse, B.T.N., Bargatze, R.F. and Butcher, E.C. (1988). A tissue-specific endothelial cell molecule involved in lymphocyte homing. *Nature*, **331**, 41–46.

Takada, Y., Elices, M.J., Crouse, C. and Hemler, M.E. (1989). The primary structure of the α_4 subunit of VLA-4: homology to other integrins and a possible cell-cell adhesion function. *EMBO J.*, **8**, 1361–1368.

Takeichi, M. (1991). Cadherin cell adhesion receptors as a morphogenetic regulator. *Science*, **251**, 1451–1455.

Takeichi, M. (1995). Morphogenetic roles of classic cadherins. *Curr. Op. Cell Biol.*, **7**: 619–627.

Tseng, J. (1984). Repopulation of the gut lamina propria with IgA-containing cells by lymphoid cells isolated from the gut lamina propria. *Eur. J. Immunol.*, **14**, 420–425.

Tseng, J. (1981). Transfer of lymphocytes of Peyer's patches between immunoglobulin allotype congenic mice: repopulation of the IgA plasma cells in the gut lamina propria. *J. Immunol.*, **127**, 2039–2043.

Williams, M.J., Hughes, P.E., O'Toole, T.E. and Ginsberg, M.H. (1994). The inner world of cell adhesion: integrin cytoplasmic domains. *Trends Cell Biol*, **4**, 109–112.

Yuan, Q., Jiang, W-M., Hollander, D., Leung, E., Watson, J.D. and Krissansen, G.W., (1991). Immunologic and structural relatedness of the integrin β_7 subunit and an antigen found highly expressed on intraepithelial lymhocytes in the small intestine. *Biochem. Biophys. Res. Commun.*, **176**, 1343–1346.

Zeitz, M., Schiefferdecker, H.L., Ullrich, R., Jahn, H-U., James, S.P. and Riecken, E-O. (1991). Phenotype and function of lamina propria T lymphocytes. *Immunol. Res.*, **10**, 199–206.

8. T CELL TRAFFICKING TO SKIN: A PARADIGM FOR TISSUE-SELECTIVE LYMPHOCYTE HOMING

LOUIS J. PICKER

Laboratories of Experimental Pathology and Clinical Immunology, Department of Pathology, University of Texas Southwestern Medical Center, Dallas, TX 75235–9072, USA

INTRODUCTION

The immune system's ability to recognize and respond to the vast spectrum of foreign antigens at all possible sites of tissue invasion depends heavily upon the efficient distribution of functionally competent lymphocytes among lymphoid tissues and effector sites [Picker and Siegelman, 1993]. To accomplish this task, evolution has developed an elaborate system of lymphocyte homing and recirculation, which acts to distribute the antigen receptor repertoire and to unify the function of the different lymphoid compartments. Most mature lymphocytes recirculate continuously – going from blood to tissue and back to blood again as often as 1–2x per day. Data gathered over the past two decades have shown conclusively that this recirculation is not random, but rather, is highly regulated so as to selectively provide "appropriate" lymphoid subsets to particular microenvironments (Butcher, 1986; Picker and Butcher, 1992; Picker, 1994; Mackay, 1993; Springer, 1994; and Butcher and Picker, 1996).

Among the most significant functions of the lymphocyte homing system is the differential distribution of "naive" and "memory/effector" (hereafter referred to as memory) populations to secondary lymphoid tissues (e.g. lymph node, Peyer's patch, tonsil, spleen) vs. extralymphoid ("tertiary" or "immune effector") sites (e.g. skin, intestinal lamina propria, pulmonary interstitium, etc.) (Butcher, 1986; Picker and Butcher, 1992; Picker and Siegelman, 1993; Mackay, 1993; Picker, 1994; and Springer, 1994). In general, naive T cells are programmed to home selectively to secondary lymphoid tissues, sites which are specialized for the antigen-induced activation and differentiation of this subset. Naive T cells do not show significant heterogeneity in their preference for the various types of secondary lymphoid tissue, and in keeping with their very limited effector capabilitites, do not migrate with any degree of efficiency to tertiary sites. In contrast, the antigen-driven naive to memory differentiation process provides the memory T cell not only with markedly enhanced effector function (Picker and Siegelman, 1993), but also with dramatically altered homing capabilities. First, memory T cells, as a group, have the ability to efficient extravasate in tertiary sites, and indeed, this population constitutes the vast majority of T cells within these sites. Second, in contrast to homogeneous virgin populations, memory T lymphocyte or immunoblast populations are heterogeneous in their ability to localize, and thus *form*

Correspondence: Tel: (214) 648–4125; Fax: (214) 648–4070

subsets with restricted, often tissue selective, homing behavior. Tissue-selective homing is thought to target memory cells and immunoblasts to sites where are they are most likely to (re)-encounter their specific antigen or are best adapted to function, and permits the segregation and specialization of immune responses in different regions of the body – e.g. mucosal vs. non-mucosal tissues (Butcher, 1986). Here, I will focus on the molecular and physiologic basis of T cell homing to skin as a paradigm for the tissue selective migration of memory lymphocytes.

THE SKIN CONSTITUTES A SELECTIVELY TARGETED MICROENVIRONMENT FOR MEMORY T CELLS

The first evidence suggesting that the skin constitutes an immunologically unique site with a distinct lymphocyte homing specificity derived from clinical observations of an unusual group of human T lineage malignancies — Mycosis Fungoides and its leukemic variant the Sezary Syndrome (collectively refered to as cutaneous T cell lymphomas or "CTCL"). These usually chronic lymphoma/leukemias show prolonged, often extensive multi-focal or diffuse involvement of epidermis, with limited involvement of other tissues (Edelson, 1980; Abel, 1985). Labeled Sezary Syndrome cells have also been shown to selectively and rapidly localize in skin upon *in vivo* re-injection (Miller *et al.*, 1980). These clinical observations were supported by accumulating evidence in animal models suggesting the ability of certain memory lymphocyte or blast subsets to home selectively to cutaneous inflammatory sites (Rose *et al.*, 1976; Issekutz *et al.*, 1980; Chin and Hay, 1980; Issekutz *et al.*, 1986), and by *in vitro* studies in the human indicating that lymphocyte binding to papillary dermal venules in inflamed (psoriatic) skin is mediated, at least in part, by molecular participants that are distinct from those involved in lymphocyte binding to the endothelium of other vascular beds involved in lymphocyte extravasation (namely the HEV of PLN and Peyer's patch; Sackstein *et al.*, 1988). However, definitive delineation of a skin-targeted memory T cell subset awaited the characterization of the Cutaneous Lymphocyte-Associated Antigen (CLA) – a lymphocyte homing receptor (HR) preferentially involved in T cell trafficking to skin.

CLA, A TISSUE-SELECTIVE LYMPHOCYTE HOMNG RECEPTOR FOR SKIN-HOMING MEMORY T CELLS

CLA was first identified as an antigenic determinant, defined by mAb HECA-452, that was selectively expressed by a subset of memory (but not naive) T cells in peripheral blood (25–30% of $CD45RA^{low}/RO^{high}$ T cells), essentially all such cells in inflamed skin (encompassing a wide variety of inflammatory conditions in skin), but very few T cells in chronically inflamed extracutaneous sites (including heart, lung, synovium, thyroid, kidney, etc.) (Picker *et al.*, 1990a and b). Examination of T cells from suction blisters placed over cutaneous delayed-type hypersensitivity (DTH) sites revealed that essentially the entire T cell population extravasating in these skin lesions is shifted positive with respect to CLA expression, with mean CLA cell surface densities an average of 23× higher than on corresponding peripheral blood T cells from the same individual (Figure 1) (Picker *et al.*, 1993a and 1994). Moreover, this striking

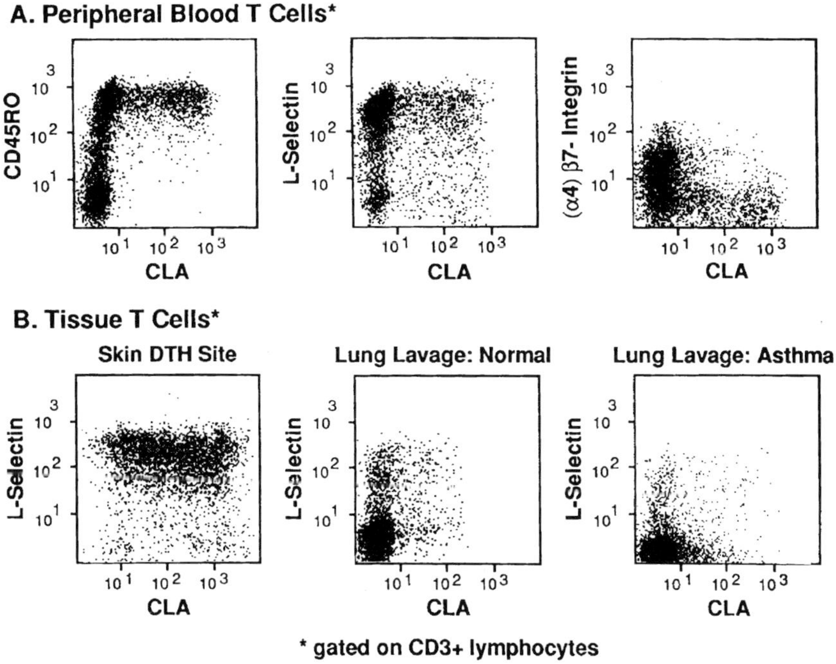

Figure 1 Expression patterns of CLA on human T cells (all profiles gated on CD3+ lymphocytes). T cells expressing CLA comprise a subset of CD45RO(+). memory T cells, which predominantly express the "peripheral lymph node" homing receptor L-selectin and lack the "mucosal" homing receptor $\alpha 4\beta 7$ integrin. Note that only a fraction of CLA (dim to intermediate). T cells express this latter homng receptor (see also Figure 2). The CLA-defined memory subset is highly enriched among cells derived from dermal-epidermal blisters placed over a cutaneous DTH sites (to mumps Ag), but not among bronchopulmonary lavage cells from either normal or inflamed (asthma). lung. Although limited CLA expression is noted on a subset of bronchopulmonary memory T cells (essentially all T cells in extralymphoid inflammatory sites are memory cells; Picker and Siegelman, 1993), both in terms of the number of CLA+ T cells and their CLA expression intensity, this expression is less than that that seen among the corresponding memory T cells in peripheral blood, consistent with a negative enrichment of the CLA+ subset in this site.

enrichment of CLA-expressing T cells in skin was observed throughout the evolution of the DTH reactions, including the earliest T cells emigrating into the blister fluid during the first few hours of the reaction (Picker *et al.*, 1993a). Thus, the profound enrichment of CLA+ T cells in skin can not be attributed to a specific upregulation of CLA expression after cutaneous extravasation, but rather appears to be due to skin-selective homing of the CLA+ T cell subset. The specificity of CLA enrichment in cutaneous inflammatory sites has been further demonstrated by the findings that, in contrast to the situation in skin, both lung and synovial memory T cells are predominantly CLA- (Picker *et al.*, 1994, Pitzalis *et al.*, 1996) (Figure 1). Moreover, CLA has been shown to be highly expressed by a series of allergen-specific T cell clones obtained from the skin of patients with atopic dermatitis, but not by non-allergen specific T cell clones obtained from the blood of the same individuals and grown under similar conditions

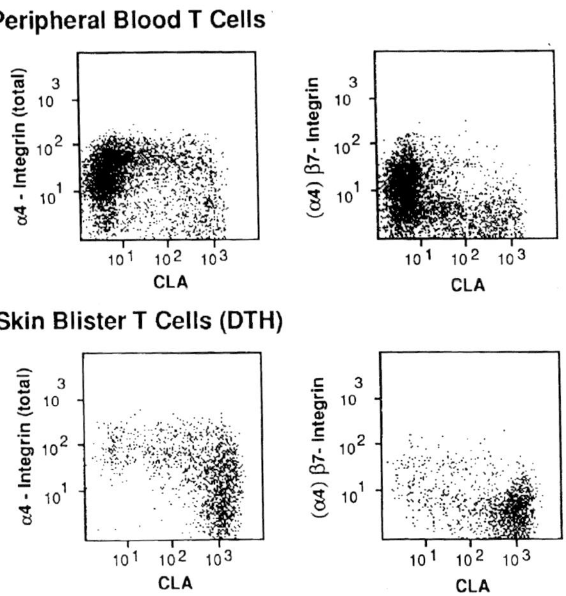

Figure 2 Comparison of CLA vs. $\alpha4$–integrin expression (total $\alpha4$–integrin – including $\alpha4\beta1$ – on the left vs. $\alpha4\beta7$ alone on the right) on cutaneous T cells taken from skin blisters overlying a mumps Ag-induced DTH site vs. peripheral blood T cells from the same individual. In this case, the DTH reaction was weak, leading to the highly selective emigration of the CLA(bright)/$\alpha4$–integrin(dim to negative). subset.

(Rossiter *et al.*, 1994). Preferential CLA expression has also been identified on the few T cells found in normal (clinically uninflamed) skin (Bos *et al.*, 1993). Finally, a recent study has demonstrated selective homing of CLA(+) human T cells to the superficial dermal vascular plexus (but not the deep vascular plexus) of human skin grafts placed on SCID mice (Kunstfeld *et al.*, 1997).

The CLA-defined memory T cell subset displays an overall pattern of HR expression consistent with its putative skin-selective homing behavior (Figure 1) (Picker *et al.*, 1994). The majority of CLA+ memory T cells also express the peripheral lymph node (PLN) HR L-selectin, in keeping with the intimate functional association between skin and its draining PLNs. In contrast, most CLA+ T cells, including essentially all the CLA(bright) cells, lack the mucosal selective homing receptor $\alpha4\beta7$–integrin.. Interestingly, the other $\alpha4$-integrin, $\alpha4\beta1$ (the primary lymphocyte receptor for VCAM-1 ligand), is strongly expressed by CLA (dim to intermediate+) T cells, but the CLA(bright+) T cell subset shows diminished expression of all $\alpha4$-integrins, including $\alpha4\beta1$ (Figure 2). The implications of this expression pattern are discussed in more detail below.

The linkage between T cell CLA expression and skin localization has been further supported by the observation that the skin localizing malignant T cells of CTCL, both Mycosis Fungoides and Sezary Syndrome cells, selectively express CLA (Picker *et al.*, 1990a; Heald *et al.*, 1993). Even Mycosis Fungoides patients without overt leukemia (e.g. Sezary Syndrome) were demonstrated to have excess numbers of circulating

CLA+ cells, the number of which correlated strongly with the extent of skin disease (Borowitz *et al.*, 1993). CLA+ T cells have also been found to predominate in the lymphoid infiltrates of non-lymphoid malignancies involving epidermis (primary or metastatic), but not extracutaneous tumors, dermal tumors, or metastatic skin-derived squamous carcinomas outside the skin (Gelb *et al.*, 1993). Besides further linking CLA expression and skin localization, these data also suggest that factors from the normal (e.g. non-neoplastic) epidermis appeared to be crucial for the skin-selective homing of the CLA+ T cell population.

CLA IS THE T CELL COUNTER-RECEPTOR FOR THE VASCULAR ADHESION MOLECULE E-SELECTIN

Taken together these data suggested a potential role for CLA in mediating tissue-selective T cell homing, perhaps as a specific receptor for an endothelial adhesion molecule. This possibility was confirmed by the discovery that CLA glycoproteins represent the major, if not exclusive, T cell ligands for the endothelial cell adhesion molecule E-selectin. It was first determined that CLA+ T cells could be greatly enriched (or depleted) by incubation with E-selectin, but not control, transfectants (Picker *et al.*, 1991). Next, highly purified CLA+ T cells were demonstrated to avidly bind E-selectin transfectants, whereas CLA- T cells bound these transfectants little, if at all (Picker *et al.*, 1991). The receptor/counter-receptor relationship of CLA and E-selectin was then formally proven by the demonstration that purified CLA glycoproteins specifically bound E-selectin transfectants (Berg *et al.*, 1991). More recently, it has been shown that CLA expression predicts E-selectin binding by a series of T cell clones (Rossiter *et al.*, 1994), and that CLA:E-selectin mediate a typical “rolling” primary interaction of T cells under physiologic conditions of flow (Jones *et al.*, 1994). Moreover, the CLA (HECA-452) determinant itself proved to be an oligosaccharide closely related to, but serologically distinct from, the prototype E-selectin ligand sialyl Lewis X (Berg *et al.*, 1991; Picker *et al.*, 1993a, and distinct from T cell ligands for the vascular adhesion molecule P-selectin; Rossiter *et al.*, 1994; Alon *et al.*, 1994; and Picker, unpublished data). CLA oligosaccharides appear to “decorate” a restricted set of core proteins, of which low M_r species of CD45 glycoproteins may be an important component (Picker, unpublished data).

The significance of the CLA/E-selectin interaction was revealed when venular E-selectin expression was examined in a wide variety of chronic inflammatory lesions (i.e. lesions with only mononuclear cell inflammatory infiltrates) of different tissues. In this setting, E-selectin was found to be preferentially expressed at high levels on the endothelial cells of superficial dermal venules (in association with CLA+ T cell infiltrates), relative to chronic inflammation-associated venules in the most other sites (Picker *et al.*, 1991). In keeping with these observations. it has also been shown that E-selectin expression is induced in the first few hours after a proinflammatory stimulus in skin, (Waldorf *et al.*, 1991), and is especially persistent on dermal microvascular endothelium (as compared to extracutaneous endothelium; Petselbauer *et al.*, 1993), likely due to alternative RNA processing (Chu *et al.*, 1994). Finally, it has recently been demonstrated that anti-E-selectin mAbs block the *in vitro* binding interaction between lymphocytes and psoriatic endothelium described above, thus, at least partially explaining this unique binding event (Cai *et al.*, 1996). Taken together, these data suggest strongly that the CLA:E-selectin interaction – although (as discussed below) by no means the only important interaction for skin homing – is required for *efficient* T cell homing to skin, but not to the vast majority of extracutaneous sites.

CLA EXPRESSION INTENSITY MAY CORRELATE WITH THE DEGREE OF SKIN HOMING SPECIFICITY AMONG MEMORY T CELLS

The pattern of $\alpha 4$-integrin expression on CLA+ memory T cells, described above and shown in Figures 1 and 2, strongly suggest that the *intensity* of CLA expression may be an important component of CLA-mediatied skin homing selectivity. Several *in vivo* observations support this view (L. Picker, unpublished data). First, whereas CLA(dim to intermediate+) cells can usually be found as a minor T cell subset in extracutaneous sites of inflammation, very few CLA(bright+) T cells are found in such sites (Figure 1). Second, CLA(bright+) T cells are usually the first T cells observed entering skin blisters overlying DTH sites, and are the "last to leave" (i.e. they persist the longest as the DTH response decreases). Finally, in some DTH reactions, usually those of lesser clinical intensity, the CLA(bright+) subset may be the predominant (or even the only) T cells to extravasate during the entire course of the reaction (Figure 2). These observations suggest two classes of skin homing T cells that are defined by their relative surface density of CLA: a CLA(bright+) T cell subset which displays strikingly skin-specific homing, and CLA(dim to intermediate+) subset which are skin-selective in their homing behavior, but retain the ability to extravasate in other sites.

The differential expression of $\alpha 4$-interins, particularly $\alpha 4\beta 1$, on CLA(dim to intermediate+) vs. CLA(bright) memory T cells has some interesting implications relative to the mechanism of tissue selective extrvasation in skin. As has been extensively reviewed (Butcher, 1991; Springer, 1994; Butcher and Picker, 1996), the extravasation process in general can be separated into 4 successive steps: 1) primary lymphocyte/endothelial cell adhesion, which is transient and reversible in seconds; 2) rapid (seconds) lymphocyte "activation"; 3) activation-dependent (secondary) "arrest" of the lymphocyte on the endothelium which is stable under shear but potentially reversible over minutes; and finally, 4) diapedesis. With regard to this model, CLA appears to function exclusively as a "primary" homing receptor (Jones *et al.*, 1994), capable of mediating initial lymphocyte/endothelial cell contact under flow and a subsequent "rolling" interaction, but not stable lymphocyte arrest on the endothelial cell surface. This functional status suggests that molecular cooperation is almost certainly essential for CLA-directed T cell homing to skin. Recent *in situ* studies of exteriorized mouse Peyer's patches (Bargatze *et al.*, 1995) provide a paradigm for how expression levels can contribute to the nature of these cooperative interactions, and thereby determine the specificity of the lymphocyte-endothelial cell interaction. In these studies it was demonstrated that molecular cooperation between the L-selectin/L-selectin ligand, $\alpha 4\beta 7$/MADCAM-1, and LFA-1/ICAM-1 or -2 adhesion molecule pairs were essential for extravasation of naive lymphocytes (L-selectin$^+$, $\alpha 4\beta 7^{\text{lo-med}}$, LFA-1$^+$) through Peyer's patch HEV (L-selectin ligand^{lo}, MAdCAM-1^{hi}, ICAM-1$^+$ and ICAM-2$^+$). In general, efficient arrest requires the sequential engagement of L-selectin to initiate contact, $\alpha 4\beta 7$ to slow rolling, and LFA-1 in conjunction with $\alpha 4\beta 7$ to mediate activation-dependent arrest. Despite the fact that $\alpha 4\beta 7$ can function as both a primary and secondary adhesion homing receptor, L-selectin dominates contact initiation because naive lymphocytes display only relatively low levels and activity of $\alpha 4\beta 7$. On the other hand, L-selectin rolling on Peyer's patch HEV (which express only low levels of L-selectin ligand) appears to be too fast to allow direct engagement of LFA-1, thus necessitating a "bridging" involvement of $\alpha 4\beta 7$. LFA-1, a exclusively secondary adhesion molecule, was also required for stable arrest, perhaps also reflecting the low levels of $\alpha 4\beta 7$ on naive cells. In contrast, $\alpha 4\beta 7$ is not required for naive cell extravasation through PLN HEV, likely because HEV in this tissue display uniquely high levels of L-selectin ligands (Michie *et al.*, 1993) that may provide for a slow enought L-selectin-mediated rolling to allow direct conversion to LFA-1-mediated arrest. Similarly, whereas $\alpha 4\beta 7$ levels on naive cells are insufficient to promote their direct binding to the MAdCAM-1$^+$, L-selectin ligand$^-$ venules in the intestinal lamina propria (an extralymphoid "effector" site), $\alpha 4\beta 7^{high}$ memory/effector lymphocytes effectively extravasate into this site in an L-selectin independent manner (Berlin *et al.*, 1995).

Thus, by varying the expression levels of key homing receptors and their endothelial counter-receptors and by allowing some of these adhesion systems to work cooperatively in sequence, the immune system can differentially target distinct lymphocyte subsets to different tissues using a relatively limited overall number of adhesion molecule pairs. By analogy, the highly variable expression of CLA may dictate a similar differential requirement for adhesion molecule cooperativity. CLA levels on the CLA (dim to intermediate) T cell subset may be insufficient to directly convert CLA-mediated rolling on E-selectin-expressing dermal venules to LFA-1 mediated arrest; however, with the "bridging" help of $\alpha 4\beta 1$ and its ligand VCAM-1 (which is also often, although not invariably, expressed by dermal venules; Rice *et al.*, 1991; Norris *et al.*, 1991; Cai *et al.*, 1996) these cell can can extravasate in skin. This cooperative interaction necessarily comes at the expense of stringent skin-homing specificity, as the high levels of $\alpha 4\beta 1$ would facililate extravasation through non-skin VCAM-1+ venules as well (VCAM-1 is widely expressed in the setting of chronic inflammation; Rice *et al.*, 1991). In contrast, the CLA(bright)/$\alpha 4\beta 1$(–) T cell subset may rely entirely on its high CLA expression to mediate the adhesive transition to LFA-1 mediated arrest, and thus this subset would be able to extravasate through E-selectin+ superficial dermal venules (including those lacking VCAM-1), but not most other vascular beds in the setting of chronic inflammation. It is important to note that this hypothesis does not rule out the participation of other factors in determing the exquisite homing specificity of CLA+ T cells. Indeed, the fact that CLA+ T cells are usually absent form acute (neutrophil-dominated) inflammatory sites with E-selectin+ vessels (L Picker, unpublished observations) suggests that the differential availability of leukocyte activating factors (step 2 of the multi-step model) may also play a key role in this process.

ANTIGEN-REACTIVE MEMORY T CELLS IN PATIENTS WITH SKIN-BASED CHRONIC INFLAMMATORY PROCESSES RECIRCULATE IN THE CLA+ SUBSET

One critical implication of the hypothesis that CLA repesents a skin-selective HR and defines a skin-homing memory subset is that most T cells involved in mediating chronic *cutaneous* inflammation would be expected to predominantly recirculate as CLA+ memory T cells in peripheral blood. This possibility has been investigated by the study of patients with cutaneous hypersensitivity to defined allergens, including patients with 1) Nickel (Ni)-induced allergic contact dermatitis, and 2) house dust mite (HDM) sensitivity manifested by either atopic dermatits alone, or asthma + an atopic dermatitis component. In these studies (Santamaria Babi *et al*,. 1995), CLA+ vs. CLA- memory T cells (CD45RO+) were physically separated, and then compared for their proliferative responses to these allergens. These experiments demonstrated that the proliferative response to Ni in the allergic contact dermatits patients was exclusively or nearly exclusively confined to the CLA+ subset. Moreover, in patients with HDM sensitivity and atopic dermatitis alone, 7 of 8 patients with measurable proliferative responses to HDM antigens showed complete or nearly complete restriction of this response to the CLA+ subset. In contrast, patients with HDM sensitivity and asthma (either alone or with an atopic dermatitis component) showed predominant (although not exclusive) HDM Ag-specific proliferative responses in the CLA- subset. Although these data are from a relatively small group of patients, they provide additional support for the association of T cell CLA expression and cutaneous immunity, and strongly suggest the existence of regulatory mechanisms capable of "sorting" individual T cell antigen specificities into homing receptor-defined memory subsets.

CLA EXPRESSION IS INDUCED DURING THE T CELL NAIVE TO MEMORY TRANSITION IN HUMAN 2° LYMPHOID TISSUES *IN VIVO* IN A TISSUE-SELECTIVE MANNER, AND CAN BE DIFFERENTIALLY REGULATED ON BOTH ACTIVATED NAIVE AND MEMORY T CELLS *IN VITRO* BY MICROENVIRONMENTAL SIGNALS

Early physiologic studies of tissue-selective homing in animal models suggested that the site of lymphocyte activation played a role in determining the subsequent homing behavior of the resultant blast or memory populations. After activation within a particular region of the body, cells tended to preferentially return to that region (e.g. mucosal activation to mucosal sites; peripheral to peripheral; Butcher, 1986). These differences in homing behavior and the subsequently characterized differences in homing recptor expression patterns between naive and memory T cells (Picker *et al.*, 1990b and 1994) suggested that the naive to memory transition in secondary lymphoid tissues may be a critical time point for this regulation. This hypothesis was investigated in the human using multiparamter flow cytometry to characterize the CLA (and L-selectin) expression of T cells undergoing the early, middle and late stages of naive to memory transition in a variety of different secondary lymphoid tissues (transition stages defined by relative levels of CD45RO, RA, and activation/proliferation markers) (Picker *et al.*, 1993a and b). First, these studies demonstrated that the heterogeneous expression of CLA and L-selectin observed among peripheral blood memory T cells is initially generated during this transition. Second, there was a clear *tissue-specificity* in this regulation: memory T cells generated in the appendix were predominantly L-selectin-/CLA-, whereas those generated in PLN were predominantly L-selectin+, and 50–60% expressed CLA. These data are consistent with the hypothesis that at least *one component* of the differential regulation of these two HRs is related to the anatomical location of the differentiation process.

The development of T cell heterogeneity in homing receptor expression and homing function during the naive to memory differentiation would appear to be analogous to (and temporally concomitant with) the much more intensively studied cytokine synthesis heterogeneity manifested by memory T cells. The differentiatial ability to produce effector cytokines (e.g. Th1 vs. Th2 cytokine synthesis subsets) has been demonstrated to be determined in large part by the microenvironmental conditions supporting the naive to memory differentiation process, particularly the presence or absence of particular immunoregulatory cytokines and/or adhesive interactions (Scott, 1993; Paul and Seder, 1994; Semnani *et al.*, 1994). *In vitro* studies with human T cells has formally demonstrated such differential regulation for CLA as well (Picker *et al.*, 1993a). When CLA- naive cells were stimulated *in vitro* with mitogen + IL-2 alone, there was little or no upregulation of CLA expression. Other selected cytokines (IL-1, IL-3, IL-4, IL-5, IL-7, IL-8, IL-10, TNF-α, GM-CSF, γ-IFN) also lacked CLA upregulatory activity when added to these cultures (Picker *et al.*, 1993a; and L. Picker, unpublished). However, two cytokines – TGF-β1, and to a lesser extent IL-6 – were able to upregulate CLA expression without substantially changing other features of the virgin to memory conversion. These cytokines also increased CLA expression on established memory T cells, both converting CLA- memory cells to CLA+ and increasing CLA expression on established CLA+ memory cells. Interestingly, L-selectin expression by activated virgin T cells was also increased by these same two cytokines – in keeping with the L-selectin+ phenotype of the vast majority of CLA-expressing memory T cells (Picker *et al.*, 1993b).

In later studies, it was observed that bacterial superAgs (including staph. enterotoxins A and B, toxic shock syndrome toxin-1 and strep. pyrogenic exotoxins A and C) induce a striking CLA upregulation, without exogenous cytokine addition (Leung *et al.*, 1995). It was further demonstrated that 1) these toxins had the ability to stimulate accessory cell production of IL-12, 2) anti-IL-12 (but not anti-TGF-β1) blocked toxin-induced CLA upregulation, and 3) adding exogenous IL-12 to mitogen-stimulated (accessory cell free) cultures also resulted in strong CLA upregulation (on both naive and memory T cells). In addition, unlike TGF-β1, IL-12 upregulated CLA without also inducing $\alpha e\beta 7$ integrin, an adhesion molecule that is absent from most circulating CLA+ memory T cells (Picker *et al.*, 1994). These data not only identified IL-12 as a potent, specific and likely physiologic upregulator of CLA, they also demonstrated that certain T cell stimuli can, by virtue of their effects on accessory cells, simultaneously activate T cells and regulate their functional differentiation.

Differential regulation of CLA expression has been demonstrated in another interesting clinical situation – that childhood milk-induced atopic dermatitis (Abernathy-Carver *et al.*, 1995). T cells from children with milk-induced dermatitis, but not children with milk-induced gastroenteritis or adult non-atopic controls, upregulated CLA strikingly after *in vitro* incubation with the milk protein casein. CLA upregulation was not observed with unrelated antigens, suggesting that these children did not suffer from a generalized overproduction of CLA+ T cells. Thus, the nature of the antigen, the genetic background of the patient, and the maturity of the immune system may all contribute to the regulation of homing receptors such as CLA. Indeed, because milk antigens are most likely delivered to the immune system via the gastrointestinal

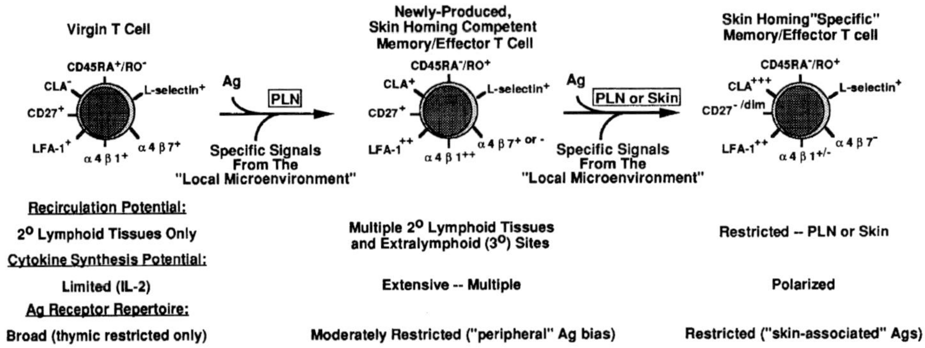

Figure 3 A model of skin-homing (CD4+). memory T cell differentiation. This model would also likely apply to CD8+ T cells with the possible exception of the CD27 phenotypes (little is known regarding the physiologic correlates of changes in CD27 expression among the CD8+ subset).

tract, these factors may, in some instances, "over-ride" the regulatory influences of that anatomical location (i.e. by producing CLA+ memory cells in mucosa-associated secondary lymphoid tissues).

A WORKING MODEL OF SKIN-HOMING MEMORY T CELL DIFFERENTIATION

Figure 3 proposes a working model of the development of skin-homing memory T cells in humans. The first basic premise of this model is that the homing characteristics of a memory T cell are differentially regulated, along with effector functions such as cytokine synthesis capabilitities, at the time of antigen-induced T cell activation by the local microenvironment that supports this activation. Important immunomodulatory components of the local microenvironment would include the nature of the antigenic stimulus itself, as well as the cytokines elaborated by and/or adhesion events supported by 1) "professional" or "semi-professional" immune accessory cells (dendritic cells, macrophages, or B cells), 2) components of the "innate" immune system (PMN, NK cells, or mast cells), 3) differentiated epithelial or mesenchymal cells of the tissue in question, 4) components of the extracellular matrix, and finally 5) T cells themselves.

The second basic premise is that memory T cell homing function can be influenced each time that cell is activated – such that the homing capabilities of that cell may reflect both its last activation and to a decreasing extent, its entire previous activation history. Analogous to recent reports examining polarized cytokine synthesis function (Openshaw *et al.*, 1995), if a cell consistently and repeatedly receives a particular set of regulatory signals, the phenotype supported by these signals would be reinforced, leading to the development of a functionally polarized memory subset. In contrast, repeated "mixed" or conflicting regulatory signals would result in the retention of a multifunctional "generalist" phenotype. In the case of skin homing, newly produced memory T cells may express low to intermediate levels of CLA, as well as other homing receptors to extracutaneous sites (e.g. $\alpha 4\beta 7$ integrin), and thus be able to access a variety of tissue sites. In contrast, "terminally differentiated" memory T cells, likely those specific for a skin-associated antigens, would have a history of consistently being activated in the skin or related microenvironments and would thus express high levels of CLA, and predominantly lack homing receptors such as $\alpha4\beta7$ integrin that are involved in homing to unrelated sites. Because of their highly skin tropic homing characeristics, these cells would functionally be skin-specific. This hypothesis is in keeping with our observations of the expression pattern of CLA vs. CD27, a proposed marker of CD4(+) memory T cell differentiation status. CD27 is thought to be permanently down-regulated on CD4+ memory T cells that are terminally differentiated (reviewed in Tortorella *et al.*, 1995), and in keeping with this, the vast majority of CLA(bright) CD4+ memory T cells have diminished expression of CD27, whereas the CLA (dim to intermediate) CD4+ memory subset is almost entirely CD27(bright) (L. Picker, unpublished observations).

From a physiologic perspective, a high degree of functional specialization would *not* appear to be appropriate for most *first-time* activated T cells as the immune system must maintain flexibility in its response to antigens that might have multiple portals of entry into the body or require a diverse functional response. On the other hand, T cells that are repeatedly stimulated under similar microenvironmental conditions would eventually become highly polarized in their functional capabilities, including both tissue-selective homing and cytokine-specific effector responses. These "fully differentiated" memory cells would be most efficient in combating antigenic challenges that are stereotyped in their mode of entry into the body or in their requirements for a specific type of effector response. Potentially pathogenic T cells would be similarly influenced by these factors, perhaps leading to functionally polarized memory cells that mediate *organ specific* immune/inflammation disease. Indeed, it is possible that inappropriate regulation of the mechanisms controlling memory T cell differentiation might, in some instances, lead to the formation of pathogenic memory populations, thereby directly contributing to the pathogenesis of specific tissue-based immune/inflammatory diseases. Further work will be required to unravel the likely complex microenvironmental regulation of homing function, to determine how homing is coordinately regulated with other effector capabilities, ane finally, to determine if and how these regulatory mechanisms contribute to immune dysfunction in disease states.

REFERENCES

Abel, E.A. Clinical features of cutaneous T cell lymphoma. (1985). *Dermatol. Clinics*, **3**, 647–664.

Abernathy-Carver, K.J., Sampson, H.A., Picker, L.J. and Leung D.Y.M. (1995). Milk-induced eczema is associated with the expansion of T cells expressing cutaneous lymphocyte antigen, *J. Clin. Invest.*, **95**, 913–918.

Bargatze, R.F., Jutila, M.A. and Butcher, E.C. (1995). Distinct roles of L-selectin and integrins alpha 4 beta 7 and LFA-1 in lymphocyte homing to Peyer's patch-HEV *in situ*: The multistep model confirmed and refined. *Immunity*, **3**, 99–108.

Berg, E.L., Yoshino, T., Rott, L.S., Robinson, M.K., Warnock, R.A., Kishimoto, T.K., Picker, L.J. and Butcher, E.C. (1991). The cutaneous lymphocyte antigen is a skin lymphocyte homing receptor for the vascular lectin endothelial cell-leukocyte adhesion molecule 1. *J. Exp. Med.*, **174**, 1461–1466.

Berlin, C., Bargatze, R.F., Campbell, J.J., von Andrian, U.H., *et al.* (1995). Alpha 4 integrins mediate lymphocyte attachment and rolling under physiologic flow. *Cell*, **80**, 413–422.

Borowitz, M.J., Weidner, A., Olsen, E.A. and Picker, L.J. (1993). Abnormalities of circulating T cell subpopulations in patients with cutaneous T cell lymphoma: Cutaneous lymphocyte-associated antigen expression on T cells correlates with extent of disease, *Leukemia*, 7, 859–863.

Bos, J.D., de Boer, O.J., Tibosch, E., Das, P.K. and Pals, S.T (1993). Skin-homing T lymphocytes: detection of cutaneous lymphocyte-associated antigen (CLA). by HECA-452 in normal human skin. *Dermatol. Res.*, **285**, 179–183.

Butcher, E.C. (1986). The regulation of lymphocyte traffic. *Curr. Topics in Microbiol. & Immunol.*, **128**, 85–122.

Butcher, E.C. (1991). Leukocyte-Endothelial Cell Recognition: Three (or More). Steps to Specificity and Diversity. *Cell*, **67**, 1033–1036.

Butcher, E.C. and Picker, L.J. (1996). Lymphocyte Homing and Homeostasis. *Science*, in press.

Cai, J.-P., Harris, K., Falanga, V., Taylor, J.R. and Chin, Y.-H. (1996). UVB therapy decreases the adhesive interaction between peripheral blood mononuclear cells and dermal microvascular endothelium and regualte the differential expresssion of CD54, VCAM-1 and E-selective in psoriatic plaques. *Br. J. Dermatol.*, **134**, 7–16.

Chin, W. and Hay, J.B. (1980). A comparison of lymphocyte migration through intestinal lymph nodes, subcutaneous lymph nodes and chronic inflammatory sites of sheep. *Gastroenterology*, 79, 1231–1242.

Chu, W., Presky, D.H., Swerlick, R.A. and Burns, D.K. (1994). Alternatively processed human E-selectin transcripts linked to chronic expression of E-selectin *in vivo. J. Immunol.*, **153**, 4179–4189.

Edelson, R.L. (1980). Cutaneous T cell lymphoma: Mycosis fungoides, Sezary Syndorme and other variants. *Am. Acad. Dermatol.*, **2**, 89–106.

Gelb, A.B., Smoller, B.R., Warnke, R.A. and Picker, L.J. (1993). Lymphocytes infiltrating primary cutaneous neoplasms selectively express the cutaneous lymphocyte-associated antigen (CLA), *Am. J. Pathol.*, **142**, 1556–1564.

Heald, P.W., Yan, S-L, Edelson, R.L„ Tigelaar, R. and Picker, L.J. (1993). Skin-Selective Lymphocyte Homing Mechanisms in the Pathogenesis of Leukemic Cutaneous T-Cell Lymphoma, *J. Invest. Dermatol.*, **101**, 222–226.

Issekutz, T.B., Chin, W. and Hay, J.B. (1980). Lymphocyte traffic throught granulomas: Differences in the recovery of indium-111-labeled lymphocytes in afferent and efferent lymph. *Cell Immunol.*, **54**, 79–86.

Issekutz, T.B., Webster, D.M. Stoltz, J.M. (1986). Lymphocyte recruitment in vaccinia virus-induced cutaneous delayed-type hypersensitivity. *Immunol.*, **58**, 647–664.

Jones, D.A., Smith, C.A., McIntire, L.V. and Picker, L.J. (1994). A two-step adhesion cascade for T cell-endothelial cell interactions under flow conditions, *J. Clin. Invest.*, **94**, 2443–2450.

Kunstfeld, R., Lechleitner, S., Gröger, M., Wolff, K. and Petzelbauer, P. (1997). HECA-452+ T cells migrate through superficial vascular plexus but not through deep vascular plexus endothelium. *J. Invest. Dermatol.*, **108**, 343–348.

Leung, D.Y.M., Gately, M., Trumble, A., Ferguson-Darnell, B., Schlievert, P.M. and Picker, L.J. (1995). Bacterial superantigens induce T cell expression of the skin-selective homing receptor, the cutaneous lymphocyte-associated antigen (CLA), via stimulation of IL-12 production, *J. Exp. Med.*, **181**, 747–753.

Mackay, C.R. (1993). Homing of Naive, Memory and Effector Lymphocytes. *Curr. Opin. Immunol.*, **5**, 423–427.

Michie, S.A., Streeter, P.R., Bolt, P.A., Butcher, E.C. and Picker, L.J. (1993). The human peripheral lymph node vascular addressin: An inducible endothelial antigen involved in lymphocyte homing, *Am. J. Pathol.*, **143**, 1688–1698.

Miller, R.A., Coleman, C.N., Fawcett, H.D., Hoppe, R.T. and McDougall, I.R. (1980). Sezary Syndrome: A model for migration of T lymphocytes to skin, *N. Engl. J. Med.*, **303**, 89–92.

Norris, P., Poston, R.N., Sian Thomas, D., Thornhill, M., Hawk, J. and Haskard, D.O. (1991). The expression of ELAM-1, ICAM-1 and VCAM-1 in experimental cutaneous inflammation: A comparison of ultraviolet B ertyhema and delayed hypersensitivity, *J. Invest. Dermatol.*, **96**, 763–770.

Openshaw, P., Murphy, E.E. Hosken, N.A., Maino, V., Davis, K., Murphy, K. and O'Garra, A. (1995). Heterogeneity of intracellular cytokine synthesis at the single-cell level in polarized T helper 1 and T helper 2 populations. *J. Exp. Med.*; **182**, 1357–1367.

Paul, W.E. and Seder, R.A. (1994). Lymphocyte Responses and Cytokines. *Cell*, **76**, 241–251.

Petzelbauer, P., Bender, J.R., Wilson, J. and Pober, J.S. (1993). Heterogeneity of dermal microvascular endothelial cell antigen expression and cytokine responsiveness *in situ*and in cell culture. *J. Immunol.*, **151**, 5062–5072.

Picker, L.J., Michie, S.A., Rott, L.S. and Butcher, E.C. (1990a). A unique phenotype of skin-associated lymphocytes in humans: Preferential expression of the HECA-452 epitope by benign and malignant T-cells at cutaneous sites. *Am. J. Pathol.*, **136**, 1053–1068.

Picker, L.J., Terstappen, L.W.M.M, Rott, L.S., Streeter, P.R., Stein, H. and Butcher, E.C. (1990b). Differential expression of homing-associated adhesion molecules by T-cell subsets in man. *J. Immunol.*, **145**, 3247–3255.

Picker, L.J., Kishimoto, T.K., Smith, C.W., Warnock, R.A. and Butcher, E.C. (1991). ELAM-1 is an adhesion molecule for skin-homing T cells. *Nature*, **349**, 796–799.

Picker, L.J. and Butcher, E.C. (1992). Physiological and Molecular Mechanisms of Lymphocyte Homing. *Annu. Rev. Immunol.*, **10**, 561–591.

Picker, L.J. and Siegelman, M.H. (1993). Chapter 6: "Lymphoid Tissues and Organs". In *Fundamental Immunology*3rd Edn. Edited by Paul WE. New York: Raven Press; pp. 145–197.

Picker, L.J., Treer, J.R., Ferguson-Darnell, B., Collins, P., Bergstresser, P. and Terstappen, L.W.M.M. (1993a). Control of Lymphocyte Recirculation in Man: II. Differential Regulation of the T Cell Skin-Selective Homing Receptor CLA. *J. Immunol.*, **150**, 1122—1136.

Picker, L.J., Treer, J.R., Ferguson-Darnell, B., Collins, P., Buck, D., Terstappen, L.W.M.M. (1993b). Control of Lymphocyte Recirculation in Man: I. Differential Regulation of the Peripheral Lymph Node Homing Receptor L-Selectin on T Cells During the Virgin to Memory Cell Transition. *J. Immunol.*, **150**, 1105–1121.

Picker, L.J., Martin, R.J., Trumble, A., Newman, L.S., Collins, P.A., Bergstresser, P.R., Leung, D.Y.M. (1994). Differential Expression of Lymphocyte Homing Receptors by Human Memory/Effector T Cells in Pulmonary vs. Cutaneous Immune Effector Sites, *Eur. J. Immunol.*, **24**, 1269–1277.

Picker, L.J. (1994). Control of Lymphocyte Homing. *Curr. Opin. Immunol.*, **6**, 394–406.

Pitzalis, C., Cauli, A., Pipitone, N., Smith, C., Barker, J., Marchesoni, A., Yanni, G., Panayi, G. (1996). Cutaneous lymphocyte antigen (CLA). positive T lymphocytes preferentially migrate to the skin but not to the joint in psoriatic arthritis. *Arthritis and Rheum.*, **39**, 137–145.

Rice, G.E., Munro, J.M., Corless, C., Bevilacqua, M.P. (1991). Vascular and non-vascular expression of INCAM-110 A target for monocnuclear adhesion in normal and inflamed human tissues. *Am. J. Pathol.*, **138**, 385–393.

Rose, M.L., Parrott, D.M.V. and Bruce, R.G. (1976). Migration of lymphoblasts to the small intestine. II. Divergent migration of mesenteric and peripheral immunoblasts to sites of inflammation in the mouse. *Cell Immunol.*, **27**, 36–46.

Rossiter, H., van Reijsen, F., Mudde, G.C., Kalthoff, F., Bruijnzeel-Koomen, C.A.F.M., Picker, L.J. and Kupper, T.S. (1994). Skin Disease-Related T Cells Bind to Endothelial Selectins: Expression of Cutaneous Lymphocyte Antigen (CLA). Predicts E-Selectin But Not P-Selectin Binding. *Eur. J. Immunol.*, **24**, 205–210.

Sackstein, R., Falange, V., Streilein, J.W. and Chin, Y-H. (1988). Lymphocyte adhesion to psoriatic dermal endothelium is mediated by a tissue-specific receptor/ligand interaction. *J. Invest. Dermatol.*, **91**, 423–428.

Santamaria Babi, L.F., Picker, L.J., Perez Soler, M.T., Drzimalla, C., Flohr, P., Blaser, K. and Hauser, C. (1995). Circulating allergen-reactive T cells from patients with atopic dermatitis and allergic contact dermatitis express the skin-selective homing receptor, the cutaneous lymphocyte-associated antigen (CLA), *J. Exp. Med.*, **181**, 1935–1940.

Scott, P. (1993). Selective Differentiation of $CD4^{+}$T Helper Cell Subsets. *Curr. Opin. Immunol.*, **5**, 391–397.

Semnani, R.T., Nutman, T.B., Hochman, P., Shaw, S. and van Seventer, G.A. (1994). Costimulation by purified intercellular adhesion molecule 1 and lymphocyte function-associated antigen 3 induces distinct proliferation, cytokine and cell surface antigen profiles in human "naive" and "memory" CD4+ T cells. *J Exp Med.*, **180**, 2125–2135.

Springer, T.A. (1994). Traffic Signals for lymphocyte recirculation and leukocyte migration: The multistep paradigm. *Cell*, **76**, 301–314.

Tortorella, C., Schulz-Koops, H., Thomas, R., Splawski, J.B., Davis, L.S., Picker, L.J. and Lipsky, P.E. (1995). Expression of CD45RB and CD27 identifies subsets of CD4+ memory T cells with different capacities to induce B cell differentiation, *J. Immunol.*, **155**, 149–162.

Waldorf, H.A., Walsh, L.J., Schechter, N.M. and Murphy, G.F. (1991). Early cellular events in evolving cutaneous delayed hypersensitivity in humans. *Am. J. Pathol.*, **138**, 477–486.

9. LYMHOCYTE TRAFFICKING THROUGH THE CENTRAL NERVOUS SYSTEM

BRITTA ENGELHARDT

Max-Planck-Institut für physiologische und klinische Forschung, W.G. Kerckhoff-Institut, 61231 Bad Nauheim, Germany

INTRODUCTION

For many years the central nervous system (CNS) was considered to be an immunologically privileged site, where normally no immunosurveillance and hence no lymphocyte traffic occurs. This theory was supported by the fact that the CNS is able to exclude components of the immune system via the vascular blood-brain barrier (BBB) and in addition, as the CNS lacks lymphatic vessels, that there is no lymphatic drainage. Furthermore, there is no constitutive expression of major histocompatibilitiy complex (MHC) — molecules in the CNS, which is a prerequisite for T cell mediated immune mechanisms. In fact, the CNS does not seem to participate in "normal" immune reactions, as various experiments show that ectopic tissue transplants survive for extended periods in the CNS, in comparison to the same tissues transplanted into other tissue sites. Thus for many years there were few reasons for investigating lymphocyte traffic in the CNS. This situation has changed recently as it has become evident that during a wide array of inflammatory diseases in the CNS, mononuclear cells readily gain access to the CNS parenchyme, and also that immunological mechanisms are intimately involved in the pathogenesis of these diseases.

It is well accepted that lymphocyte migration through the body is not random and that lymphocyte subpopulations may vary in their recirculation patterns. The specificity of lymphocyte migration is in part mediated by the expression pattern of cell adhesion molecules (CAMs) on lymphocytes as well as on endothelial cells. Thus endothelial cells actively participate in the regulation of lymphocyte entry into various tissues.

This chapter will summarize the current knowledge about the interaction between the immune system and the CNS, as provided by the migration of lymphocytes through the CNS. The molecular mechanisms controlling the entry of lymphocytes across the highly specialized vascular BBB as well as possible pathways for lymphocyte exit from the CNS will be discussed.

This will lead us to the central issue, that is, whether the observed peculiarities of lymphocyte trafficking through the CNS are due to specialized immune mechanisms applied within the CNS or, whether the unique cellular environment of the CNS is responsible for the observed alterations in immune reactions, and thus in lymphocyte trafficking.

THE BLOOD-BRAIN BARRIER

In order to gain access to the CNS, circulating lymphocytes have to migrate across a highly specialized vascular wall — the BBB. The BBB separates the CNS tissue from the blood by preventing the free exchange of molecules between the blood and neuropil. This fact was discovered almost 100 years ago by Paul Ehrlich (Ehrlich, 1904), when he injected dyes into the vascular system which were rapidly taken up by all organs with the exception of the brain. The same dyes, when injected into the cerebrospinal fluid, readily stained neural tissue, but were incapable of staining all the other organs. Obviously, these dyes were prevented from penetrating the wall of cerebral blood vessels. The concept of the BBB which also serves as a brain-blood barrier, was born (Bradbury, 1979; Bradbury, 1984). It took another 70 years after Ehrlich's initial observations to ultrastructurally locate the barrier to brain endothelial cells. Only then it was shown that horseradish peroxidase, a small molecule of 40 kDa, was prevented from passing between endothelial cells by tight junctions, which connect most brain endothelial cells (Reese and Karnovsky, 1967; Brightman and Reese, 1969). How cerebral vessels achieve their tightness is still not completely understood, and is the focus of current research.

Development

The vascular system within the CNS develops via a mechanism called angiogenesis (Noden, 1991; Risau and Wolburg, 1990). Vascular sprouts originating from the perineural vascular plexus, which covers the entire surface of the neural tube, radially invade the proliferating neuroectoderm. These sprouts elongate, give rise to manifold branches and finally anastomose with adjacent sprouts to form a plexus of undifferentiated capillaries in the developing brain (Bär, 1980; Roncali *et al.*, 1986). Whereas the factors, which regulate the pattern formation of the cerebral vasculature remain unknown, vascular endothelial cell growth factor (VEGF) probably plays a key role in brain angiogenesis (reviewed in Plate *et al.*, 1994). VEGF is highly expressed in the ventricular layer of the developing CNS, and cells within this layer probably direct the ingrowth of vascular sprouts, which express the VEGF-receptors (Breier *et al.*, 1992; Breier *et al.*, 1995). VEGF and its receptors are highly expressed throughout the process of brain angiogenesis, and are downregualted in the adult brain when angiogenesis has ceased (Millauer *et al.*, 1993). Brain endothelial cells start to differ from endothelial cells at other sites at this early timepoint during embryogenesis by expressing specific enzymes, such as P-glycoprotein (Qin and Sato, 1995) or the non-receptor tyrosine kinase lyn (Achen *et al.*, 1995), which are not detectable in other endothelial cells in the body. As the VEGF/VEGF-receptor interaction is not specific to the process of brain angiogenesis, the interaction of VEGF with its receptors can not be responsible for the expression of these specific molecules by brain endothelial cells. Therefore alternative signals, probably provided by the developing neuroectoderm, must induce the expression of brain endothelial specific enzymes.

Differentiation and Maintenance

The differentiation of the BBB into a tight barrier is accompanied by further changes in the phenotype of cerebral endothelial cells, leading to the unique barrier characteristics of the BBB. Differentiated brain endothelial cells build a physical barrier for hydrophilic substances. The endothelial cells of the brain lack pinocytotic vesicles and thus transcellular bulk flux and additionally, they form a complex network of tight junctions between them, which prevents paracellular diffusion. Most important in the context of this chapter is that by structural analysis tight junctions between endothelial cells of the fully differentiated BBB have been shown to be more complex than those in other endothelial cells of the body (Nagy *et al.*, 1984; Nico *et al.*, 1992). This complex network of tight junctions is thought to be primarily responsible for the high electrical resistance (about 2000 Ohm $\times$ cm^2) across the cerebral vascular wall (Crone and Olesen, 1982). In addition BBB tight junction particles of brain capillary endothelial cells, in contrast to tight junctions of other endothelial cells, are associated with the P-face (the cytoplasmic leaflet; Wolburg *et al.*, 1994). This P-face association of the tight junction particles in BBB endothelium probably reflects the strong connection to the cytoskeleton of the endothelial cell which is mediated by tight junction associated proteins. To date, none of the tight junction associated proteins such as ZO-1, ZO-2 or cingulin have been proven sufficient to establish the "tightness" of junctions as measured by electrical resistance over cellular monolayers *in vitro* (Phelps, 1972; Stevenson *et al.*, 1988).

The barrier properties of the BBB are not merely achieved by building a physical barrier but also by establishing a metabolic barrier. The polar expression of specific transport systems at either the luminal or abluminal surfaces of the brain endothelium, ensure the transport of sufficient metabolites from the blood into the brain and guarantee quick removal of toxic substances from the brain (for review see Engelhardt and Risau, 1995). In addition, certain enzymes which are present in the cytoplasm of cerebral endothelial cells are able to degrade metabolites, which are otherwise toxic to the CNS. As the expression of specific enzymes correlates with BBB differentiation, they have generally been used as specific BBB markers. Besides enzymes there are many more molecules which are highly associated with the differentiated BBB, however, their functions are generally unknown (reviewed in Engelhardt and Risau, 1995). Another peculiarity of the BBB endothelium is that several endothelial cell specific molecules are downregulated following BBB differentiation; these include the endothelial specific VE-cadherin and the as yet undefined endothelial antigen, which is recognized by the monoclonal antibody (mAb) MECA-32. Both molecules are expressed by all other endothelial cells in the adult mouse (Breier *et al.*, 1995; Hallmann *et al.*, 1995). The significance of downregulation of certain molecules following BBB differentiation is not yet understood. It should also be noted that, to date, no BBB specific cell adhesion molecules (CAMs) have been characterized.

The molecular mechanisms which are responsible for the induction of specific endothelial characteristics acquired during BBB differentiation are not yet defined. Although it is well established that development of BBB characteristics is a function of the neuroectodermal environment, how capillaries recognize these brain-specific environmental clues, which lead to the development and maintenance of their unique BBB phenotype, is still uncertain. Stewart and Wiley (1981) were the first to demonstrate that the specific characteristics of brain endothelial cells are not pre-determined but can be induced in non-brain endothelial cells by the neural microenvironment following transplantation. Many of the BBB characteristics of brain endothelial cells are lost when these cells are isolated and cultured *in vitro*. Thus differentiation of the BBB, that is "barriergenesis", is a reversible event that requires permanent maintenance, which is probably provided via signals originating from the neuroectoderm. This current state of knowledge led us to the proposal that there are 2 phases of endothelial development which lead to BBB differentiation (Engelhardt and Risau, 1995). The first irreversible differentiation step is exemplified by the early induction of certain enzymes and/or markers which are specifically expressed by brain endothelial cells and the second differentiation step is characterized by the expression of additional markers and the development of high electrical resistance (for review see Dermietzel and Krause, 1991; Risau, 1995; Risau and Wolburg, 1990). The second differentiation phase of cerebral endothelium is reversible and is only maintained within the healthy CNS. This could explain why under pathological conditions, such as inflammation or trauma in the CNS, specific characteristics of the BBB are lost and the BBB becomes leaky.

The question remains, however, which cell type or combination of cells within the neuroectoderm contributes to the induction and maintenance of BBB characteristics in the cerebral endothelium. In the adult CNS, astrocytic end-feet almost completely cover the abluminal surface of cerebral vessels (Figure 1). Since this astroglial perivasular collar is a unique feature of the CNS microvasculature, and forms at around the same time as BBB differentiation occurs (Phelps, 1972) astrocytes and their precursors have been implicated in the induction of the BBB (Goldstein, 1988; Janzer and Raff, 1987). There is experimental evidence for a role of type I astrocytes in barrier differentiation (Arthur *et al.*, 1987; Dehouck *et al.*, 1990), however no factors, capable of inducing the BBB phenotype in endothelial cells, have been identified from astrocytes (Holash *et al.*, 1993). Therefore, whether there is a role of astrocytes or other cells intimately associated with the BBB in the maintenance of the BBB remains to be determined. A large number of pericytes are located in the basal lamina of the cerebral microvasculature (Sims, 1986). As pericytes express smooth muscle cell actin they are hypothesized to function as contractile cells within the microvasculature. Additionally, there is a subpopulation of perivascular microglial cells, which are specifically associated with cerebral vessels (Hickey and Kimura, 1988). Both, pericytes and perivascular microglial cells can be induced to express MHC class II molecules and are capable of becoming actively phagocytic. They are therefore regarded as the "second line of defence" at the interface between blood vessels and the neuropil (Farrell and Risau, 1994).

In summary, in order to get access to the CNS, circulating lymphocytes have to penetrate a highly specialized endothelial cell barrier. While the endothelial cells form the barrier proper, the complex interaction of endothelial cells with adjacent cells like pericytes, perivascular microglial cells and astrocytes (Figure 1) seems to be a prerequisite for the maintenance of the barrier function.

IMMUNE PRIVILEGE OF THE CNS

The presence of the BBB as well as additional observations regarding the special immune status of the CNS led Barker and Billingham (Barker and Billingham, 1977) to coin the term "immune privilege" which describes a deficit of immune reactivity or immune surveillance seen in the brain and other organs. The peculiar immune status of the CNS was first noted by Sir Peter Medawar (Medawar, 1948), who discovered immunological tolerance and the process of tissue rejection. In a series of experiments Medawar and colleagues determined that tissue ectopically transplanted into the brain

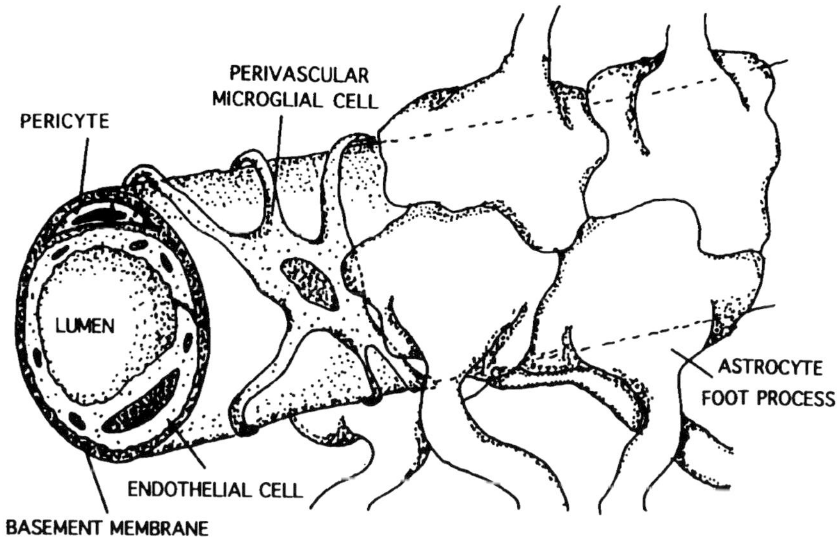

Figure 1 The cellular elements of the differentiated blood-brain barrier.

survived for extended periods, unlike transplants in other tissue sites, which were rapidly rejected (Medawar, 1948). Many subsequent experiments confirmed that immune recognition of foreign transplants is altered in the CNS. This inability to reject transplanted tissue seems to be mainly due to the lack of classical antigen-presenting cells (APCs) within the CNS. Indeed, dendritic cells, which constitutively express both classes of MHC molecules, and serve as professional APCs, are found in almost any tissue other than the CNS (Hart and Fabre, 1981). This poses the question, whether antigen presentation takes place within the CNS and which cells are responsible for presentation of the immune complex. One possibility is that under physiological conditions there is no antigen presentation within the CNS at all. Indeed, very recently the electrical activity of CNS neurons has been correlated to the active inhibition of MHC expression in the CNS (Neumann *et al.*, 1995). Although the CNS lacks a fully developed lymphatic vasculature there is an accumulating body of evidence, which supports the hypothesis that the CNS extracellular fluid drains into the deep cervical lymph nodes (Cserr and Knopf, 1992). Thus it is possible that all CNS antigen-presenting events could take place peripherally, leaving the fragile CNS microenvironment unaltered.

However, at least under pathological conditions, such as inflammatory reactions, neoplasm or brain injury the expression of MHC class I and/or II molecules can be induced on several cells including astrocytes, endothelial cells, microglial cells, pericytes and choroid plexus epithelium and antigen-presentation could take place within the CNS parenchyme (reviewed by Fabry *et al.*, 1994). Indeed, it has been shown at least for astrocytes and microglial cells that they are capable to present antigen *in vitro* (Fontana *et al.*, 1984; Hickey and Kimura, 1988).

Taken together these observations have challenged the idea that the CNS is completely secluded from the immune system.

EXPERIMENTAL AUTOIMMUNE ENCEPHALOMYELITIS (EAE) - A MODEL OF T CELL TRAFFICKING THROUGH THE CNS

Most of our current knowledge about cellular immune reactivity within the CNS comes from studies performed in the animal model known as experimental autoimmune encephalomyelitis (EAE). EAE is a T cell mediated inflammatory demyelinating disease of the CNS, which because of its clinical and histopathological similarities is considered the best animal model for human inflammatory demyelinating diseases of the CNS like multiple sclerosis (MS).

Encephalomyelitis resulting from immunization was first observed in humans vaccinated with rabies virus extracted from infected rabbit CNS tissue (Jochowed, 1925). Two decades later Rivers and colleagues could show that most experimental animals, when repeatedly injected subcutaneously with heterologous brain extracts over a prolonged period of time, developed inflammatory CNS lesions accompanied by demyelination (Rivers *et al.*, 1933). Pathoanatomically, the perivenular inflammatory lesions, so called inflammatory cuffs, and the demyelinating foci resembled those described in MS. Following these initial studies, this experimental demyelinating disease was termed experimental autoimmune encephalomyelitis (EAE). The development of Freund's adjuvants accelerated research in this field (Freund *et al.*, 1947). A single subcutaneous injection of CNS homogenate in complete Freund's adjuvant (CFA) was enough to induce severe disseminated encephalomyelitis in a relatively short period of time in rabbits, guinea pigs and monkeys and later also in certain inbred rodent strains susceptible for the disease. Genes of the major histocompatibility complex (MHC) have been demonstrated to be most important factors in determining susceptibility to EAE. The following research focused on identifying the specific molecular fraction within the CNS homogenate responsible for inducing EAE. Probably because of ease in isolation, myelin basic protein (MBP) was the first encephalitogenic, that is disease inducing, protein examined in detail. Since the isolation of MBP several other proteins have been demonstrated to elicit EAE upon injection in CFA. Other myelin components such as proteinlipid protein (PLP), myelin oligodendrocyte glycoprotein (MOG) and also non-myelin components such as S100β can serve as auto-antigenic agents for the induction of EAE in susceptible animals (Kojima *et al.*, 1994; and reviewed by Martin and McFarland, 1995). Paterson was first to note that EAE is mediated by autoaggressive lymphocytes rather than by autoantibodies reacting against myelin components (Paterson, 1960). Formal proof for the pathogenic role of T cells in EAE was provided when Ben-Nun and colleagues succeded in growing MBP-monospecific T cell lines from peripheral lymph nodes of rats afflicted with EAE and were able to transfer EAE to syngeneic rats with those MBP-specific T cells (Ben-Nun *et al.*, 1981). It is important to note that the transfer of EAE only succeded with T cells that were freshly activated *in vitro* either by presentation of the auto-antigen MBP or by the polyclonal T cell mitogen Concanavalin A. Extensive research has been undertaken to immunologically characterize the encephalitogenic T cell lines. A quite unexpected finding is that all the encephalitogenic T cell lines characterized to date are members of the $CD4^+$ T- helper 1 (T_{h1}) subset. It was equally unexpected to find that MBP-specific encephalitogenic T cells use a very restricted repertoire of T cell receptor (TCR) variable chains (reviewed by Martin and McFarland, 1995). Finally, it is noteworthy to mention, that although from the CD4 subpopulation, encephalitogenic T cells have been shown to efficiently lyse target cells in a MHC class II restricted manner (Engelhardt, *et al.*, 1994, Sun and Wekerle, 1986).

THE BBB IS NOT A STRICT BARRIER TO LYMPHOID CELLS

Immigration of Lymphoid Cells into the Healthy CNS

The fact that intravenously injected T cells are able to induce an inflammatory response in the CNS contradicted the hypothesis of the BBB being an impermeable barrier for immune cells. The first observations that activated, antigen-specific T line cells are capable of penetrating the healthy BBB were obtained from the group of Wekerle (Wekerle *et al.*, 1986) and extended by Hickey and co-workers (Hickey *et al.*, 1991). Both groups used the experimental approach of tracing the migration pattern of intravenously injected ^{14}C-Thymidine-labeled $CD4^{+}$ MBP-specific T line cells in Lewis rats. The T cells used in these studies were freshly activated, either by presentation of their specific antigen or by polyclonal stimulation with the T cell mitogen concanavalin A. The results of both groups indicated that T cell entry into the CNS following intravenous (i.v.) injection occured in 2 phases. The first phase of entry occurred within the first 24 hours after injection, where T cells were detected already 6 hours after injection in every organ examined including the CNS (Meyermann *et al.*, 1987). The second phase of entry occured about 96 hours post-injection (p.I.) which was coincidental to the onset of clinical symptoms of T cell transferred EAE (tEAE). Whereas the first phase of T cell immigration must have occured across the "normal" BBB, the second phase of immigration took place at a time when the BBB was probably pre-stimulated by the previous entry of auto-agressive T cells into the CNS. The finding that autoantigen-specific T cells, which mediate EAE, get access into the CNS was not completely unexpected. It was surprising, however, that T cell lines, which are specific for an antigen that is not present in the CNS and which do not cause EAE also migrated into the CNS. Ovalbumin-specific T cells could be demonstrated to show a monophasic immigration into the CNS which was indistinguishable from the first immigration phase of MBP-specific T cells. Furthermore, MBP-specific T cells, when injected into allogeneic MHC-incompatible rat strains (where they cannot recognize MBP in the correct MHC context within the CNS) could also readily be detected within the CNS (Hickey *et al.*, 1991). In the allogeneic situation, immigration of MBP-specific T cells into the CNS followed a monophasic pattern of immigration into the CNS, suggesting that local antigen-recognition and T cell stimulation is necessary to induce the secondary phase of lymphocyte immigration. Another difference which is dependent on antigen-recognition was noted with regard to the disappearance of T cells from the CNS. Whereas MBP-specific T cells seemed to be cleared from the CNS only gradually in the syngeneic host, ovalbumin-specific T cell clearance took place more rapidly and so did T cell clearance from the allogeneic CNS. It should be noted that in both studies resting T line cells, irrespective of their antigen-specificity, were never demonstrated to pass through the normal BBB. The observation that only activated but not resting auto-aggressive T cells can be used to transfer EAE to syngeneic recipients can then be explained by the failure of resting T cells to penetrate the BBB.

Immigration of Lymphoid Cells into the Inflamed CNS

The immigation of mononuclear cells into the CNS during EAE does not seem to be random but rather a selective process, as the inflammatory infiltrate is characterized by the presence of T-cells and macrophages, whereas B-lymphocytes or neutrophils are rarely detected. During the acute phase of clinical EAE there is evidence for BBB leakyness, as documented by the presence of serum albumin in the perivascular space. Nevertheless, there is no evidence for passive "leakage" of cellular elements across the inflamed BBB. This is suggested by the well defined inflammatory population and further stressed by the absence of erythrocytes in the infiltrate. Furthermore, similar to the pathophysiology of MS, the inflammatory cell subsets in EAE change during the timecourse of the disease. As in early lesions of MS, T cells in EAE are predominantly from the $CD4^+$ subpopulation, whereas $CD8^+$ T cells are more prominent in later stages of MS or EAE, when clinical relapses occur. Within actively demyelinating lesions, the T cells are outnumbered by macrophage-like cells. Thus, although during inflammatory diseases of the CNS there is evidence for BBB leakage, the mechanism for immune cell entry into the CNS do not seem to be random at any time.

MECHANISMS OF T CELL ENTRY INTO THE CNS

Auto-antigen-presentation by BBB Endothelium: The First Step of T Cell Entry into the CNS?

The question of whether BBB endothelium is capable of presenting antigen to T cells has been the subject of extensive research. The evidence that only activated T cells can penetrate the BBB raised the question where T cells are activated *in vivo*. One possibility is that localized presentation of auto-antigens at the luminal surface of the BBB can lead to T cell activation and subsequent transmigration into the CNS. In order to present antigen to $CD4^+$ T cells, endothelial cells must express MHC class II antigens. As we learned before, MHC expression is low or absent in the healthy CNS. Many investigations were performed to determine if there is MHC class II expression in EAE. The results of these investigations were quite controversal and seemed to dependend on the animal and the EAE model used (reviewed by Wekerle *et al.*, 1990). Although there were several independent claims of MHC class II expression by cerebral endothelial cells during CNS inflammation, demonstration of MHC class II expression at the electron microscopic level has only been described in the guinea pig model of EAE (Sobel *et al.*, 1987; Wilcox *et al.*, 1989). After the establishment of pure cultures of BBB vasculature, it became possible to address the inducibility of MHC class II products on brain endothelium *in vitro*. Mediators such as TNF-α and γ-IFN were shown to induce MHC class II expression by those cultured endothelial cells *in vitro* (Male and Pryce, 1988; Risau *et al.*, 1990). Direct studies of the antigen-presenting potential of MHC class II induced cerebral endothelial cells *in vitro*, however, showed an unexpected result. T cells, which were cocultured with syngeneic MHC class II induced endothelial cells in the presence of their specific antigen were not induced to undergo blastogenesis and proliferation (Risau *et al.*, 1990; Sedgwick *et al.*, 1990). This failure to undergo blastogenesis was not due to the fact that the endothelial cells were unable to take up antigen or to process and present them in a suitable MHC class II restricted manner to the T cells. Instead, MHC class II induced rat brain endothelial cells became targets for lysis mediated by $CD4^+$ MBP-specific T cells (Risau *et al.*, 1990; Sedgwick, *et al.*, 1990; Engelhardt *et al.*, 1994). The lysis of brain endothelial cells was strictly dependent on the presence of antigen and the appropriate MHC class II molecule. These findings show that rat BBB endothelial cells qualify as incompletely competent antigen-presenting cells, which lack the "costimulatory factors" necessary to drive T cells into the full activation program.

In summary, antigen-presentation at the BBB is unlikely to be the first step mediating T cell migration across the BBB. This is further substantiated by the experiments described above which showed that activated T cells obtain access to the CNS, irrespective of their anigen-specificity or MHC-compatibility.

Antigen-specificity of T Cells Present in CNS Inflammatory Infiltrates

Further evidence for antigen-independent mechanisms of T cell immigration into the CNS have come from analyzing the T cell populations present within the inflammatory cuffs in EAE and also within MS. It was quite striking to note, that when inflammatory cells are isolated from the CNS, only a low number seems to specifically recognize CNS-antigens presented *in vitro* (reviewed by Martin and McFarland, 1995; Zamwil and Steinman, 1990)

MBP-specific T cell clones have been shown to rearrange a restricted number of TCR $V\alpha$ and $V\beta$ genes in different species. This is quite striking, as the MHC complex and the encephalitogenic determinant recognized on MBP are different between different species (Burns *et al.*, 1989). When inflammatory cell populations present in the CNS of rodents afflicted with EAE and also of MS patients were characterized with regard to their TCR expression, a diversity of rearranged TCR genes was often detected (reviewed by Martin and McFarland, 1994). In a kinetic EAE study, restricted TCR usage was only demonstrated 48 hours after injection of autoaggressive T cells (Karin *et al.*, 1993). During the clinical progression of EAE, when the secondary phase of inflammatory cells was recruited, no such restriction could be detected. After the clinical disease ablated there seemed to be again a more restricted inflammatory T cell population, with regard to TCR usage, present in the CNS.

A different approach, which addressed the antigen-specificity of inflammatory T cells present within the CNS, was an investigation, which localized ^{14}C-Thymidine labeled T-line cells during chronic relapsing EAE in the SJL/J mouse (Cross *et al.*, 1990). Labeled cells were detected in association with cerebral blood vessels 24 hours prior to disease onset. Quantitaion of the numbers of inflammatory cells surrounding blood vessels during acute EAE, revealed that labeled cells represented a minority (1 - 4%) of the inflammatory infiltrate (Cross *et al.*, 1990). These different experimental approaches provide evidence, which taken together, refutes the hypothesis of an antigen-specific mechanism allowing T cell entry into the CNS.

Role of Cell Adhesion Molecules (CAMs) in T Cell Migration into the CNS

Expression of CAMs at the BBB

It has become evident that CAMs are actively involved in mediating the recruitment of specific lymphocyte subsets into different tissues (Picker and Butcher, 1992).

Table 1 Expression of cell adhesion molecules at the BBB.

CAM	species	normal BBB	inflamed BBB	Reference#
ICAM-1	rat	+	++	Archelos *et al.*, 1993
	mouse	+	++	Canella *et al.*, 1990
		+	++	Steffen *et al.*, 1994
		+	++	Deckert-Schlüter *et al.*, 1994
	human	+	++	Sobel *et al.*, 1990
ICAM-2	mouse	+	+	Engelhardt, unpublished
VCAM-1	mouse	nd	++	Baron *et al.*, 1993
		+/–	++	Steffen *et al.*, 1994
		+	++	Deckert-Schlüter *et al.*, 1994
	human	+/–	++	Verbeek *et al.*, 1994
	primates	+/–	++	Sasseville *et al.*, 1992
MAdCAM-1	mouse	-	+	O'Neill *et al.*, 1991
PECAM-1/CD31	mouse	+	+	Engelhardt, unpublished
LFA-3	rat	+	+	Jung *et al.*, 1995
	human	+	+	Rossler et al, 1992
CD44	human	+	+	Rossler *et al.*, 1992
HECA-452-Ag*	human	-	+/-	Raine *et al.*, 1990
E-selectin	human	nd	+/-	Washington *et al.*, 1994

*Ag= antigen; #The table does not provide a complete list of all references. – = negative; +/– = low expression; + = positive; ++ = increased expression

With respect to the structural uniqueness of the BBB endothelium, the major question for lymphocyte traffick into the CNS has been whether the differentiation of BBB endothelium extends to the presence of BBB-specific CAMs. There is no evidence for BBB specific CAMs, although the role of CAMs during the progression of EAE has recently been the subject of extensive research.

In the healthy CNS cerebral endothelial cells constitutively express ICAM-2, PECAM-1 and CD34 like all other endothelial cells in the body (Schulz and Engelhardt, unpublished results; Table 1). The expression of these molecules does not seem to be altered during inflammation in the CNS. In the rat and in man, expression of LFA-3 has been reported on the BBB endothelium (Jung *et al.*, 1995; Rossler *et al.*, 1992). There is low level of constitutive expression of ICAM-1, and possibly of VCAM-1, on some larger venules within the healthy CNS (Engelhardt *et al.*, 1994; Steffen *et al.*, 1994); Figure 2). After induction of EAE in susceptible mice there is a dramatic upregulation of ICAM-1 and VCAM-1 on CNS venules and in the case of ICAM-1 also on capillaries (Figure 2). This upregulation precedes the perivascular inflammatory cell cuffing and the onset of clinical disease (Cannella *et al.*, 1990; and Schulz and Engelhardt, unpublished results). During the first clinical

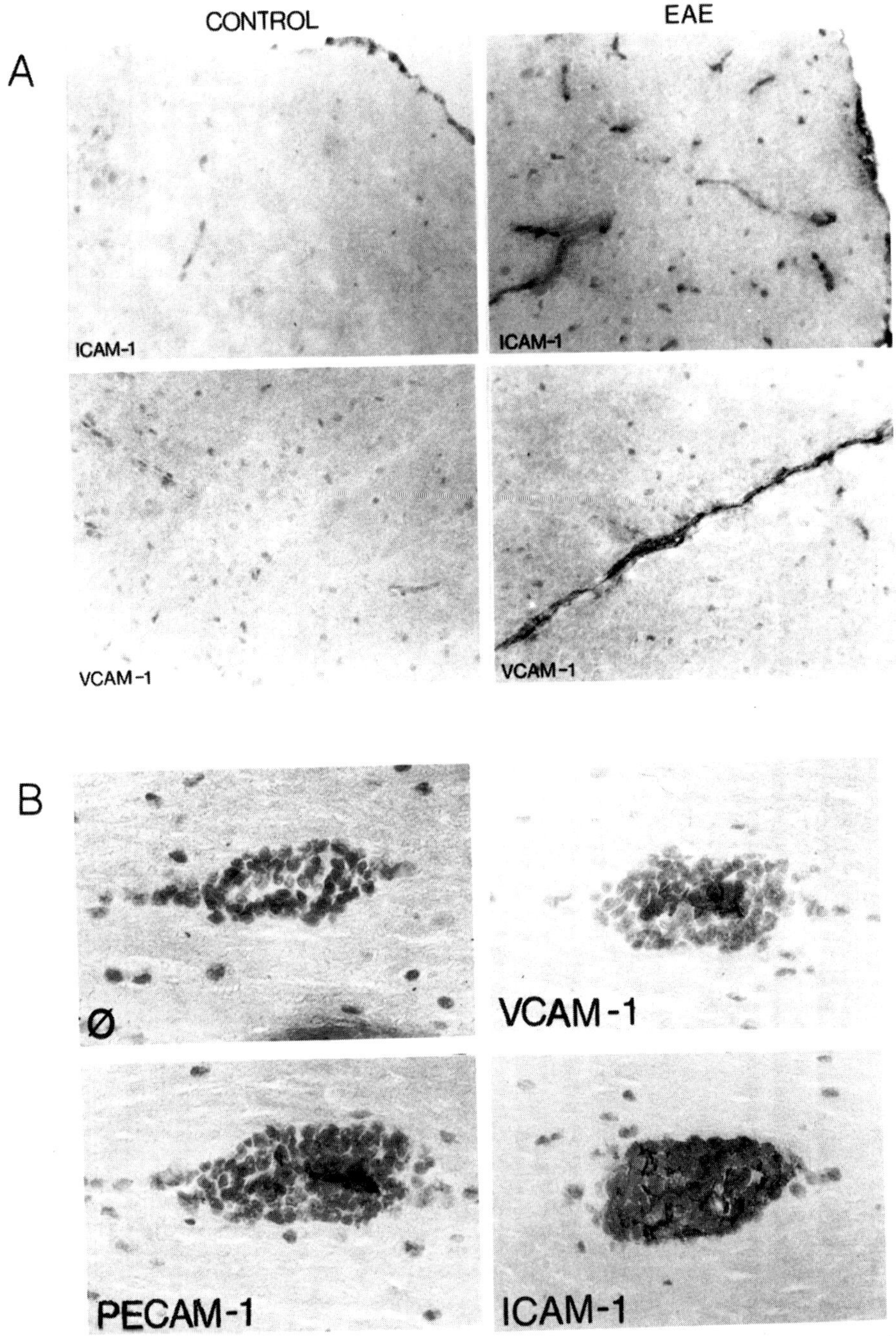

Figure 2 Expression of adhesion molecules in EAE. The low constitutive expression of ICAM-1 and VCAM-1 in the brain of healthy controls is dramatically upregulated in the brain of mice afflicted with EAE. Upregulation of ICAM-1 and VCAM-1 is not restricted to vessels surrounded by an inflammatory cuff (A). B shows an inflammatory cuff in the spinal cord of a mouse afflicted with EAE. The PECAM-1 (CD31) positive vessel also stains positive for ICAM-1 and VCAM-1. ∅ = isotype matched control.

episode, expression of ICAM-1 and VCAM-1 remains upregulated (Steffen *et al.*, 1994). Very importantly, upregulation of ICAM-1 and VCAM-1 on cerebral vessels is not restricted to vessels surrounded by an inflammatory infiltrate (Figure 2), suggesting that expression of these molecules is a prerequisite of inflammatory cell entry but is not sufficient to mediate BBB transmigration. In addition to EAE, ICAM-1 and VCAM-1 have been demonstrated to be upregulated on cerebral endothelial cells in a wide array of inflammatory conditions in the CNS like virus (Sasseville *et al.*, 1992) and parasite (Deckert-Schlüter, *et al.*, 1994) induced encephalitis, as well as in response to tissue injury and bacterial stimulation (Engelhardt, *et al.*, 1994) and also in neoplasia (Lassmann, *et al.*, 1991). Endothelial ICAM-1 expression has also been demonstrated in MS (Sobel *et al.*, 1990). In the case of EAE, it seems that once an autoaggressive T cell encounters its antigen in the CNS parenchyme, inflammatory mediators are secreted that alter the phenotype of the BBB endothelium, thus paving the way for massive cellular infiltrations. This has also been suggested by a study where the modulation of endothelial ICAM-1 expression in the CNS could be shown to correlate with the pathogenesis of chronically relapsing remitting EAE. In this study the upregulated levels of ICAM-1 were shown to be downregulated after the first clinical episode of EAE and again upregulated prior to the clinical relapse and development of new inflammatory infiltrates (Cannella *et al.*, 1990).

Evidence for the functional importance of cerebral endothelial expression of ICAM-1 and VCAM-1 is provided by several studies using a modified version of the Stamper-Woodruff frozen section adhesion assay (Stamper and Woodruff, 1976). In these assays, frozen sections derived from brain tissue of animals afflicted with EAE (Steffen *et al.*, 1994; Yednock *et al.*, 1992) or virus induced encephalitis (Sasseville *et al.*, 1994) were coincubated with immune cells. These studies clearly demonstrate that lymphocytes or macrophages can bind to inflamed cerebral vessels with increased levels of ICAM-1 and VCAM-1 via the interaction of their known ligands LFA-1/Mac-1 and α4-integrins (Figure 3).

In addition to the upregulation of ICAM-1 and VCAM-1 the altered phenotype of the inflamed BBB-endothelium is reflected by the modulation of cellular levels of other molecules with no known adhesive functions (reviewed by Engelhardt and Risau, 1996). One such example is the re-expression of the MECA-32 antigen, which is observed on cerebral vessels surrounded by large inflammatory cuffs (Engelhardt *et al.*, 1994; and Engelhardt, unpublished results). It is tempting to speculate that soluble inflammatory factors induce an inflamed BBB phenotype which is characterized by the up-regulation of ICAM-1 and VCAM-1, whereas the presence of a cellular infiltrate around the vessel leads to an interruption of the BBB maintaining signals provided by the neuropil and thus eventually to BBB leakage as often seen at inflammatory sites. Thus the embryonic marker MECA-32 may reflect the leaky BBB phenotype seen in severe CNS inflammation (Engelhardt *et al.*, 1994).

During chronic inflammation, which occurs in the relapsing/remitting EAE models, further changes of cerebral vessels have been reported that include a change in morphology of cerebral endothelial cells. Cerebral vessels resembling high endothelial venules (HEVs) in lymph nodes have been described. The induction of a not as yet defined HEV-specific molecule, recognized by the monoclonal antibody (mAb) MECA-325, on these vessels seems thus not too surprising (Cannella *et al.*, 1990). With regard to HEV specific CAMs, the so called addressins, only the expression of MAdCAM-1 has been reported on cerebral vessels during chronic EAE

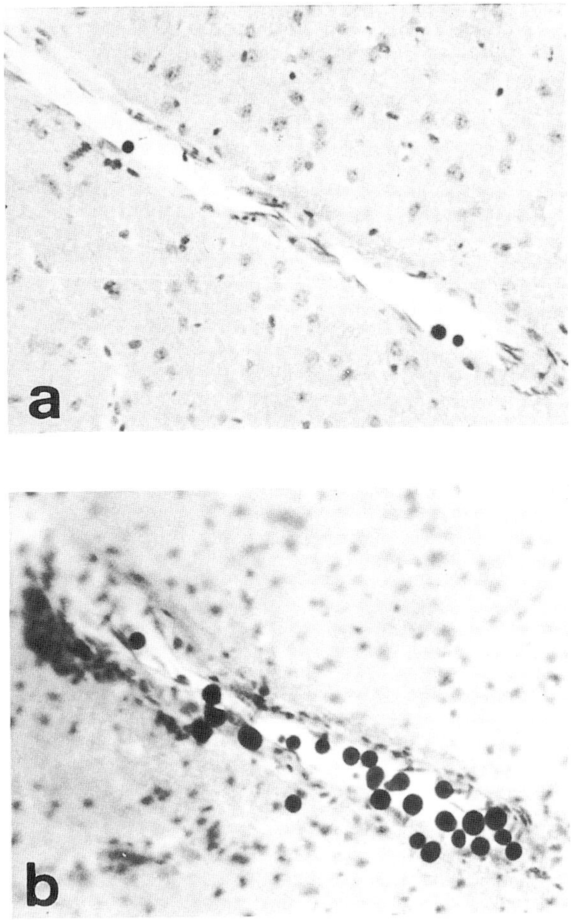

Figure 3 Frozens section adhesion assay. Lymphocytes specifically bind to venules on a frozen section from an EAE brain (b), but not to venules on a brain section form a healthy control (a).

(O'Neill *et al.*, 1991). No vessels reactive with the mAb MECA-79 recognizing the peripheral lymph node addressin (PNAd) has been described to date. In addition to these observations, in MS, there has been a single report describing a reactivity of inflamed vessels with the monoclonal antibody HECA-452 (Raine *et al.*, 1990). HECA-452 has been reported to recognize the E-selectin ligand cutaneous lymphocyte antigen (CLA) on a sub-population of T memory cells but also a carbohydrate moiety on inflamed vessels in tonsils. E-Selectin expression has been reported on some vessels isloated from MS brains (Washington *et al.*, 1994)

Are there brain seeking lymphocytes?

As there is no evidence for antigen-specific mechanisms driving T cell recruitment into the CNS, the question arose whether there is a population of lymphocytes which preferentially migrates through the CNS as this has been shown to be true for other tissues. Several studies characterized the phenotype of infiltrating T lymphocytes using immunohistochemistry or by performing FACS analysis on inflammatory cells isolated from inflamed CNS tissue. There is general agreement that CNS infiltrating T cells are "recently activated/memory" lymphocytes as characterized by their high expression of CD44, LFA-1 and ICAM-1 and a low expression of CD45RB in the mouse (Engelhardt *et al.*, 1995; Hurwitz *et al.*, 1992; Renno *et al.*, 1994). This subset of inflammatory T cells is very similar to the inflammatory T cell infiltrate in other non-lymphoid tissue, where T memory cells are also preferentially found (Mackay, 1991). However, inflammatory T cells in the CNS differ from T cells present in other inflammatory sites by their distinct expression of certain integrins (Engelhardt *et al.* submitted). Whereas inflammatory T cells present in the CNS express $\alpha 4$-integrin at the same level as circulating T cells, they express low to undetectable levels of the $\alpha 4\beta 7$–heterodimer (Engelhardt *et al.*, 1995). This implies that inflammatory T cells present in the CNS only express the $\alpha 4\beta 1$-heterodimer. Besides, T cells present in the inflamed CNS lack expression of $\alpha 6$-integrin and the αE-integrin chain (Engelhardt *et al.*, 1995; Steffen *et al.*, 1994; and Engelhardt, unpublished results). Whether lymphocytes, after having entered the CNS, express L-selectin is still controversal (Dopp *et al.*, 1994; Engelhardt, *et al.*, 1995). As it is known that L-selectin can be downregulated upon lymphocyte contact with endothelium the role of L-selectin in T cell immigration into the CNS during EAE remains speculative.

The phenotype of the inflammatory cells isolated from inflamed brains could be confirmed by their selective ability to bind to purified ICAM-1 and VCAM-1, but not to MAdCAM-1 (Engelhardt *et al.*, 1995). As there is no detectable expression of the $\alpha 4\beta 7$-ligand MAdCAM-1 on cerebral endothelial cells during the early phase of inflammation in the CNS, the lack of $\beta 7$-integrin on the surface of inflammatory T cells within the CNS is not completely unexpected. However, when investigating the adhesion of lymphocytes to inflamed cerebral vessels in the frozen section assay *in vitro*, antibody inhibition studies clearly showed that T cells can bind via $\alpha 4\beta 1$-integrin but also via $\alpha 4\beta 7$ to VCAM-1 on inflamed cerebral vessels. Thus, the reason for the lack of $\beta 7$-integrin expression on inflammatory T cells within the CNS must be due to selective interactions involved later in T cell/BBB interaction during T cell migration across the BBB. Taken together, the current data provide evidence for an involvement of ICAM-1 and VCAM-1 in lymphocyte interactions with cerebral endothelium in EAE.

Migration of Lymphocytes Across the BBB

Given the distinct cell surface integrin expression of inflammatory T cells present in the CNS, and the modulation of the BBB endothelium with regard to CAM expression, one of the principal questions to address has been to define the molecular mechanisms mediating the interaction between inflammatory cells and the cells composing the BBB *in vivo*. This has been widely addressed in models of EAE by investigating the effect of monoclonal antibodies directed against CAMs on the evolution of EAE. A summary of these observations is given in Table 2. Most of these studies fail to provide definitive evidence that the inhibitory effect of the applied antibody is truly mediated by inhibiting T cell traffic into the CNS, rather than by interfering

Table 2 Anti-adhesion molecule therapy in EAE.

anti-	Host	Model+	Prevention*/Treatment*	Reference°
α4-integrin	rat	tEAE	yes/nd#	Yednock *et al.*, 1992
	guinea pig	tEAE	yes/yes	Kent *et al.*, 1995
	guinea pig	aEAE	yes/yes	Kent *et al.*, 1995
LFA-1a	mouse	tEAE	↑severity of disease	Welsh *et al.*, 1993
	mouse	tEAE	dto.	Cannella *et al.*, 1993
Mac-1/CR3	rat	aEAE	yes/nd	Huitinga *et al.*, 1993
L-selectin	mouse	tEAE	no/no	Verooma *et al.*, 1993
	mouse	aEAE	yes/nd	Engelhardt; unpublished
CD44	mouse	tEAE	yes/nd	Verooma *et al.*, 1993
CD2	rat	tEAE	yes/yes	Jung *et al.*, 1995
	rat	aEAE	yes/yes	Jung *et al.*, 1995
ICAM-1	rat	aEAE	yes/nd	Archelos *et al.*, 1993
	rat	tEAE	no/nd	Archelos *et al.*, 1993
	mouse	tEAE	no/little	Canella *et al.*, 1993

*For prevention of EAE the antibody was given before induction of disease; in case of treatment the antibody was given after disease induction. #nd = not done. +aEAE = actively induced EAE, tEAE = passively transferred EAE. °The table does not provide a complete list of all references.

with antigen-recognition, which is a key event for the initiation and development of EAE. Interaction of ICAM-1 and VCAM-1 with their respective ligands plays an important role in antigen-presentation and recognition, and hence T cell activation and proliferation (reviewed in Carlos and Harlan, 1994). These findings might explain why antibodies directed against the ICAM-1 ligand LFA-1 can increase the severity of EAE (Welsh *et al.*, 1993), whereas treatment with anti-ICAM-1 antibodies in other studies either inhibited the development of EAE (Archelos, *et al.*, 1993) or did not have any effect (Willenborg *et al.*, 1993). The effective LFA-1 and ICAM-1 antibodies used in these studies were shown to interfere with the process of antigen-specific stimulation of T cells *in vitro* thus complicating the interpretation of their effects *in vivo*. Some light can be shed on this problem by comparing the results of administration of anti-CAM antibodies in actively induced EAE (aEAE) versus EAE which is initiated by the transfer of activated encephalitogenic T cells (tEAE). It has been shown, although in independent studies that repeated injections of the antibody Mel-14, which is directed against L-selectin, had no influence on the development of tEAE (Veromaa *et al.*, 1993), however, this antibody readily inhibits aEAE (Schulz and Engelhardt, unpublished observations). This result lends credence to the suggestion that the inhibitory effect on disease progression, which is seen following the administration of Mel-14 in aEAE, is due to the interference with antigen recognition by T cells rather than with T cell trafficking into the CNS.

A key role for α4-integrin mediated lymphocyte recruitment across the BBB during EAE is supported by the results of several experiments. Intravenous administration of monoclonal antibodies directed against α4-integrin inhibits EAE disease progression in the rat and most importantly is capable of reversing the disease process in the guinea pig (Kent *et al.*, 1995; Yednock *et al.*, 1992). The absence of an inflammatory cell infiltrate within the CNS of treated animals supports the hypothesis of a dual role for the α4-integrin, namely in the initial phase of T cell immigration across the intact BBB into the CNS as well as in the later recruitment of inflammatory cells across the inflamed BBB. Additionally, antibodies directed against α4-integrin can block lymphocyte adhesion to inflamed cerebral vessels as assessed in the frozen section assay (Steffen *et al.*, 1994; Yednock *et al.*, 1992). Furthermore, the expression of the α4-integrin chain on MBP-specific T cell lines has been shown to correlate with the encephalitogenic potential of these T cells (Baron *et al.*, 1993). It is a well known phenomenon that when culturing murine MBP-specific T cell lines over an extended period of time they eventually loose their encephalitogenic potential without changing their antigen-specificity. Baron and colleagues provided the first experimental evidence that loss of encephalitogenicity correlates with loss of α4-integrin expression on the surface of MBP-specific T cells. As there is no experimental evidence to date which describes the functional importance of selectins in the recruitment of lymphocyte into the CNS, the importance of α4-integrin becomes even more compelling, because α4-integrins are integrins with a "dual adhesive function". $\alpha4\beta7$–integrin can in addition to the classical activation-dependent integrin mediated adhesion mediate the activation independent interaction with its ligand MAdCAM-1 under shear stress, which is visualized as "rolling" (Berlin *et al.*, 1995). The same "dual adhesive function" has been reported for $\alpha4\beta1$-integrin interaction with its ligand VCAM-1 (Alon *et al.*, 1995). Therefore it is possible that $\alpha4\beta1$-integrin could initiate T cell interaction with the vascular BBB as well as mediate the secondary firm adhesion of T cells to the BBB wall (Figure 4).

The Pathway of Transmigration Through the BBB

A key issue is the question by which route lymphocytes traverse the BBB. Using transmission electron microscopy, T lymphocytes were shown to pass the BBB via a transendothelial route, preferentially in parajunctional areas, the precise mechanisms, however, remaining unclear (Calder and Greenwood, 1995; Lossinski *et al.*, 1989; Raine *et al.*, 1990). Lassmann and coworkers were able to show by immunelectron-microscopic studies that VCAM-1 seems to be upregulated over the entire surface of inflamed BBB endothelial cells, in contrast to ICAM-1, which is at a particularly high density in parajunctional areas (Lassmann *et al.*, 1991). ICAM-1 is a key regulator in T lymphocyte transmigration across endothelial monolayers *in vitro* (Oppenheimer-Marks *et al.*, 1994) Thus the VCAM-1/α4-integrin pathway may play a major role in T cell adhesion to BBB endothelium, whereas the parajunctionally located ICAM-1 guides the LFA-1 expressing T lymphocytes and Mac-1 expressing monocytes through the parajunctional area of BBB endothelial cells.

Lymphocyte-BBB Interaction: The Multi-step Paradigm

Lymphocyte extravasation into peripheral tissue is exquisitely regulated *in vivo* by mechanisms of selective lymphocyte-endothelial cell recognition, which can display

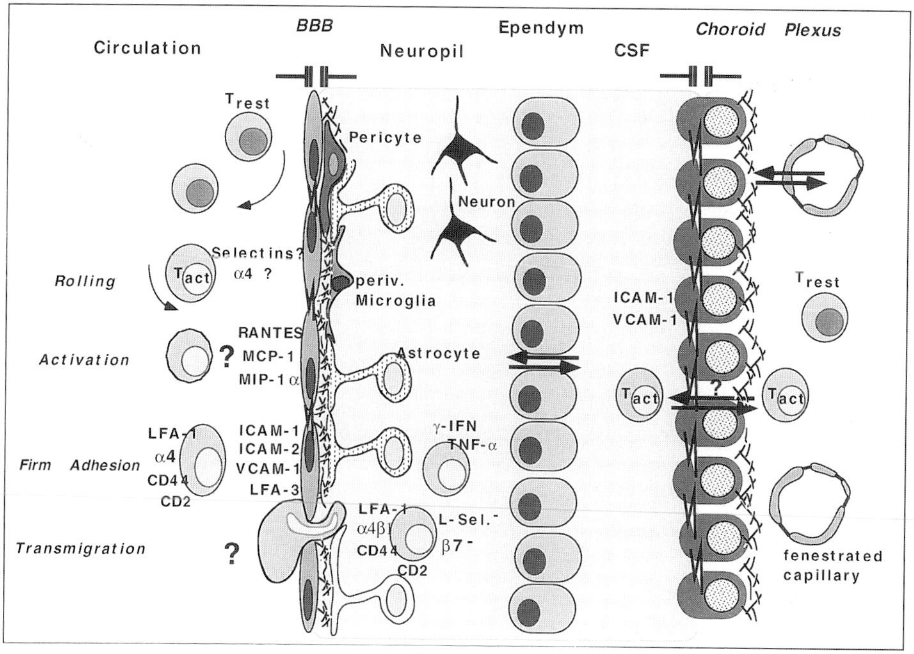

Figure 4 Lymphocyte interaction with the BBB: The multistep-paradigm. The cartoon summarizes the current knowledge of the signals involved in lymphocyte migration across the BBB. The function of CAMs on choroid plexus epithelium is not yet defined. =| |= = physical and metabolic barrier; ⇔ = free exchange of molecules. Tact act = subscript activated T cell, T_{rest} = resting T cell.

extraordinary specificity depending on the tissue site or organ involved, the nature of an inflammatory stimulus and also the timepoint during the inflammatory response. The observation that the specificity of lymphocyte recruitment can not be explained by simple lock-and-key models of lymphocyte-endothelial interaction led E. Butcher (1991) to propose a model which views lymphocyte-endothelial interaction as an active process requiring multiple steps. The multi-step paradigm postulates that traffic signals for lymphocytes function in a sequence of steps, allowing multiple molecular choices at each step, thus providing great combinatorial diversity in signals. The selectins have been shown to regulate the first step of this process, the tethering of lymphocytes from the flowing blood. The subsequent firm adhesion of lymphocytes to the endothelial cell surface and migration across the vessel wall (diapedesis) are regulated independently of the inital binding. These latter stages are thought to be mediated by functional activation of integrins on the lymphocyte by chemoattractants located in the vessel wall (Butcher, 1991; Springer, 1994).

The molecules mediating the tethering of circulating T cells to the BBB endothelium are unknown. It remains to be investigated, whether α4-integrin or selectins, the classical CAMs mediating lymphocyte rolling in extraneural tissue, are invloved in lymphocyte rolling in the CNS. The chemotactic signals leading to the induction of firm adhesion and transmigration of lymphocytes across the BBB remain also largely unknown. Chemokines have recently been identified as candidate molecules, which deliver activation signals to rolling cells, as they induce leukocyte accumulation in tissue sites of inflammation. The protein levels of chemokines such as MIP-1α, RANTES, MCP-1 and IP-10 are specifically upregulated in the CNS during EAE, and can possibly mediate the specific recruitment of T lymphocytes and monocytes across the BBB (Karpus *et al.*, 1995; Ransohoff *et al.*, 1993; and Engelhardt, unpublished results). Synthesis of chemokines is likely to take place in local astrocytes *in vivo* as data from *in vitro* studies demonstrate that astrocytes express mRNA for these chemokines after stimulation with either TNF-α or γ-IFN (Ransohoff *et al.*, 1993). Both, TNF-α and γ-IFN, which are produced by T cells present in conditions of CNS inflammation, play an important role in EAE as antibodies directed against these cytokines were capable of abolishing EAE following i. v. administration (Ruddle *et al.*, 1990). Most of our current knowledge about the traffic signals involved in lymphocyte interaction with the BBB concerns the third step of interaction, namely the stabile binding of lymphocytes to the BBB endothelium. There is good evidence (see above) that endothelial VCAM-1 and ICAM-1 can mediate the firm adhesion of lymphocytes to the cerebral vascular wall by binding their integrin ligands α4-integrin or LFA-1 and Mac-1. Figure 4 summarizes the known traffic signals for lymphocyte recruitment into the CNS based on the multi-step paradigm.

TRAFFICKING THROUGH AND OUT OF THE CNS?

Evidence supporting the hypothesis that there are migratory pathways for lymphocytes through and out of the CNS are sparse and contradictory. By the combination of immunohistochemistry, electron microscopy and *in situ* nick translation it was shown that most auto-antigen specific T cells present in the CNS undergo apoptosis (Schmied *et al.*, 1993). In addition, other studies concluded that no proliferation of T cells could be detected within the inflamed CNS (Ohmori *et al.*, 1992). Conversely, morphological and functional studies have provided evidence that at least some parts of the brain may be linked to the deep cervical lymph nodes (Cserr and Knopf, 1992). Lymphatic connections for solute movement between the CNS interstitial spaces and the local lymph nodes have been described in the rat. Recently, Weller and coworkers provided evidence for a direct route between the cerebrospinal fluid (CSF) and the deep cervical lymph nodes, by introduction of particulate tracers into the CSF of rats (Kida *et al.*, 1993). In these studies, when indian ink was injected into the CSF of rats, it drained from the subarachnoid space directly into nasal lymphatics via the paravascular channels, through the cibriform plate and deposited in the nasal mucosa (Kida *et al.*, 1993). Whether this pathway of CNS lymphatic drainage into the nasal mucosae is found to be important in humans remains to be shown. Due to the size differences between rodents and humans it is possible that the high drainage volume of CSF produced by the choroid plexus within the ventricles, which travels via the subarachnoid space and arachnoid granulations into the venous sinuses, overshadows such a direct connection to the nasal lymphatics as seen in rodents. Within the CNS there is — at least in the rat — evidence that interstitial fluid spreads diffusely throughout the white matter and then drains into the ventricular cerebrospinal fluid (Weller *et al.*, 1992). In the grey matter, however, tracers pass along the perivascular spaces to the surface ot the brain and then into CSF (Kida *et al.*, 1993).

That cells might exit the CNS via the same pathways as used by particulate tracers is suggested by results provided by Oehmichen and co-workers (Oehmichen *et al.*,

1979). Their studies and more recent investigations (Kruse *et al.*, 1994) showed that activated cells when injected into the cannulated CNS can be found to preferentially follow white matter tracts, and on the occasions when these cells were detected in the grey matter, they obviously tracked along the perivascualar spaces (Kruse *et al.*, 1994). ^{14}C-labeled MBP-specific T cells were found to remain exclusively in the perivascular spaces throughout the timecourse of chronic relapsing EAE (Cross *et al.*, 1990). It may be inferred from this data that in the CNS the perivascular space is a site for antigen-presentation to MBP-specific T cells. After active immunization, antigen-specific T cells and lymphocytes expressing a recently activated/memory phenotype "CD44^{high}, LFA-1^{high}/ICAM-1^{+}, CD45RB^{low}" can be found in the CSF. Unless these recently activated/memory T cells are derived from other sources, like the meninges or the choroid plexus, they are likely to have reached their compartment after having crossed the CNS parenchyme via the perivascular spaces in the grey matter or alternatively by following white matter tracts. A certain "affinity" of lymphocytes for CNS white matter has been deduced from the results of *in vitro* adhesion assays on frozen brain sections which demonstrated the L-selectin mediated binding of lymphocytes to an unknown myelin ligand (Huang *et al.*, 1994). As data concerning the L-selectin expression of CNS infiltrating lymphocytes are still contradictory, making interpretation of these findings is difficult (Dopp *et al.*, 1994; Engelhardt *et al.*, 1995).

That afferent connections from the CNS to the immune system indeed exist has been supported by investigations, which demonstrated that injecting antigen into CNS tissue elicits a systemic antigen-specific humoral response with concomitant production of antigen-specific antibody in the deep cervial lymph nodes and the spleen (Harling-Berg *et al.*, 1989). Cellular immune reactivity has also been shown to be affected via lymphatic afferences from the CNS, as MBP injected into the CSF effectively suppresses EAE (Gordon *et al.*, 1992; Harling-Berg *et al.*, 1991). Additionally, when encephalitogenic T cells were injected intrathecally, they failed to elicit clinical EAE (Wekerle *et al.*, 1986). Detailed morphological analysis of intrathecally injected rodents revealed that the encephalitogenic T cells rather than directly invading the CNS, initially migrated out via the leptomeningeal vessels and entered the bloodstream. After entering the bloodstream some encephalitogenic T cells re-entered the brain inducing small, clinically silent, perivascular cuffs.

All aforementioned studies are complicated by the fact that local tissue trauma is induced by the application of the antigen or the cells. Intrathecal injection of antigen or cells may open connections between the CNS parenchyme and the periphery, which are disturbing the physiological properties of the CNS.

Lassmann and coworkers detected macrophages in lumbar lymph nodes isolated from animals suffering from EAE which stained positively for the CNS-specific myelin protein PLP (Lassmann, personal communication). This provides further evidence for an afferent arm of the immune system from the CNS. Taken together many observations support the concept of a strictly controlled one-way lymphocyte traffick through the CNS.

CHOROID PLEXUS AND CIRCUMVENTRICULAR ORGANS

In certain areas of the brain, endothelial cells do not form tight junctions; thus allowing a free exchange of molecules between the blood and the neuropil. Most of these

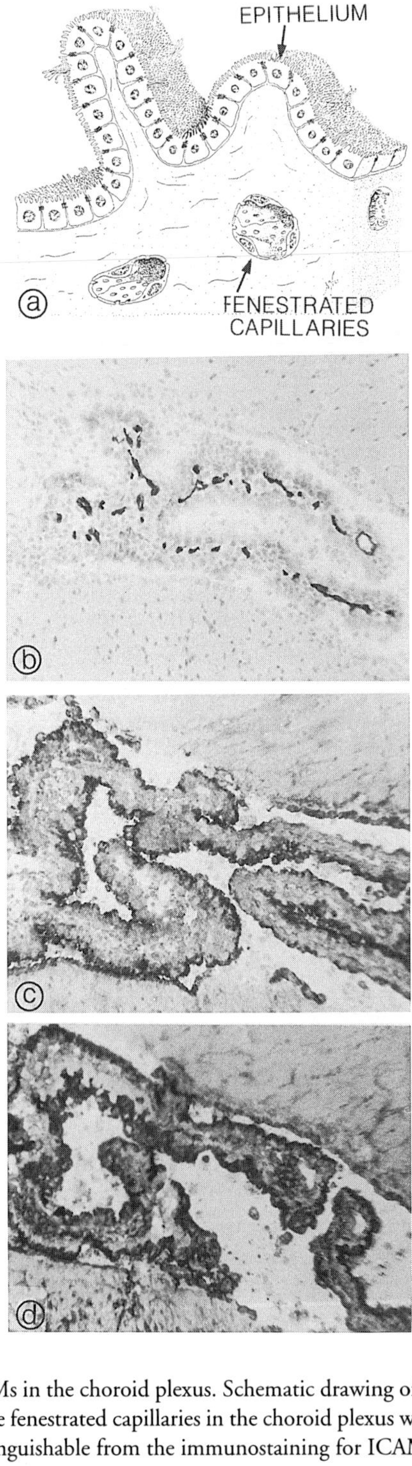

Figure 5 Expression of CAMs in the choroid plexus. Schematic drawing of the choroid plexus is shown in (a). Immunostaining of the fenestrated capillaries in the choroid plexus with the monoclonal antibody MECA-32 (b) is clearly distinguishable from the immunostaining for ICAM-1 (c) and VCAM-1 (d) on the choroid plexus epithelium.

specialized areas, in which the endothelial cells lack tight junctions, are situated close to the cerebral ventricles and are therefore called circumventricular organs (CVOs). They include morphologically and functionally distinct areas known as the pituitary, median eminence, subfornical organ, pineal gand, organum vasculosum lamina terminalis, the area postrema and the choroid plexus.(for review see Leonhardt, 1980). Within the CVOs, the neurons are directly exposed to the blood compartment. This morphological arrangement is of functional importance, as several neuroendocrine hormones are released from specialized nerve terminals within the CVOs directly into the blood stream. In other CVOs, neurons monitor the level of circulating hormones in the bloodstream. CVOs are sealed from the adjacent nervous tissue by barriers, which are built by specialized ependymal cells. It is not known whether lymphocytes get access into the CNS or leave the CNS via these organs during inflammation.

Recently several independent groups have described the intense expression of CAMs at the blood-cerebrospinal fluid (CSF)-barrier located within the choroid plexus (Deckert-Schlüter, *et al.*, 1994; H. Lassmann and R. Sobel, personal communication). The choroid plexus (CP) is a villous structure extending intraluminally from the wall of all cerebral ventricles. The CP forms a direct barrier between the blood compartment and the CSF. Histochemical and ultrastructural investigations have revealed that the choroid plexus epithelial cells have morphological features similar to these of secretory cells and there is a wealth of physiological data indicating that the choroid plexus is the major site of CSF production (reviewed in Betz *et al.*, 1989; Bradbury, 1979). The capillaries in the choroid plexus differ in several respects from those of the brain; the most significant of which is the free movement of molecules across the endothelial cells via the fenestrations and intercellular gaps (Figure 5; reviewed by Betz *et al.*, 1989; Bradbury, 1979). Thus, in the choroid plexus, the blood-CSF barrier is located at the level of the epithelial cells, which are joined together by tight junctions, located close to their apical surface (Cserr and Bundgaard, 1984). Although the expression of CAMs in the choroid plexus has been descibed (Deckert-Schlüter *et al.*, 1994; Sasseville *et al.*, 1992) there is no information regarding their function and regulation. Recently, our laboratory showed that there is constitutive expression of ICAM-1 and VCAM-1, at both the mRNA and the protein level, in the choroid plexus and that during inflammatory conditions — at least in the mouse CNS — there is pronounced increase of ICAM-1 and VCAM-1 at this site (Figure 5; Steffen *et al.*, 1996). In EAE the expression of epithelial MAdCAM-1 is also observed at the mRNA and protein level. CAM expression could not be detected on choroid plexus endothelium, although adjacent cerebral endothelial cells clearly upregulate ICAM-1 and VCAM-1 at the same timepoint during EAE (Steffen *et al.*, 1995). Induction of CAM expression in choroid plexus epithelial cell cultures can be achieved with pro-inflammatory cytokines as shown for endothelial cells. CAMs expressed on the choroid plexus could be shown to bind lymphocytes via their known ligands *in vitro*; it remains unknown whether choroid plexus CAMs serve to regulate lymphocyte traffic across this surface *in vivo*. In the light of observations showing that choroid plexus epithelial cells can express MHC class I and MHC class II molecules and probably present antigen to T cell lines as well as the observations described above it is likely that the choroid plexus plays a critical role in the communication of the immune system with the CNS.

CONCLUSION

There is abundant evidence that the CNS is not secluded from immunosurveillance, although the pathways and molecular mechanisms of lymphocyte trafficking through the CNS are only partially understood. Thus the term "immune privilege" of the CNS has to be reconsidered. It is not the immune system that actively provides a "special service" of immunosurveillance for the CNS, but it is rather the CNS that actively regulates and modulates the "service" provided by the immune system. Thus the CNS ensures its immunosurveillance without risking to disturb its fragile homeostasis.

As the CNS is difficult to access for direct investigations, the definition of the molecular mechanisms mediating lymphocyte traffic through the CNS will also rely on the development of suitable methods, which will allow investigations *in vitro*. To understand the molecular mechanisms that direct lymphocyte migration across the BBB, the development of specialized BBB endothelial cell cultures *in vitro* will be essential. These efforts will have to be paralleled by investigations designed to delineate the neuroectodermal signals that maintain specialized structures like the BBB.

The identification of the endothelial signals mediating the entry of immune cells into the CNS is likely to be of great clinical importance. Blocking of these signals would be desirable in order to treat CNS autoimmune disease, whereas induction or enhancement of lymphocyte traffic into the CNS could have a beneficial effect in infection or neoplasia.

ACKNOWLEDGEMENT

I owe special thanks to Dr. Christopher Mitchell for his devoted discussion of this book chapter. I would also like to thank Gregor Fachinger, Dr. Werner Risau and Dr. Urban Deutsch for critically reading the manuscript.

REFERENCES

Achen, M.G., Clauss, M., Schnürch, H. and Risau, W. (1995). The non- receptor tyrosine kinase Lyn is localized in the developing blood- brain barrier. *Differentation*, **59**, 15–24.

Alon, R., Kassner, P.D., Carr, M.W., Finger, E.B., Hemler, M.E. and Springer, T.A. (1995). The integrin VLA-4 supports tethering and rolling in flow on VCAM-1. *J. Cell Biol.*, **128**, 1243–1253.

Archelos, J.J., Jung, S., Maurer, M., Schmied, M., Lassmann, H., Tamatani, T., Miyasaka, M., Toyka, K.V. and Hartung, H.P. (1993). Inhibition of experimental autoimmune encephalomyelitis by an antibody to the intercellular adhesion molecule ICAM-1. **34**, 145–154.

Arthur, F.E., Shivers, R.R. and Bowman, P.D. (1987). Astrocyte-mediated induction of tight junctions in brain capillary endothelium: an efficient in vitro model. *Brain Res.*, **433**, 155–159.

Bär, T. (1980). The vascular system of the cerebral cortex. *Adv. Anat. Embryol. Cell Biol.*, **59**, 1–62.

Barker, C.F. and Billingham, R.E. (1977). Immunologically priviledged sites. *Adv. Immunol.*, **25**, 1–54.

Baron, J.L., Madri, J.A., Ruddle, N.H., Hashim, G. and Janeway, C.A., Jr. (1993). Surface expression of alpha 4 integrin by CD4 T cells is required for their entry into brain parenchyma. *J. Exp. Med.*, 177, 57– 68.

Ben-Nun, A., Wekerle, H. and Cohen, I. (1981). The rapid isolation of clonable antigen-specific T cell lines capable of mediating autoimmune encephalomyelitis. *Eur. J. Immunol.*, **11**, 195.

Berlin, C., Bargatze, R.F., Campbell, J.J., von Andrian, U.H., Szabo, M.C., Hasslen, S.R., Nelson, R.D., Berg, E.L., Erlandsen, S.L. and Butcher, E.C. (1995). a4 integrins mediate lymphocyte attachment and rolling under physiologic flow. *Cell*, **80**, 413–422.

Betz, L.A., Goldstein, G.W. and Katzman, R. (1989). Blood-Brain- Cerebrospinal Fluid Barriers. In *Basic Neurochemistry: Molecular, cellular and medical aspects, vol. 4th Ed.* G.J. Siegel, ed. Raven Press, New York, pp. 591–606.

Bradbury, M. (1979). *The Concept of a Blood-Brain Barrier.* John Wiley & Sons, Chichester.

Bradbury, M.W.B. (1984). The structure and function of the blood-brain barrier. *Fed. Proc.*, **43**, 186–189.

Breier, G., Albrecht, U., Sterrer, S. and Risau, W. (1992). Expression of vascular endothelial growth factor during embryonic angiogenesis and endothelial cell differentiation. *Development*, **114**, 521–532.

Breier, G., Breviaro, F., Caveda, L., Berthier, R., Schnürch, H., Gotsch, U., Vestweber, D., Risau, W. and Dejana, E. (1995). Molecular cloning and expression of murine VE-cadherin in early developing cardiovascular system. *Blood*, in press.

Breier, G., Clauss, M. and Risau, W. (1995). Coordinate expression of VEGF-receptor 1 (flt-1). and its ligand suggests a paracrine regulation of murine vascular development. *Devel. Dynamics*, **204**, 228–239.

Brightman, M.W. and Reese, T.S. (1969). Junctions between intimately apposed cell membranes in the vertebrate brain. *J.Cell Biol.*, **40**, 648–677.

Burns, R.R., Li, X., Shen, H., *et al.* (1989). Both rat and mouse T cell receptors specific for the encephalitogenic determinant of myelin basic protein use similar Va and Vb chain genes even though the major histocompatiblility complex and encephalitogenic determiants being recognized are different. *J. Exp. Med.*, **169**, 27–40.

Butcher, E.C. (1991). Leukocyte-endothelial cell recognition: three (or more). steps to specificity and diversity. *Cell*, **67**, 1033–1036.

Calder, V. and Greenwood, J. (1995). Role of the vascular endothelium in immunologically mediated neurological disease. In *Immunological aspects of the vascular endothelium.* C.O.S. Savage and J.D. Pearson, eds. Cambridge University Press, Cambridge, pp. 96–123.

Cannella, B., Cross, A.H. and Raine, C.S. (1990). Upregulation and coexpression of adhesion molecules correlate with relapsing autoimmune demyelination in the central nervous system. *J. Exp. Med.*, **172**, 1521–1524.

Carlos, T.M. and Harlan, J.M. (1994). Leukocyte-endothelial adhesion molecules. *Blood*, 7, 2068–2101.

Crone, C. and Olesen, S.P. (1982). Electrical resistance of brain microvascular endothelium. *Brain Res.*, **241**, 49–55.

Cross, A.H., Cannella, B., Brosnan, C.F. and Raine, C.S. (1990). Homing to central nervous system vasculature by antigen-specific lymphocytes. I. Localization of 14C-labeled cells during acute, chronic and relapsing experimental allergic encephalomyelitis. *Lab. Invest.*, **63**, 162–170.

Cserr, H.F. and Bundgaard, M. (1984). Blood-brain interfaces in vertebrates: comparative approach. *Am. J. Physiol.*, **246**, 277–288.

Cserr, H.F. and Knopf, P.M. (1992). Cervical lymphatics, the blood-brain barrier and the immunoreactivity of the brain: a new view. *Immunol. Today*, **13**, 507–512.

Deckert-Schlüter, M., Schlüter, D., Hof, H., Wiestler, O.D. and Lassmann, H. (1994). Differential expression of ICAM-1, VCAM-1 and their ligands LFA-1, Mac-1, CD43, VLA-4 and MHC class II antigens in murine Toxoplasma encephalitis: a light microscopic and ultrastructural immunohistochemical study. *J. Neuropathol. Exp. Neurol.*, **53**, 457–468.

Dehouck, M.P., Meresse, S., Delorme, P., Fruchart, J.C. and Cecchelli, R. (1990). An easier, reproducible and mass-production method to study the blood-brain barrier *in vitro*. *J. Neurochem.*, **54**, 1798–1801.

Dermietzel, R. and Krause, D. (1991). Molecular anatomy of the blood- brain-barrier as defined by immunocytochemistry. *Int. Review Cytol.*, **127**, 57–109.

Dopp, J.M., Breneman, S.M. and Olshcowka, J.A. (1994). Expression of ICAM-1, VCAM-1, L-selectin and leukosialin in the mouse central nervous system during the induction and remission stages of experimental allergic encephalomyelitis. *J. Neuroimmunol.*, **54**, 129–144.

Ehrlich, P. (1904). *Über die Beziehung chemischer Constitution, Vertheilung, und pharmakologischer Wirkung.* Berlin.

Engelhardt, B., Conley, F.C., Kilshaw, P.J. and Burcher, E.C. (1995). Lymphocytes infiltrating the CNS during inflammation display a distinctive phenotype and bind to VCAM-1 but not to MAdCAM-1. *Int. Immunol.*, 7, 481–491.

Engelhardt, B., Conley, F.K. and Butcher, E.C. (1994). Cell adhesion molecules on vessels during inflammation in the mouse central nervous system. *J. Neuroimmunol.*, **51**, 199–208.

Engelhardt, B., Gorlach, H., Risau, W. and Wekerle, H. (1994). Lysis of rat brain microvascular endothelial cells mediated by resting but not activated MBP-specific CD4+ T cell lines. *J. Neuroimmunol.*, **55**, 69–80.

Engelhardt, B. and Risau, W. 1995. The development of the blood-brain barrier. In *New concepts of a blood-brain barrier.* J. Greenwood, D. Begley and M. Segal, eds. Plenum Press, London.

Fabry, Z., Raine, C.S. and Hart, M.N. (1994). Nervous tissue as an immune compartment: teh dialect ot the immune response in the CNS. *Immunol. Today*, **15**, 218–224.

Farrell, C.L. and Risau, W. (1994). Normal and abnormal development of the blood-brain barrier. *Microsc. Res. Tech.*, 27, 495–506 .

Fontana, A., Fierz, W. and Wekerle, H. (1984). Astrocytes present myelin basic protein to encephalitogenic T cell lines. *Nature*, **307**, 273–275.

Freund, J., Stern, E.R. and Pisani, T.M. (1947). *J. Immunol.*, 57, 179.

Goldstein, G.W. (1988). Endothelial cell-astrocyte interactions. A cellular model of the blood-brain barrier. *Ann. N.Y. Acad. Sci.*, **529**, 31–39.

Gordon, L.B., Knopf, P.M. and Cserr, H.F. (1992). Ovalbumin is more immunogenic when introduced into brain or cerebrospinal fluid that into extracerebral sites. *J. Neuroimmunol.*, **40**, 81–88.

Hallmann, R., Mayer, D.N., Berg, E.L., Broermann, R. and Butcher, E.C. (1995). Novel mouse endothelial cell surface marker is suppressed during differentation of the blood-brain barrier. *Development. Dyn.*, **202**, 325–332.

Harling-Berg, C., Knopf, P.M. and Cserr, H.F. (1991). Myelin basic protein infusee into cerebrospinal fluid suppresses experimeental autoimmune encephalomyelitis. *J. Neuroimmunol.*, **35**, 45–51.

Harling-Berg, C., Knopf, P.M., Merriam, J. and Cserr, H. (1989). Role of cervical lymph nodes in the systemic humoral immune response to human serum albumin microinfused into rat cerebrospinal fluid. *J. Neuroimmunol.*, **25**, 185–193.

Hart, D.N.J. and Fabre, J.W. (1981). Demonstration and characterization of Ia-positive dendritic cells in the interstitial connective tissues of rat heart and other tissues, but not brain. *J. Exp. Med.*, **154**, 347–361.

Hickey, W.F., Hsu, B.L. and Kimura, H. (1991). T-lymphocyte entry into the central nervous system. *J. Neurosci. Res.*, **28**, 254–260.

Hickey, W.F. and Kimura, H. (1988). Perivascular microglial cells of the CNS are bone marrow-derived and present antigen *in vivo*. *Science*, **239**, 290–292.

Holash, H.A., Noden, D.M. and Stewart, P.A. (1993). Re-evaluating the role of astrocytes in blood-brain barrier induction. *Dev. Dyn.*, **197**, 14–25, ISSN: 1058–8388.

Huang, K., Kikuta, A. and Rosen, S. (1994). Myelin Localization of a central nervous system ligand for L-Selectin. *J. Neuroimmunol.*, **53**, 133–141.

Hurwitz, A.A., Lyman, W.D., Guida, M.P., Calderon, T.M., Berman, J.W., Jensen, M.A., Arnason, B.G., Toscas, A. and Noronha, A. (1992). Tumor necrosis factor alpha induces adhesion molecule expression on human fetal astrocytes: Preferential increase of IL-2R+ CD4+ T cells and CD45RB- CD4+ T cells in the central nervous system in experimental allergic encephalomyelitis. *J. Exp. Med.*, **176**, 1631–1636.

Janzer, R.C. and Raff, M.C. (1987). Astrocytes induce blood-brain barrier properties in endothelial cells. *Nature*, **325**, 253–257.

Jochowed, B. (1925). Ein Fall einer Erkrankung des Nervensystems im Verlaufe der Schutzimpfung gegen Tollwut. *Z. Gesamt. Neurol. Psychiatr.*, **41**, 583.

Jung, S., Toyka, K. and Hartung, H.-P. (1995). Suppression of experimental autoimmune encephalomyelitis in Lewis rats by antibodies against CD2. *Eur. J. Immunol.*, **25**, 1391–1398.

Karin, N., Szafer, F., Mitchell, D., Gold, D.P. and Steinman, L. (1993). Selective and non-selective stages in homing of T lymphocytes to teh central nervous system during experimental allergic encephalomyelitis. *J. Immunol.*, **150**, 4116–4124.

Karpus, J.W., Lukacs, N., W.„ McRae, B.L., Strieter, R.M., Kunkel, S.L. and Miller, S.D. (1995). An important role for the chemokine macrophage inflammatroy protein-1a in the pathogenesis of teh T cell mediated autoimmune disease, experimental autoimmune encephalomyelitis. *J. Immunol.*, **155**, 5003–5010.

Kent, S.J., Karlik, S.J., Cannon, C., Hines, D.K., Yednock, T.A., Fritz, L.C. and Horner, H.C. (1995). A monoclonal antibody to alpha 4 integrin suppresses and reverses active experimental allergic encephalomyelitis. *Journal of Neuroimmunology*, **58**, 1–10.

Kida, S., Pantazis, A. and O., W.R. (1993). CSF drain sdirectly from the subarachnoid space into nasal lymphatics in the rat. Anatomy, histology and immunological significance. *Neuropathol. App. Neurobiol.*, **19**, 480–488.

Kojima, K., Berger, T., Lassmann, H., Hinze-Selch, D., Zhang, Y., Gehrmann, J., Reske, K., Wekerle, H. and Linington, C. (1994). Experimental autoimmune panencephalitis and uveoretinitis transferred to the Lewis rat by T lymphocytes specific for the S100 beta molecule, a calcium binding protein of astroglia. *J. Exp. Med.*, **180**, 817–829.

Kruse, C.A., Kong, Q., Schiltz, P.M. and Kleinschmidt-DeMasters, B.K. (1994). Migration of activated lymphocytes when adoptively trasnferrd into cannulated rat brain. *J. Neuroimmunol.*, **55**, 11–21.

Lassmann, H., Rossler, K., Zimprich, F. and Vass, K. (1991). Expression of adhesion molecules and histocompatibility antigens at the blood-brain barrier. *Brain Pathol.*, **1**, 115–123.

Leonhardt, H. (1980). *Ependym und zirkumventrikuläre Organe.* Springr Verlag, Berlin.

Lossinski, A.S., Badmajew, V., Robson, J., A.„ Moretz, R.C. and Wisniewski, H., M. (1989). Sites of egress of inflammatory cells and horseradish peroxidase transport across the blood-brain barrier in a murine model of chronic relapsing experimental allergic encephalomyelitis. *Acta Neuropathol.*, **78**, 359–371.

Mackay, C.R. (1991). T-cell memory: the connection between function, phenotype and migration pathways. *Immunol. Today*, **12**, 189–192.

Male, D. and Pryce, G. (1988). Kinetics of MHC gene expression and mRNA synthesis in brain endothelium. *Immunology*, **63**, 37–42.

Martin, R. and MaFarland, H.F. (1995). Immunological aspects of experimental allergic encephalomyelitis and multiple sclerosis. *Crit. Rev. Clin. Lab. Sci.*, **32**, 121–182.

Medawar, P.B. (1948). Immunity to homologous grafted skin. III. The fate of skin homografts transplanted to the brain, to subcutaneous tissue and to anterior chamber of teh eye. *Br. J. Exp. Pathol.*, **29**, 58–69.

Meyermann, R., Lampert, P.W., Korr, H. and Wekerle, H. (1987). The blood- brain barrier — a strict border to lymphoid cells? In *Stroke and Microcirculation.* J. Cervas-Navarro and P. Ferrst, eds. Raven Press, New York, pp. 289–296.

Millauer, B., Wizigmann, V.S., Schnurch, H., Martinez, R., Moller, N.P., Risau, W. and Ullrich, A. (1993). High affinity VEGF binding and developmental expression suggest Flk-1 as a major regulator of vasculogenesis and angiogenesis. *Cell*, 72, 835–846.

Nagy, Z., Peters, H. and Hüttner, I. (1984). Fracture faces of cell junctions in cerebral endothelium during normal and hyperosmotic conditions. *Lab. Invest.*, **50**, 313–322.

Neumann, H., Cavalie, A., Jenne, D.E., Wekerle, H. (1995). Induction of MHC class I genes in neurons. *Science*, **269**, 549–552.

Nico, B., Cantino, D., Bertossi, M., Ribatti, D., Sassoe, M. and Roncali, L. (1992). Tight endothelial junctions in the developing microvasculature — a thin-section and freeze-fracture study in the chick-embryo optic tectum. *J. Submicr. Cytol. P*, **24**, 85–95.

Noden, D.M. (1991). Development of craniofacial blood vessels. In *The development of the vascular system*. R.N. Feinberg, G.K. Sherer and R. Auerbach, eds. Karger, Basel, pp. 1–24.

O'Neill, J.K., Butter, C., Baker, D., Gschmeissner, S.E., Kraal, G., Butcher, E.C. and Turk, J.L. (1991). Expression of vascular addressins and ICAM-1 by endothelial cells in the spinal cord during chronic relapsing experimental allergic encephalomyelitis in the Biozzi AB/H mouse. *Immunology*, 72, 520–525.

Oehmichen, M., Güninger, H., Withölter, H. and Genic, M. (1979). Lymphatic efflux of intracerebrally injected cells. *Acta Neuropathol.*, **45**, 61–65.

Ohmori, K., Hong, Y., Fujiwara, M. and Matsumoto, Y. (1992). In situ demonstration of proliferating cells in the rat central nervous system during experimental autoimmune encphalomyelitis. Evidence suggesting that most infiltrating T cells do not proliferate in the target organ. *Lab. Invest.*, **66**, 54–62.

Oppenheimer-Marks, N., Kavanaugh, A.F. and Lipsky, P.E. (1994). Inhibition of the transendothelial migration of human T-lymphocytes by prostaglandin e(2). *J. Immunol.*, **152**(N12), 5703–5713.

Paterson, P.Y. (1960). Transfer of allergic encephalomyelitis in rats by means of lymph node cells. *J. Exp. Med.*, **111**, 119.

Phelps, C.H. (1972). The development of glio-vascular relationships in the rat spinal cord. *Z. Zellforsch.*, **128**, 555–563.

Picker, L.J. and Butcher, E.C. (1992). Physiological and molecular mechanisms of lymphocyte homing. *Annu. Rev. Immunol.*, **10**, 561–591.

Plate, K.H., Breier, G. and Risau, W. (1994). Molecular mechanisms of developmental and tumor angiogenesis. *Brain Pathol.*, 4, 207–218.

Qin, Y. and Sato, T.N. (1995). Mouse multidrug resistance 1a/3 gene is the earliest known endothelial cell differentiation marker during blood- brain barrier development. *Develop. Dyn.*, **202**, 172–180.

Raine, C.S., Cannella, B., Duijvestijn, A.M. and Cross, A.H. (1990). Homing to central nervous system vasculature by antigen-specific lymphocytes. II. Lymphocyte/endothelial cell adhesion during the initial stages of autoimmune demyelination. *Lab. Invest.*, **63**, 476–489.

Raine, C.S., Lee, S.C., Scheinberg, L.C., Duijvestin, A.M. and Cross, A.H. (1990). Adhesion molecules on endothelial cells in the central nervous system: an emerging area in the neuroimmunology of multiple sclerosis. *Clin. Immunol. Immunopathol.*, **57**, 173–187.

Ransohoff, R., Hamilton, T., Tani, M., Stoler, M., Shick, H., Major, M., Estes, M., Thomas, D. and Tuohy, V.K. (1993). Astrocyte expression of mRNA encoding cytokines IP-10 and JE/MCP-1 in experimental autoimmune encephalomyelitis. *FASEB J.*, 7, 592.

Reese, T.S. and Karnovsky, M.J. (1967). Fine structural localization of a blood-brain barrier to exogenous peroxidase. *J. Cell Biol.*, **34**, 207–217.

Renno, T., Zeine, R., Girard, J.M., Gillani, S., Dodelet, V. and Owens, T. (1994). Selective enrichment of Th1 CD45RBlow CD4+ T cells in autoimmune infiltrates in experimental allergic encephalomyelitis. *Int. Immunol.*, **6**, 347–354.

Risau, W. (1995). Differentation of endothelium. *FASEB J.*, **9**, 926–933.

Risau, W., Engelhardt, B. and Wekerle, H. (1990). Immune function of the blood-brain barrier: Incomplete presentation of protein (auto-) antigens by rat brain microvascular endothelium in vitro. *J. Cell Biol.*, **110**, 1757–1766.

Risau, W. and Wolburg, H. (1990). Development of the blood-brain barrier. *Trends Neurosci.*, **13**, 174–178.

Risau, W. and Wolburg, H. (1990). Development of the blood-brain-barrier. *TINS*, **13**, 174–178.

Rivers, T.M., Sprunt, D.H. and Berry, G.P. (1933). Observations on attempts to produce acute disseminated encephalomyelitis in monkeys. *J. Exp. Med.*, **58**, 39.

Roncali, L., Nico, B., Ribatti, D., Bertossi, M. and Mancini, L. (1986). Microscopical and ultrastructural investigations on the development of the blood-brain barrier in the chick embryo optic tectum. *Acta Neuropathol. (Berl.)*, **70**, 193–201.

Rossler, K., Neuchrist, C., Kitz, K., Scheiner, O., Kraft, D. and Lassmann, H. (1992). Expression of leucocyte adhesion molecules at the human blood-brain barrier (BBB). *J. Neurosci. Res.*, **31**, 365–374.

Ruddle, N.H., Bergman, C.M., McGrath, K.M., Lingenheld, E.G., Grunnet, M.L., Padula, S.J. and Clark, R.B. (1990). An antibody to lymphotoxin and tumor necrosis factor prevents transfer of experimental allergic encephalomyelitis. *J. Exp. Med.*, **172**, 1193–1200.

Sasseville, V.G., Newman, W., Brodie, S.J., Hesterberg, P., Pauley, D. and Ringler, D.J. (1994). Monocyte adhesion to endothelium in simian immunodeficiency virus-induced aids encephalitis is mediated by vascular cell-adhesion molecule-1/alpha-4-beta-1 integrin interactions. *Am. J. Pathol.*, **144**(N1), 27–40.

Sasseville, V.G., Newman, W.A., Lackner, A.A., Smith, M.O., Lausen, N.C., Beall, D. and Ringler, D.J. (1992). Elevated vascular cell adhesion molecule-1 in AIDS encephalitis induced by simian immunodeficiency virus. *Am. J. Pathol.*, **141**, 1021–1030.

Schmied, M., Breitschopf, H., Gold, R., Zischler, H., Rothe, G., Wekerle, H. and Lassmann, H. (1993). Apoptosis of T lymphocytes in experimental autoimmune encephalomyelitis. Evidence for programmed cell death as a mechanism to control inflammation in the brain. *Am J. Pathol.*, **143**, 446–452.

Sedgwick, J.D., Hughes, C.C., Male, D.K., MacPhee, I.A. and ter, M.V. (1990). Antigen-specific damage to brain vascular endothelial cells mediated by encephalitogenic and nonencephalitogenic CD4+ T cell lines *in vitro*. *J. Immunol.*, **145**, 2474–2481.

Sims, D.E. (1986). The Pericyte – A review. *Tissue and Cell*, **18**, 153–174.

Sobel, R.A., Mitchell, M.E. and Fondren, G. (1990). Intercellular Adheison Molecule-1 (ICAM-1). in cellular immune reactions in the human central nervous system. *Am. J. Pathol.*, **136**, 1309–1316.

Sobel, R.A., Natale, J.M. and Schneeberger, E.E. (1987). The immunopathology of acute experimental allergic encephalomyelitis. IV. An ultrastructural immunocytochemical study of class II major histocompatiblity comples (Ia). expression. *J. Neuropathol. Exp. Neurol.*, **46**, 239–249.

Springer, T.A. (1994). Traffic signals for lymphocyte recirculation and leukocyte emigration: the multistep paradigm. *Cell*, **76**, 301–314.

Stamper, H.B.J. and Woodruff, J.J. (1976). Lymphocyte homing into lymph nodes: in vitro demonstration of the selective affinity of recirculating lymphocytes for high endothelial venules. *J. Exp. Med.*, **144**, 828.

Steffen, B.J., Breier, G., Butcher, E.C., Schulz, M. and Engelhardt, B. (1996). ICAM-1, VCAM-1 and MAdCAM-1 are expressed on choroid plexus epithelium but not endothelium and mediate binding of lymphocytes *in vitro*. *Am. J. Pathol.*, **148**, 1819–1838.

Steffen, B.J., Butcher, E.C. and Engelhardt, B. (1994). Evidence for involvement of ICAM-1 and VCAM-1 in lymphocyte interaction with endothelium in experimental autoimmune encephalomyelitis in the central nervous system in the SJL/J mouse. *Am. J. Pathol.*, **145**(N1), 189–201.

Stevenson, B.R., Anderson, J.M., Godenough, D.A. and Mooseker, M.S. (1988). Tight junction structure and ZO-1 content are identical in two strains of Madin-Darby canine kidney cells which differ in transepithelial resistance. *J. Cell Biol.*, **107**, 2401–2408.

Stewart, P.A. and Wiley, M.J. (1981). Developing nervous tissue induces formation of blood-brain barrier characteristics in invading endothelial cells: A study using quail-chick transplantation chimeras. *Dev. Biol.*, **84**, 183–192.

Sun, D. and Wekerle, H. (1986). Ia-restricted encephalitogenic T lymphocytes mediating EAE lyse autoantigen-presentign astrocytes. *Nature*, **320**, 70.

Verbeek, M. M., Westphal, J. R., Ruiter, D. J., de Waal, R. M. W. (1994). T lymphocyte adhesion to human brain pericytes is mediated via VLA- 4/VCAM-1 interactions. *J. Immunol.*, **154**, 5876–5884.

Veromaa, T., Weissman, I.L., Steinman, L. and Bocke, S. (1993). A critical role for integrin $\alpha 4$ and CD44 in the induction of experimental autoimmune encephalomyelitis (EAE). in mice. *J. Biol. Chem.*, **17A**, 356.

Washington, R., Burton, J., Todd, R.F.R., Newman, W., Dragovic, L. and Dore-Duffy, P. (1994). Expression of immunologically relevant endothelial cell activation antigens on isolated central nervous system microvessels from patients with multiple sclerosis. *Ann. Neurol.*, **35**, 89–97.

Wekerle, H., Engelhardt, B., Risau, W. and Meyermann, R. 1990. Passage of lymphocytes across the blood-brain barrier in health and disease. In *Pathophysiology of the blood-brain barrier.* B.B. Johansson, C. Owman and H. Widner, eds, Elsevier, pp. 439–445.

Wekerle, H., Linington, C., Lassmann, H. and Meyermann, R. (1986). Cellular immune reactivity within the CNS. *TINS*, **9**, 271–277.

Weller, R.O., Shinya, K. and Zhang, E.-T. (1992). Pathways of fluid drainage from the brain-morphological aspects and immunological significance in rat and man. *Brain Pathol.*, **2**, 277–284.

Welsh, C.T., Rose, J.W., Hill, K.E. and Townsend, J.J. (1993). Augmentation of adoptively transferred experimental allergic encephalomyelitis by administration of a monoclonal antibody specific for LFA-1a. *J. Neuroimmunol.*, **43**, 161–168.

Wilcox, C.E., Healey, D.G., Baker, D., Willoughby, D.A. and Turk, J.L. (1989). Presentation of myelin basic-protein by normal guinea-pig brain endothelial-cells and its relevance to experimental allergic encephalomyelitis. *Immunol.*, **67**, 435–440.

Willenborg, D. O., Simmons, R. D., Tamatani, T., Miyasaka, M. (1993). ICAM-1-dependent pathway is not critically involved in the inflammatory process of autoimmune encephgalomyelitis or in cytokine-induced inflammation of the central nervous system. *J. Neuroimmunol.*, **45**, 147–154.

Wolburg, H., Neuhaus, J., Kniesel, U., Krauss, B., Schmid, E.-M., Öcalan, M., Farrell, C. and Risau, W. (1994). Modulation of tight junction structure in blood-brain barrier endothelial cells. *J. Cell Sci.*, **107**, 1347–1357.

Yednock, T.A., Cannon, C., Fritz, L.C., Sanchez Madrid, F., Steinman, L. and Karin, N. (1992). Prevention of experimental autoimmune encephalomyelitis by antibodies against alpha 4 beta 1 integrin. *Nature*, **356**, 63–66.

Zamwil, S.S. and Steinman, L. (1990). The T lymphocyte in experimental allergic encephalomyelitis. *Ann. Rev. Immunol.*, **8**, 579– 621.

10. MOLECULES INVOLVED IN TRAFFICKING OF NK CELLS AND DENDRITIC CELLS: IMPLICATIONS FOR TUMOUR IMMUNOTHERAPY

PAOLA ALLAVENA[1], SILVANO SOZZANI[1] and ALBERTO MANTOVANI[1,2,*]

[1] *Department Immunology and Cell Biology, "Mario Negri" Institute, 20157 Milan, Italy*
[2] *Division General Pathololog*y*, University of Brescia, 25123 Brescia, Italy*

INTRODUCTION

Leukocyte infiltration is a common feature of experimental and human tumors. Cells of the monocyte-macrophage and lymphoid lineage are most frequently represented. Being strategically located at the very interface between tumor and host, they represent a target for therapeutic interventions (Allavena and Mantovani, 1994). Natural killer cells (NK) and dendritic cells (DC) usually do not represent a major component of the lymphoreticular infiltrate of tumors. Yet these cells have the potential to elicit and orient anti-tumor immunity. DC are the most effective antigen presenting cells (APC) and are particularly equipped to activate naive T cells (Steinman, 1991). NK cells are a prime target of IL-12 and, by producing IFNγ, orient immunity to macrophage-dependent responses (Trinchieri, 1993). Thus the recruitment and functional orientation of these cells may bear relevance to tumor immunotherapy.

Here we will review current understanding of the molecules involved in the trafficking of NK cells and DC and discuss how this information may relate to targeting anti-tumor immune responses.

NK CELLS

5–15% of lymphoid cells are large granular lymphocytes, a distinct population also called natural killer (NK) cells for their spontaneous cytotoxic activity against various target cells. NK cells represent a third lymphoid population which is distinct from T and B lymphocytes as they lack a T cell receptor and surface immunoglobulins and express specific markers such as CD56 and CD16. NK cells spontaneously lyse a variety of parasites, bacteria, fungi, virus-infected cells and some neoplastic cells, and thus are considered to play a role in the early protection against microbial infections and – possibly – tumors, prior to the development of specific immunity (Herberman and Ortaldo, 1981; Trinchieri, 1989; Robertson and Ritz, 1990). Another major

*Correspondence: Alberto Mantovani, Department Immunology and Cell Biology, "Mario Negri" Institute, 20157 Milan, Italy.

function of NK cells is to secrete cytokines, including IFNγ, TNF and GM-CSF, which activate other immune effectors and shift the immune response towards a Th1 type (Trinchieri, 1993).

Most NK cells reside in blood and spleen but it is well established that they can be found, under physiological conditions, at extra-hematic sites, such as lung and liver parenchyma, uterine and intestinal mucosa (Trinchieri, 1989). Moreover, in certain pathological conditions NK cells accumulate at sites of viral and bacterial infections (Natuk and Welsh, 1987; Holmberg, Springer and Ault, 1981) and are found in the allograft infiltrate during the early phase of rejection (McIntyre and Welsh, 1986; Inverardi *et al.* 1992).

In this chapter we will review present knowledge of the mechanisms of NK cells extravasation and entry into tissues: in particular we will focus on their interaction with the vascular endothelium, and their migratory response to chemotactic factors.

Expression of Adhesive Molecules Relevant for NK Cell/Endothelial Cell Interaction

The adhesive equipment of NK cells has been studied by several groups; a summary of the adhesion molecules is given here in Table 1. About 50% of NK cells express L-selectin (CD62L) (Maenpaa *et al.*, 1993) as well as the carbohydrates Sialyl-Lewis X, a ligand for E- and P-selectin (Ohmori *et al.*, 1989; Moore and Thompson, 1992; Pinola *et al.*, 1994a). Moreover an NK cell surface molecule, CD57 (HNK1), binds to L- and P-selectin (Needham and Schnaar, 1993). As for $\beta2$ integrins, all NK cells express LFA-1 (CD11a/CD18) and about 70–80% express Mac-1 (CD11b/CD18), while only a sub-population (20–30%) have detectable p150,95 (CD11c/CD18) (Maenpaa *et al.*, 1993; Timonen, Gahmberg and Patarroyo, 1990; Robertson *et al.*, 1990). Among $\beta1$ integrins, VLA-4 and VLA-5 are expressed by freshly isolated NK cells (Maenpaa *et al.*, 1993; Gismondi *et al.*, 1991; Morrone *et al.*, 1991); in addition, cultured NK cells propagated with IL-2 also express VLA-1, VLA-2 and VLA-6 (Morrone *et al.*, 1991; Maenpaa *et al.*, 1993). A sub-population of NK cells also express $\alpha4\beta7$, a ligand for MAdCAM-1, VCAM-1 and fibronectin (Pinola *et al.*, 1994b).

NK cells are positive for CD31, a member of the Ig superfamily also expressed by endothelial cells (EC) and involved in transmigration as well as in cell activation (Berman, Xie and Muller, 1996). Another transmembrane protein is CD44, which binds to hyaluronic acid and to extracellular matrix proteins. In T cells, triggering of CD44 transduces activation signals through $\beta2$ integrins (Koopman *et al.*, 1990). NK cells express CD44, and although its role in NK cell adhesion and transmigration has not been studied, it is known that cross-linking of CD44 increases NK cell cytotoxicity (Galandrini *et al.*, 1994).

Adhesion to the Vascular Endothelium

NK cells leave the blood stream through the flat endothelium of post-capillary venules. Unlike other lymphoid cells they do not recirculate, and are in fact virtually absent in the lymph and lymph-nodes. NK cells spontaneously adhere to monolayers of vascular endothelium *in vitro* (Allavena *et al.*, 1991; Aronson *et al.*, 1988). In a large series of experiments we estimated that, in the absence of stimuli, 10–20% of NK cells bind to EC, a considerable proportion, considering that T cells and neutrophils bind

Table 1 NK cell adhesion molecules interacting with vascular endothellum.

Molecule	CD	Ligand	% positive	Role in EC interaction
Selectins				
L-selectin	CD62L	sLe^x-related	50	Rolling
sLe^x		E-selectin P-selectin	50-100	Rolling
HNK1	CD57	E- P-selectin	50–100	Rolling
β_2 Integrins				
LFA-1	CD11a	ICAM-1,2	100	Firm adhesion/ Transmigration
Mac-1	CD11b	ICAM-1, C3	70–80	not involved
p 150,95	CD11c	C3	20–30	not involved
β_1 Integrins				
VLA-4	CD49d	VCAM-I/FN	>90	Firm adhesion/ Transmigration
β_1 Superfamily				
PECAM-1	CD31	PECAM-1	>90	Transmigration

very poorly (less than 5%). In contrast, monocytes show a much higher frequency of up to 40%. The adhesion to IL-1-stimulated EC is usually increased by two to three fold, as might be expected from the appearance on the endothelium of newly expressed adhesive ligands.

The adhesive ability of NK cells is also stimulated by certain cytokines. Even a short incubation with IL-2 and IL-12 significantly increases NK cell binding to resting and "inflammed" EC (Allavena *et al.*, 1994b). However, IL-4, a cytokine showing inhibitory functions on NK cell activities, decreases their adhesion to EC (Paganin *et al.*, 1993).

One of the mechanisms through which cytokines might mediate increased NK cell adhesion to endothelium is an effect on adhesion molecules. It has been reported that IL-2 and IL-12 augment the expression of adhesion molecules on NK cells cultured *in vitro* for several days (Robertson *et al.*, 1992; Rabinovich, Herberman and Whiteside, 1993). An increase in EC binding, in contrast, is already observable after a few hours of stimulation. The expression of LFA-1 after a short incubation with cytokines was not changed when conventional antibodies were used. However, we have found an increased expression of an activated epitope of LFA-1 detected by the mAb NKI L16 after stimulation with IL-2 and IL-12 (Allavena *et al.*, 1994b).

Thus IL-2 and IL-12 rapidly increase NK cell binding to EC by functionally activating already expressed adhesion molecules, in particular LFA-1, which will bind to its ligand more efficiently.

Adhesive Molecules Involved in NK Cell Adhesion to EC

The role of selectins in the adhesion of NK cells to EC monolayers has been extensively studied by Pinola *et al.* An increase of NK cell binding to EC treated for 4h with TNF is inhibited by anti-E selectin mAbs (Pinola *et al.*, 1994a).

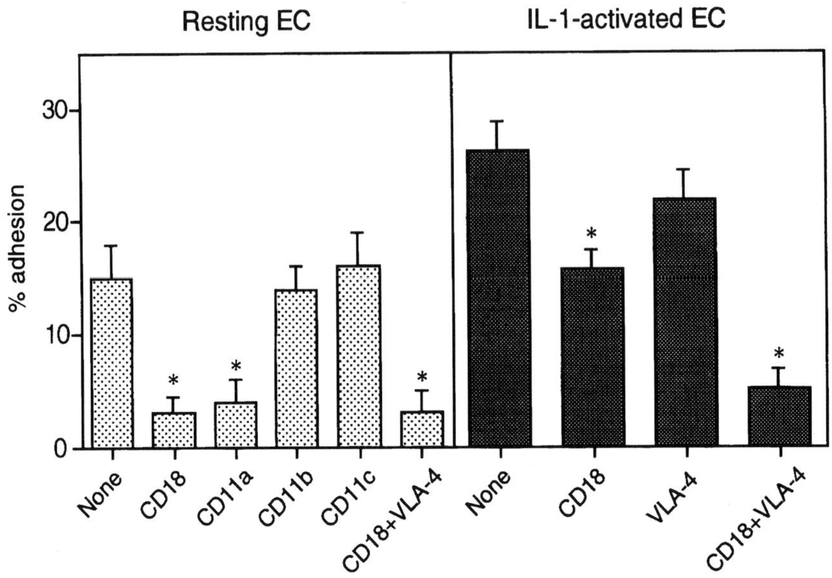

Figure 1 Adhesion molecules involved in NK cell binding to resting and to IL-1 activated EC. ^{51}Cr-labelled NK cells were pre-incubated with saturating concentrations of mAbs. NK cells and EC were coincubated for 30 min. Data are expressed as percentage of adherent cells referred to input cells seeded. Values are the mean ± SD of 4 replicates. * Statistically different from control cells incubated with an irrelevant Mab. p < O.O1.

We have investigated the role of $\beta 1$ and $\beta 2$ integrins, a representative experiment is shown in Figure 1. When resting EC are used, mAbs anti-CD18 and anti-CD11a inhibit most of the binding (up to 80%), while CD11b and CD11c have no effect. When IL-1-activated EC (24 h treatment) are used, anti-CD18 alone is less inhibitory <40%, but when anti-CD18 is applied together with anti-VLA-4, which alone has a small effect, they combined to give a much greater inhibition (up to 70–80%). Thus, the most important adhesive molecule involved in NK cell/resting EC interaction is LFA-1(CD18/CD11a), the only $\beta 2$ integrin utilized by NK cells. CD11b, highly expressed in these cells and involved in the binding of neutrophils and monocytes to EC (Springer, 1994), is not apparently used by NK cells for EC adhesion, though we had found that CD11b is highly enriched at intercellular contacts among IL-2-activated NK cells (Allavena *et al.*, 1991). It is interesting to note that CD11b, rather than CD11a, is the most important molecule involved in the migration of resting NK cells through polycarbonate filters for chemotaxis assays (Somersalo *et al.*, 1992).

When VCAM-1 is expressed on EC after cytokine activation, VLA-4 is also involved in binding. It is of interest that mAb anti-VLA-4 alone causes only a minor inhibition of NK cell binding. The role of VLA-4 is evident only when anti-VLA-4 are used in concert with anti-CD18, and results in a greater inhibition compared to that resulting from anti-CD18 alone.

This suggests that LFA-1 is more important than VLA-4 for EC binding, and that a hierarchy exists among adhesion molecules. It is interesting to note that LFA-1 and

VLA-4 have a distinct distribution on the cell surface. We have reported that NK cells adhere to a substratum (including EC) by organizing cytoskeletal protrusions from the ventral membrane, called podosomes, which are also present in other cell types (Marchisio *et al.*, 1984). These podosomes are formed by a core of F-actin and a surrounding ring of vinculin. These protrusions are the very first portion of the cell membrane which comes into contact with the substratum. By immunofluorescence we have demonstrated that LFA-1 is localized mainly in podosomes, while VLA-4 is diffused on the membrane (Allavena *et al.*, 1991). This finding may provide an explanation of the primary role of LFA-1 in NK cell binding to EC.

Adhesive Molecules Involved in NK Cell Transmigration

In a transendothelial migration assay in which a monolayer of EC is allowed to grow on a porous polycarbonate filter, a substantial proportion of NK cells has been shown to migrate across the endothelial barrier (Bianchi *et al.*, 1993). Within the 60 min assay about 5% of NK cells transmigrated across resting endothelium and 8–10% across IL-1–activated endothelium. In comparison, under the same experimental conditions, transmigration of T lymphocytes was less than 2% and 4%, respectively. The percentage of transmigrating cells is usually one third of the adherent population; this ratio is unchanged by cytokine treatment of EC or of NK cells. As successful migration implies first binding to EC, mAbs anti-CD18 and anti-VLA-4 were also inhibitory on transmigration. In order to asses a more direct role of CD18 and VLA-4 in migration, we have performed experiments in which mAbs were added onto already adherent NK cells. Under these conditions, the combination of anti-CD18+ anti-VLA-4 completely blocked transmigration (>95% inhibition) (Bianchi *et al.*, 1993).

It is of interest that in a large series of experiments we found that the combination anti-VLA4+CD18 inhibited transmigration significantly more than anti-VCAM-1+CD18; indeed, in the representative experiment reported in Figure 2, inhibition was 68% and 58% respectively ($p< 0.05$). The two mAb combinations, however, were similarly inhibitory to adhesion. Therefore we investigated whether during transmigration VLA-4, in addition to VCAM-1, could interact with its other ligand, fibronectin. In two experiments, a CS1 peptide, recognized by VLA-4, and an irrelevant control peptide were used as inhibitors. As shown in Figure 6, when the CS1 peptide was combined with anti-VCAM-1+CD18 mAb, a small but significant increase was observed in the blocking of migration, with 69 % inhibition compared to 58% for anti-VCAM-1+CD18 and 55% for anti-VCAM-1+CD18+, a scrambled peptide. In contrast, the CS1 peptide, alone or combined with anti-VLA4+CD18, had no additive effect on binding (not shown) and migration, compared to VLA-4+CD18 (Bianchi *et al.*, 1993).

In conclusion, not only LFA-1 but also VLA-4 are involved in the process of transendothelial migration. In addition, during sub-endothelial migration, NK cells use VLA-4 to interact with fibronectin, a protein of the extra-cellular matrix (ECM), as demonstrated by the inhibition of transmigration (but not adhesion) caused by fibronectin peptides.

Anti-CD31 mAbs inhibit by 50% transmigration of NK cells through resting endothelium. It is of interest that ligation of CD31 on NK cells causes a 4 to 5 fold increase of NK cell binding by a mechanism which is $\beta 2$ integrin dependent, as it is inhibited by anti-CD18 (Berman, Xie and Muller, 1996).

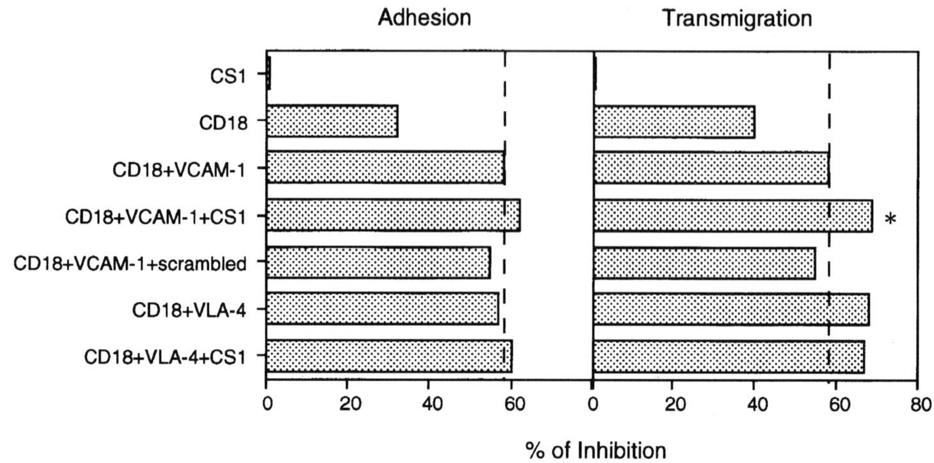

Figure 2 Adhesion molecules involved in the adhesion and transmigration of NK cells through IL-1 activated EC. ^{51}Cr-labelled NK cells were preincubated with saturating concentrations of mAbs. The CS1 peptide of 25 aa with the sequence of the IIICS region of fibronectin or a scrambled peptide were used at the concentration of 1 mg/ml. NK cells and EC were coincubated for 60 min. Data are expressed as percentage of inhibition referred to cells incubated with an irrelevant Mab. Values are the mean ± SD of 4 replicates. * Statistically different from the appropriate control group (CD18+VCAM-1) $p<0.01$.

Regulation of Motility: The Role of Chemotactic Factors

In addition to the above mentioned molecules, migration into tissues is regulated by chemotactic signals, to which NK cells may respond (Table 2). In the mid '80s we and others described NK cell migration through nitrocellulose filters in response to activated serum , fMLP and casein (Bottazzi *et al.*, 1985; Polentarutti *et al.*, 1986; Pohajdak *et al.*, 1986). Several studies agreed that the migratory capacity of NK cells is higher than that of T lymphocytes (Bottazzi *et al.*, 1985; Somersalo and Saksela, 1991). The fact that NK cells are more rapidly mobilizable is in line with the concept that they act as a first line of natural defense. Several *in vivo* experiments demonstrated that accumulation of NK cells is detected at sites of virus inoculation. In these studies too NK cells accumulated within the first two days, before the arrival of specific T cells (Natuk and Welsh, 1987).

Whether NK cells produce chemotactic factors which later attract T lymphocytes is not fully established. Of interest is a report showing that supernatants from NK cells attract T lymphocytes and contain IL-8 (Somersalo, Carpen and Saksela, 1994). We have looked at the production of Monocyte Chemotactic Protein 1 (MCP-1) by NK cells. While freshly isolated NK cells do not produce MCP-1, cultured NK , propagated with IL-2, produced substantial amounts of MCP-1 (Allavena, unpublished).

Migration of NK cells through chemotaxis filters is increased after cytokine activation: we and others have demonstrated that IL-2 and IFNγ increase chemotaxis and chemokinesis (Bottazzi *et al.*, 1985; Polentarutti *et al.*, 1986; Pohajdak *et al.*, 1986; Somersalo and Saksela, 1991; Sebok *et al.*, 1993). In addition, TNF (Maghazachi, 1991) and IL-12 (Allavena *et al.*, 1994b) attract NK cells primarily in a chemotactic

Table 2 Spectrum of Action of Chemokines.

Family	Chemokine	Neu	Eo	Ba	Mo	T	B	NK	DC
C-X-C	IL-8	+	–	+/–	–	+	–	+	–
	NAP-2	+	NT	+	–	NT	NT	NT	NT
	MGSA/gro α	+	NT	NT	–	NT	NT	NT	–
	IP10	–	NT	NT	+	+	–	+	–
C-C	MCP-1	–	–	+	+	+	–	+	–
	MCP-2	–	+/–	NT	+	+	NT	+	–
	MCP-3	+	+	+	+	+	–	+	+
	RANTES	–	+	+	+	+	+	+	+
	MIP-1α	–	+	–	+	+	+	+	+
	MIP-1β	–	–	–	+	+	–	–	NT
C	Lymphotactin	–	NT	NT	–	+	NT	+	NT

Neu, neutrophils; Eo, eosinophils; Ba, basophils; Mo, monocytes; T, T lymphocytes; B, B lymphocytes; NK, natural killer cells; DC, dendritic cells.

manner. Fibronectin increases NK cell migration (Somersalo and Saksela, 1991). Coating of chemotaxis filters with fibronectin and other ECM proteins is currently used to facilitate cell attachment to the filter surface, as NK cells are poorly adherent in resting conditions.

The study of NK cell migration in response to chemotactic factors received renewed interest with the discovery of a new family of chemotactic cytokines (chemokines) (Baggiolini, Dewald and Moser, 1994; Oppenheim *et al.*, 1991). Chemokines are characterized by four Cys residues, the first two in tandem, whose relative position is conserved among all members. On the basis of the relative position of the first Cys tandem, a Cys-X-Cys family (or α) and a Cys-Cys family (or β) can be identified. Chemokines have distinct but overlapping spectra of action (Table 2). For instance IL-8 and other a chemokines are active primarily on neutrophils whereas MCP-1 and other members of the β family (Rantes, MIP-1α) are active on monocytes. However in the last two years there have been many reports on the effect of chemokines on other cell types, including T lymphocytes, NK cells, eosinophils, basophils and dendritic cells (see below), and it is now evident that the spectrum of action of chemokines has become broader and less distinct than initially thought.

In 1993 Maghazachi *et al.*, reported that IL-2–activated adherent NK cells migrate in response to IL-8 (Sebok *et al.*, 1993). We and others examined the ability of NK cells to respond to MCP-1 and related proteins (Maghazachi, Alaoukaty and Schall, 1994; Allavena *et al.*, 1994a). We showed that MCP-1 induces migration across polycarbonate filters of IL-2-activated NK cells, whereas it is a weak attractant for unstimulated cells. The MCP-1-related chemokines MCP-2 and MCP-3 are also active. IL-2-activated NK cells show specific binding sites for labeled MCP-1 and cell migration was inhibited by both Cholera and Bordetella pertussis toxins (Allavena *et al.*, 1994a).

The migration of NK cells in response to C-X-C and C-C chemokines has also been thoroughly investigated by other groups. Freshly isolated NK cells migrate through fibronectin-coated polycarbonate filters in response to IP10, MIP-1α, IL-8, MCP-1

(Taub *et al.*, 1995). These results have recently been confirmed by the use of NK cell clones (Loetscher *et al.*, 1996).

Finally, a third member of the chemokine family showing only one Cys, lymphotactin, has been recently cloned and shown to attract T lymphocytes (Kelner *et al.*, 1994). Our recent results indicate that lymphotactin is also chemotactic for freshly isolated and IL-2-cultured NK cells (Bianchi *et al.*, 1996). Thus it is now well established that NK cells respond to a variety of chemotactic signals, including classical chemotactic factors, cytokines and chemokines.

It is of interest that in our study (Allavena *et al.*, 1994a) as well as in another report (Maghazachi, Alaoukaty and Schall, 1994) NK cells optimally responded to MCP-1 only after IL-2 activation. Although an overnight activation with IL-2 was sufficient to induce significant migration to the chemokine, better responses were obtained with long-term (15 days) IL-2-propagated NK cells. These results are in agreement with Northern blot analysis of MCP-1 R (CCR2) transcripts which are expressed at substantial levels only in NK cells cultured with IL-2 for two weeks (Polentarutti, *et al.*, 1997).

Migration of freshly isolated NK cells requires the pre-coating of filters with fibronectin to obtain successful migration. Thus the activation of NK cells is a pre-requisite for optimal migration. On the other hand it is of interest that resting T cells have been shown to respond to MCP-1 in a transendothelial migration system (Woldemar Carr *et al.*, 1994). It is possible that the interaction of T cell adhesive molecules with their corresponding ligands on EC provides an initial signal that enables T cells to respond to chemotactic signals.

The fact that NK cells are highly motile only when activated may bear some relevance *in vivo*. NK cells are found at sites of viral or bacterial infections but are rarely seen in chronic infiltrates or within most solid tumors (Trinchieri, 1989; Introna and Mantovani, 1983).There is evidence that several human and murine tumors express MCP-1 (Van Damme *et al.*, 1992) and that this chemokine is an important determinant in reflecting the levels of tumor-associated macrophages (Mantovani *et al.*, 1992). The requirement for an activation step, which distinguishes the migratory response of NK cells to MCP-1 from that of monocytes, may account for the apparent paradox of a substantial accumulation of mononuclear phagocytes, but not NK cells, in MCP-1-producing tumors (Negus *et al.*, 1995).

TARGETING OF LEUKOCYTES INTO TUMORS

By immunohistochemistry it has been demonstrated that the leukocyte infiltrate of human tumors is composed mainly of macrophages and T lymphocytes. CD4+ cells usually predominate over CD8+. Minor components are granulocytes (<10%), B cells (<5%) and very a few eosinophils and NK cells (Ruco, 1994). When CD56+ cells are found they are usually T cells co-expressing CD3+. As for CD16, it has been demonstrated that many tumor-associated macrophages express CD16+ (van Ravenswaay Claasen, Kluin and Fleuren, 1992). The poor representation of NK cells within the tumor infiltrate has raised doubts about their role as anti-tumor effector cells. It should be noted, however, that established tumors probably do not represent the optimal condition for the detection of tissue-infiltrating NK cells. A report indicated that NK cells infiltrate the lesion of an experimental murine tumor only at a very early stage (3 days after transplant; Kurosawa *et al.*, 1993). In allograft rejection (Nemlander, Saksela and Hayrj, 1983) and viral infection (Natuk and Welsh, 1987), tissue infiltrating NK cells preceded the entry of T cells and were not found at later times.

Thus evidence supports the hypothesis that the unsuccessful search for NK cells within the tumor infiltrate is due to a change in cell composition over time. Moreover it should be considered that NK cells have a shorter life span than other lymphoid cells and macrophages. The estimated half life of an NK cell has not been clearly identified but is approximately a few weeks (Trinchieri, 1989). In *in vitro* experiments we have demonstrated that in the absence of exogenous stimuli, NK cells are rapidly lost within 3–5 days of culture, while T lymphocytes survive much longer (>7 days) (Allavena, unpublished). As a matter of fact, neutrophils, another cell population with a very short half life, but much more represented in the circulation, is also poorly represented in the leukocyte infiltrate of tumors (Ruco, 1994).

Immunohistological studies on tumor samples from patients receiving immunotherapy have demonstrated that the systemic infusion of cytokines induced quantitative and qualitative changes in the tumor infiltrate . For instance after IL-2 treatment the number of intratumoral leukocytes is increased, with more macrophages and lymphocytes, but NK cells are rarely seen in these samples (Cohen *et al.*, 1987; Velotti *et al.*, 1991).

Adoptive transfer of IL-2-activated cytotoxic cells has been performed since the mid ’80s in a large series of tumor patients and resulted in a low but considerable proportion of clinical response, especially in renal cell carcinoma and melanoma patients (Rosenberg, 1992). A crucial aspect in the adoptive transfer is the localization of the systemically infused cells to the tumor deposits. Studies with ^{111}In-labelled cells or cells transduced with a “marker” gene (resistance to neomycin) have demonstrated that a slow accumulation of infused effector cells occurs at the tumor site over a period of a few days. Nevertheless the proportion of LAK cells or TIL able to penetrate tumor tissues is very small (Rosenberg, 1992).

The systemic administration of cytokines ignores the paracrine nature of their action. An alternative approach to attract and activate leukocytes within the tumor mass is the transduction of cytokine genes into tumor cells. Under these circumstances high concentrations of cytokines are secreted locally in the tumor and induce a strong local inflammatory response. In gene-transduced tumor cells the characteristics of the infiltrate are quite distinct and depend both on the lymphokine being expressed and on the amount of the lymphokine secreted (reviewed in Pardoll, 1995). For example IL-2-transduced tumors are characterized by extensive lymphocyte infiltrates consisting of CD4+ and CD8+ lymphocytes, as well as of cells with NK markers. Rejection of IL-2–transduced tumors is dependent on CD8+ and NK cells but not CD4+ cells (Gansbacher *et al.*, 1990b). In contrast IL-4-transduced tumors are characterized by infiltrating macrophages and eosinophils and some lymphocytes, mainly CD8+ (Tepper, Pattengale and Leder, 1989). The striking eosinophil infiltrate in these tumors has been attributed to the induction of VCAM 1 expression on local vascular EC by IL-4. In fact eosinophils express high levels of VLA-4, which binds to VCAM-1 (Springer, 1994).

Introduction of the IFNγ gene into tumor cells generally induces or upregulates both MHC Class I and Class II antigens and lymphocyte infiltration (Gansbacher *et al.*, 1990a). In some tumors, IFN gamma expression results in rejection of the tumor and induction of systemic immunity (Restifo *et al.*, 1992). Tumor cells transduced with IL-7 elicited an anti-tumor response that was independent of CD8+ cells but dependent on macrophages and CD4+ lymphocytes (Hock, Dorsch and Diamantstein, 1991).

Using a poorly immunogenic murine melanoma, Dranoff *et al.*, demonstrated that vaccination with tumor cells transduced with the gene of GM-CSF potently protected mice from a second challenge with a wild type tumor. The cellular infiltrate in GM-CSF-transduced tumors was predominantly composed of mononuclear cells and few T cells. The striking result is that vaccination with GM-CSF-transduced tumors induced long-lasting systemic immunity which depended on both CD4+ and CD8+ cells and not on macrophages or NK cells (Dranoff *et al.*, 1993). The authors proposed that the paracrine secretion of GM-CSF induces the differentiation of bone marrow-derived antigen presenting cells (APC). It is now well established that GM-CSF is a key cytokine for the differentiation of dendritic cells (DC) from blood or bone marrow-derived precursors (Caux *et al.*, 1992). APC such as tissue DC are stimulated to pick up antigens released by tumor disintegration. DC then stimulate T lymphocytes with tumor antigens, thus inducing the effector mechanisms for tumor rejection and immunological memory.

These experiments suggest that successful immunotherapy can be achieved not only by transducing cytokines directly acting on the effector cells but also by transducing cytokines acting on professional APC, which potently stimulates an anti-tumor response.

The interest in APC and in particular in DC has increased greatly during the last two years, due to the availability of new methods to differentiate *in vitro* large numbers of DC from circulating precursors or from CD34+ stem cells (Caux, Liu and Banchereau, 1995). As mentioned above, GM-CSF play a crucial role in this differentiation, together with other cytokines such as TNF and IL-4 (Caux *et al.*, 1992; Sallusto and Lanzavecchia, 1994). The feasibility of exploiting the potent ability of DC to capture, process and present antigens to T lymphocytes is now a reality and is currently being pursued by many investigators.

DENDRITIC CELLS

DC precursors originate in the bone marrow and subsequently migrate into peripheral tissues and primary lymphoid organs, where they efficiently take up and process soluble antigens (Ag). After processing, DC migrate via the afferent lymphatics to draining lymph nodes, where they stimulate T cells (Caux, Liu and Banchereau, 1995). Thus attraction of DC precursors to peripheral tissues and migration of Ag-carrying DC to lymph nodes is essential for the induction of immune responses. As the molecules controlling these events are unknown, we have investigated which chemotactic factors could attract this cell population and regulate their trafficking in tissues (Sozzani *et al.*, 1995).

For these experiments, DC were obtained from peripheral blood precursor cells cultured in the presence of GM-CSF and IL-4 or IL-13 (Sallusto and Lanzavecchia, 1994; Piemonti *et al.*, 1995). These cells were $CD1a^{+}$, MHC class II^{++}, $CD14^{-}$, $CD3^{-}$ and $CD20^{-}$ and behaved as classical immature DC, being active in eliciting the proliferation of naive T lymphocytes and showing a strong ability to uptake FITC-dextran (Piemonti *et al.*, 1995). In a classical microwell chemotaxis assay,

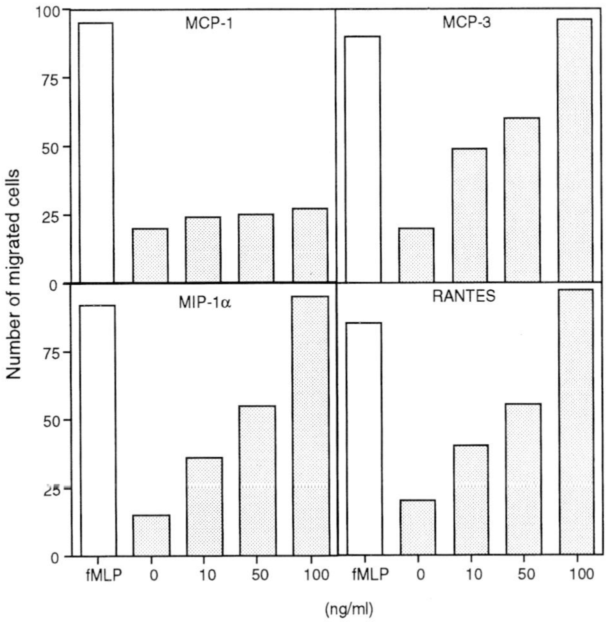

Figure 3 Migration of dendritic cells to C-C chemokines. DC were tested for their ability to migrate across a 5 μm pore-size polycarbonate filter in response to different concentrations of chemokines or to an optimal concentration (10^{-8}M) of fMLP, a classical chemoattractant. At the end of the incubation (90 min, at 37°C), the number of cells in five high power-immersion fields was evaluated.

a representative experiment is shown in Figure 3, these cells migrated in response to a selected pattern of C-C chemokines, such as MCP-3, RANTES and MIP-1α and to two prototypical chemotactic factors, fMLP and C5a (Sozzani *et al.*, 1995). MCP-1 and MCP-2, two other C-C chemokines and all of the C-X-C chemokines tested (IL-8, IP-10 and Groß) were inactive. Peak active concentrations and the percentage of input cells migrating in response to C-C chemokines were comparable to those observed with monocytes.

MCP-3, MIP-1α and RANTES showed a complex pattern of cross-desensitization, suggesting the presence of multiple promiscuous receptors on these cells. In preliminary studies of Northern blot and receptor binding, it has been found that DC express the C-C CKR1 receptor (Sozzani, unpublished). The expression of other chemokine receptors is currently under investigation. Thus, C-C chemokines appear to function as pure chemotactic signals for these cells, at least when tested *in vitro*.

Recently it has also been reported that chemokines can induce tissue accumulation of DC *in vivo*. Transgenic mice for MCP-1, in which MCP-1 is constitutively expressed in the epidermis under the control of human keratin 14 promoter, responded to a contact hypersensitivity challenge with an increased lichenoid infiltration of monocytes, T lymphocytes and CD45⁺, I-A⁺ cells that assumed a dendritic morphology in situ (Nakamura, Williams and Kupper, 1995). We also have observed that a melanoma cell line transfected with murine MCP-3 caused DC accumulation when inoculated *in vivo* (Vecchi, unpublished). The reason for the apparent discrepancy between the accumulation of Langerhans cells in the skin, observed in MCP-1 transgenic mice (Nakamura, Williams and Kupper, 1995), and the lack of chemotactic response to MCP-1 reported for human *in vitro*-generated DC (Sozzani *et al.*, 1995) is at present unknown. It is possible that the culture conditions used to generate DC *in vitro*, the possible different bone marrow precursor (i.e. bone marrow cells vs. circulating monocytes) and the stage of maturation could affect the expression of chemotactic receptors and the relative response to chemokines.

These data, taken together, show that the spectrum of action of C-C chemokines, initially restricted to monocytes, now encompasses several leukocyte populations including DC (Table 2). The exact role of chemokines in the onset of the specific immune responses is still unclear. However, the data so far available strongly suggest that these proteins could be one important determinant in directing the trafficking of DC *in vivo*.

CONCLUDING REMARKS

NK cells and DC have considerable interest as tools for immunotherapy for neoplastic diseases. DC are most efficient APC and may therefore act as effective initiators of antitumor immunity. NK cells are involved in orienting immunity to Th1 type responses. The results summarized herein identify molecules which are likely to be involved in trafficking of NK cells and DC. While, at least for NK cells, a considerable amount of information is available *in vitro*, the actual *in vivo* relevance of these observations and the fate of recruited cells remain largely unexplored and information is scanty and fragmentary. This is particularly true for DC which after being recruited in tissues and encountering antigen travel to regional lymphnodes. It is probable that a better definition *in vitro* and *in vivo* of the molecular basis of the trafficking of DC and NK cells may provide tools to generate and orient effective antitumor immunity.

ACKNOWLEDGMENTS

Supported in part by the Italian Association for Cancer Research (AIRC) and Istituto Superiore di Sanitá (project AIDS) and National Research Council (CNR).

REFERENCES

Allavena, P., C. Paganin, I. Martin Padura, G. Peri, M. Gaboli, E. Dejana, P.C. Marchisio and A. Mantovani (1991). Molecules and structures involved in the adhesion of natural killer cells to vascular endothelium, *J. Exp. Med.*, **173**, 439.

Allavena, P., G. Bianchi, D. Zhou, J. Van Damme, P. Jìlek, S. Sozzani and A. Mantovani (1994a) Induction of natural killer cell migration by monocyte chemotactic protein-1, -2 and -3, *Eur. J. Immunol.*, **24**, 3233.

Allavena, P., C. Paganin, D. Zhou, G. Bianchi, S. Sozzani and A. Mantovani (1994b). Interleukin-12 is chemotactic for natural killer cells and stimulates their interaction with vascular endothelium, *Blood*, **84**, 2261.

Allavena, P. and A. Mantovani (1994). Therapeutic strategies targeted to tumor-associated leukocytes in human tumors, in: Tumor-associated leukocytes: Pathophysiology and therapeutic applications, ed. A. Mantovani (R.G. Landes Publisher, Austin, USA) p. 90.

Aronson, F.R., P. Libby, E.P. Brandon, M.W. Janicka and J.W. Mier (1988). IL-2 rapidly induces natural killer cell adhesion to human endothelial cells. A potential mechanism for endothelial injury, *J. Immunol.*, **141**, 158.

Baggiolini, M., B. Dewald and B. Moser (1994). Interleukin-8 and related chemotactic cytokines – CXC and CC chemokines, *Adv. Immunol.*, **55**, 99.

Berman, M.E., Y. Xie and W.A. Muller (1996). Roles of platelet/endothelial cell adhesion molecule-1 (PECAM-1,CD31) in natural killer cell transendothelial migration and $\beta 2$ integrin activation, *J. Immunol.*, **156**, 1515.

Bianchi, G., M. Sironi, E. Ghibaudi, C. Selvaggini, M. Elices, P. Allavena and A. Mantovani (1993). Migration of natural killer cells across endothelial cell monolayers, *J. Immunol.*, **151**, 5135.

Bianchi *et al.* (1996). Migratory response of human natural killer cells to lymphotactin, *Eur. J. Immynol.*, **26**, 3238–3241.

Bottazzi, B., M. Introna, P. Allavena, A. Villa and A. Mantovani (1985). *In vitro* migration of human large granular lymphocytes, *J. Immunol.*, **134**, 2316.

Caux, C., C. Dezutter-Dambuyant, D. Schmitt and J. Banchereau (1992). GM-CSF and TNF-α cooperate in the generation of dendritic Langerhans cells, *Nature*, **360**, 258.

Caux, C., Y.J. Liu and J. Banchereau (1995). Recent advances in the study of dendritic cells and follicular dendritic cells, *Immunol. Today*, **16**, 2.

Cohen, P.J., M.T. Lotze, J.R. Roberts, S.A. Rosenberg and E.S. Jaffe (1987). The immunopathology of sequential tumor biopsies in patients treated with interleukin-2. Correlation of response with T-cell infiltration and HLA-DR expression, *Am. J. Pathol.*, **129**, 208.

Dranoff, G., E. Jaffee, A. Lazenby, P. Golumbek, H. Levitsky, K. Brose, V. Jackson, H. Hamada, D. Pardoll and R.C. Mulligan (1993). Vaccination with irradiated tumor cells engineered to secrete murine granulocyte-macrophage colony-stimulating factor stimulates potent, specific, and long-lasting anti-tumor immunity, *Proc. Natl. Acad. Sci. USA*, **90**, 3539.

Galandrini, R., R. De Maria, M. Piccoli, L. Frati and A. Santoni (1994). CD44 triggering enhances human NK cell cytotoxic functions, *J. Immunol.*, **153**, 4399.

Gansbacher, B., R. Bannerji, B. Daniels, K. Zier, K. Cronin and E. Gilboa (1990a), Retroviral vector-mediated gamma-interferon gene transfer into tumor cells generates potent and long lasting antitumor immunity, *Cancer Res.*, **50**, 7820.

Gansbacher, B., K. Zier, B. Daniels, K. Cronin, R. Bannerji and E. Gilboa (1990b, Interleukin 2 gene transfer into tumor cells abrogates tumorigenicity and induces protective immunity, *J. Exp. Med.*, **172**, 1217.

Gismondi, A., S. Morrone, M.J. Humphries, M. Piccoli, L. Frati and A. Santoni (1991). Human natural killer cells express VLA-4 and VLA-5, which mediate their adhesion to fibronectin, *J. Immunol.*, **146**, 384.

Herberman, R.B. and J.R. Ortaldo (1981). NK cells: their role in defence against disease, *Science*, **214**, 24.

Hock, H., M. Dorsch and T. Diamantstein (1991). Interleukin 7 induces $CD42^+$ T cell-dependent tumor rejection, *J. Exp. Med.*, **174**, 1291.

Holmberg, L.A., K.A. Springer and K.A. Ault (1981). Natural Killer activity in the peritoneal exudate of mice infected with listeria monocitogenes, *J. Immunol.*, **127**, 1792.

Introna, M. and A. Mantovani (1983). Natural killer cells in human solid tumors, *Cancer Metastasis. Rev.*, **2**, 337.

Inverardi, L., M. Samaja, R. Motterlini, F. Mangili, J.R. Bender and R. Pardi (1992). Early recognition of a discordant xenogeneic organ by human circulating lymphocytes, *J. Immunol.*, **149**, 1416.

Kelner, G.S., J. Kennedy, K.B. Bacon, S. Kleyensteuber, D.A. Largaespada, N.A. Jenkins, N.G. Copeland, J.F. Bazan, K.W. Moore, T.J. Schall and A. Zlotnik (1994). Lymphotactin: A cytokine that represents a new class of chemokine, *Science*, **266**, 1395.

Koopman, G., Y. van Kooyk, M. de Graaff, C.J. Meyer, C.G. Figdor and S.T. Pals (1990). Triggering of the CD44 antigen on T lymphocytes promotes T cell adhesion through the LFA-1 pathway, *J. Immunol.*, **145**, 3589.

Kurosawa, S., G. Matsuzaki, M. Harada, T. Ando and K. Nomoto (1993). Early appearance and activation of natural killer cells in tumor-infiltrating lymphoid cells during tumor development, *Eur. J. Immunol.*, **23**, 1029.

Loetscher, P., M. Seitz, I. Clarklewis, M. Baggiolini and B. Moser (1996). Activation of NK cells by CC chemokines – Chemotaxis, Ca2+ mobilization, and enzyme release, *J. Immunol.*, **156**, 322.

Maenpaa, A., J. Jaaskelainen, O. Carpen, M. Patarroyo and T. Timonen (1993). Expression of integrins and other adhesion molecules on NK cells; impact of IL-2 on short- and long-term cultures, *Int. J. Cancer*, **53**, 850.

Maghazachi, A.A. (1991). Tumor necrosis factor-alpha is chemokinetic for lymphokine-activated killer cells: regulation by cyclic adenosine monophosphate, *J. Leukoc. Biol.*, **49**, 302.

Maghazachi, A.A., A. Alaoukaty and T.J. Schall (1994). C-C chemokines induce the chemotaxis of NK and IL-2-activated NK cells — Role for G proteins, *J. Immunol.*, **153**, 4969.

Mantovani, A., B. Bottazzi, F. Colotta, S. Sozzani and L. Ruco (1992). The origin and function of tumor-associated macrophages, *Immunol. Today*, **13**, 265.

Marchisio, P.C., D. Cirillo, L. Naldini, M.V. Primavera, A. Teti and A. Zambonin-Zallone (1984). Cell-substratum interaction of cultured avian osteoclasts is mediated by specific adhesion structures, *J. Cell Biol.*, **99**, 1696.

McIntyre, K.W. and R.M. Welsh (1986). Accumulation of natural killer and cytotoxic T large granular lymphocytes in the liver during virus infection, *J. Exp. Med.*, **164**, 1667.

Moore, K.L. and L.F. Thompson (1992). P-selectin (CD62) binds to subpopulations of human memory T lymphocytes and natural killer cells, *Biochem. Biophys. Res. Commun.*, **186**, 173.

Morrone, S., A. Gismondi, F. Mainiero, M. Milella, G. Santoni, R. Falcioni, M. Piccoli, L. Frati and A. Santoni (1991). Expression of beta1 integrin laminin receptors by fresh and cultured human NK cells, *Nat. Immun. Cell Growth. Regul.*, **10**, 135.

Nakamura, K., I.R. Williams and T.S. Kupper (1995). Keratinocyte-derived monocyte chemoattractant protein 1 (MCP-1): Analysis in a transgenic model demonstrates MCP-1 can recruit dendritic and langerhans cells to skin, *J. Invest. Dermatol.*, **105**, 635.

Natuk, R.J. and R.M. Welsh (1987). Accumulation and chemotaxis of natural killer/large granular lymphocytes at sites of virus replication, *J. Immunol.*, **138**, 877.

Needham, L.K. and R.L. Schnaar (1993). The HNK-1 reactive sulfoglucuronyl glycolipids are ligands for L-selectin and P-selectin but not E-selectin, *Proc. Natl. Acad. Sci. USA*, **90**, 1359.

Negus, R.P., G.W. Stamp, M.G. Relf, F. Burke, S.T. Malik, S. Bernasconi, P. Allavena, S. Sozzani, A. Mantovani and F.R. Balkwill (1995). The detection and localization of monocyte chemoattractant protein-1 (MCP-1) in human ovarian cancer, *J. Clin. Invest.*, **95**, 2391.

Nemlander, A., E. Saksela and P. Hayrj (1983). Are natural killer cells involved in allograft rejection? *Eur. J. Immunol.*, **13**, 348.

Ohmori, K., T. Yoneda, G. Ishihara, K. Shigeta, K. Hirashima, M. Kanai, S. Itai, T. Sasaoki, S. Arii, H. Arita, *et al.* (1989). Sialyl SSEA-1 antigen as a carbohydrate marker of human natural killer cells and immature lymphoid cells, *Blood*, 74, 255.

Oppenheim, J.J., C.O. Zachariae, N. Mukaida and K. Matsushima (1991). Properties of the novel proinflammatory supergene "intercrine" cytokine family, *Annu. Rev. Immunol.*, **9**, 617.

Paganin, C., C. Matteucci, S. Censuales, A. Mantovani and P. Allavena (1993). IL-4 inhibits binding and cytotoxicity of NK cells to vascular endothelium, *Cytokine* (in press).

Pardoll, D.M. (1995). Paracrine cytokine adjuvants in cancer immunotherapy, *Annu. Rev. Immunol.*, **13**, 399.

Piemonti, L., S. Bernasconi, W. Luini, Z. Trobonjaca, A. Minty, P. Allavena and A. Mantovani (1995). IL-13 supports differentiation of dendritic cells from circulating precursors in concert with GM-CSF, *Eur. Cytokine Netw*, **6**, 245.

Pinola, M., R. Renkonen, M.L. Majuri, S. Tiisala and E. Saksela (1994a, Characterization of the E-selectin ligand on NK cells, *J. Immunol.*, **152**, 3586.

Pinola, M., E. Saksela, S. Tiisala and R. Renkonen (1994b, Human NK cells expressing alpha 4 beta 1/beta 7 adhere to VCAM-1 without preactivation, *Scand. J. Immunol.*, **39**, 131.

Pohajdak, B., J. Gomez, F.W. Orr, N. Khalil, M. Talgoy and A.H. Greenberg (1986). Chemotaxis of large granular lymphocytes, *J. Immunol.*, **136**, 278.

Polentarutti, N., B. Bottazzi, C. Balotta, A. Erroi and A. Mantovani (1986). Modulation of the locomotory capacity of human large granular lymphocytes, *Cell Immunol.*, **101**, 204.

Polentarutti, N. *et al.* (1997). Il-2 Reulated expression of the monocyte chemotactic protein-1 receptor (CC CKR2) in human NK cells: Characterization of a predominant 3.4 Kb transcript containing CC CKR2A and CC CKR2B sequences, *J. Immunol.*, **158**, 2689–2694.

Rabinovich, H., R.B. Herberman and T. Whiteside (1993). Differential effects of IL-12 and IL-2 in expression and function of cellular adhesion molecules on purified NK cells, *Cell Immunol.*, **152**, 481.

Restifo, N.P., P.J. Spiess, S.E. Karp, J.J. Mule and S.A. Rosenberg (1992). A nonimmunogenic sarcoma transduced with the cDNA for interferon gamma elicits CD8+ T cells against the wild-type tumor: correlation with antigen presentation capability, *J. Exp. Med.*, **175**, 1423.

Robertson, M.J., M.A. Caligiuri, T.J. Manley, H. Levine and J. Ritz (1990). Human natural killer cell adhesion molecules. Differential expression after activation and participation in cytolysis, *J. Immunol.*, **145**, 3194.

Robertson, M.J., R.J. Soiffer, S.F. Wolf, T.J. Manley, C. Donahue, D. Young, S.H. Herrmann and J. Ritz (1992). Response of human natural killer (NK) cells to NK cell stimulatory factor (NKSF): cytolytic activity and proliferation of NK cells are differentially regulated by NKSF, *J. Exp. Med.*, **175**, 779.

Robertson, M.J. and J. Ritz (1990). Biology and clinical relevance of human natural killer cells, *Blood*, **76**, 2421.

Rosenberg, S.A. (1992). Karnofsky Memorial Lecture. The immunotherapy and gene therapy of cancer, *J. Clin. Oncol.*, **10**, 180.

Ruco, L. (1994). Tumor-infiltrating leukocytes: histology, immunohistochemistry and in situ hybridization, in: Tumor-associated leukocytes: pathophysiology and therapeutic applications, ed. A. Mantovani (R.G. Landes Co., Austin, USA) p. 15.

Sallusto, F. and A. Lanzavecchia (1994). Efficient Presentation of Soluble Antigen by Cultured Human Dendritic Cells Is Maintained by Granulocyte/ Macrophage Colony-Stimulating Factor Plus Interleukin-4 and Downregulated by Tumor Necrosis Factor-α, *J. Exp. Med.*, **179**, 1109.

Sebok, K., D. Woodside, A. al Aoukaty, A.D. Ho, S. Gluck and A.A. Maghazachi (1993). IL-8 induces the locomotion of human IL-2-activated natural killer cells. Involvement of a guanine nucleotide binding (Go) protein, *J. Immunol.*, **150**, 1524.

Somersalo, K. and E. Saksela (1991). Fibronectin facilitates the migration of human natural killer cells, *Eur. J. Immunol.*, **21**, 35.

Somersalo, K., J. Tarkkanen, M. Patarroyo and E. Saksela (1992). Involvement of beta 2–integrins in the migration of human natural killer cells, *J. Immunol.*, **149**, 590.

Somersalo, K., O. Carpen and E. Saksela (1994). Stimulated natural killer cells secrete factors with chemotactic activity, including NAP-1 IL-8, which supports VLA-4- and VLA-5-mediated migration of T lymphocytes, *Eur. J. Immunol.*, **24**, 2957.

Sozzani, S., F. Sallusto, W. Luini, D. Zhou, L. Piemonti, P. Allavena, J. Van Damme, S. Valitutti, A. Lanzavecchia and A. Mantovani (1995). Migration of dendritic cells in response to formyl peptides, C5α and a distinct set of chemokines, *J. Immunol.*, **155**, 3292.

Springer, T.A. (1994). Traffic signal for lymphocyte recirculation and leukocyte emigration: the mutistep paradigm, *Cell*, **76**, 301.

Steinman, R.M. (1991). The dendritic cell system and its role in immunogenicity, *Annu. Rev. Immunol.*, **9**, 271.

Taub, D.D., T.J. Sayers, C.R.D. Carter and J.R. Ortaldo (1995). α and β chemokines induce NK cell migration and enhance NK-mediated cytolysis, *J. Immunol.*, **155**, 3877.

Tepper, R.I., P.K. Pattengale and P. Leder (1989). Murine interleukin-4 displays potent anti-tumor activity *in vivo*, *Cell*, **57**, 503.

Timonen, T., C.G. Gahmberg and M. Patarroyo (1990). Participation of CD11a-c/CD18, CD2 and RGD-binding receptors in endogenous and interleukin-2-stimulated NK activity of CD3-negative large granular lymphocytes, *Int. J. Cancer*, **46**, 1035.

Trinchieri, G. (1989). Biology of natural killer cells, *Adv. Immunol.*, **47**, 187.

Trinchieri, G. (1993). Interleukin-12 and its role in the generation of TH1 cells, *Immunol. Today*, **14**, 335.

Van Damme, J., P. Proost, J.P. Lenaerts and G. Opdenakker (1992). Structural and functional identification of two human, tumor-derived monocyte chemotactic proteins (MCP-2 and MCP-3) belonging to the chemokine family, *J. Exp. Med.*, **176**, 59.

van Ravenswaay Claasen, H.H., P.M. Kluin and G.J. Fleuren (1992). Tumor infiltrating cells in human cancer. On the possible role of CD16+ macrophages in antitumor cytotoxicity, *Lab. Invest.*, **67**, 166.

Velotti, F., A. Stoppacciaro, L. Ruco, A. Tubaro, A. Pettinato, S. Morrone, T. Napolitano, P.C. Bossola, C.R. Franks, P. Palmer and et al (1991). Local activation of immune response in bladder cancer patients treated with intraarterial infusion of recombinant interleukin-2, *Cancer Res.*, **51**, 2456.

Woldemar Carr, M., S.J. Roth, E. Luther, S.S. Rose and T.A. Springer (1994). Monocyte chemoattractant protein-1 acts as a T-lymphocyte chemoattractant, *Proc. Natl. Acad. Sci. USA.*, **91**, 3652.

11. ENDOTHELIUM AND INFLAMMATION: MOLECULES AND THERAPEUTIC PROSPECTS

LAURELEE OSBORN*

Biogen, Inc., Cambridge, MA 02142, USA

INTRODUCTION

Recruitment of leukocytes from the bloodstream to inflamed or infected tissue is a key event in the inflammatory/immune response. Inflammatory mediators generated by resident leukocytes or by the cells composing the injured or infected tissue itself cause endothelial cells lining the nearby vasculature to present adhesion molecules and chemokines, which bind and activate circulating leukocytes. A healthy response to injury or infection gives rise to carefully orchestrated expression and then down-regulation of these endothelial recruiting molecules, resulting in quick neutralization of the insult while causing minimal damage to host tissues. Disorders involving the inflammatory cascade are, however, wide-spread, perhaps reflecting the extrememly subtle balance of pro- and anti-inflammatory influences that must be operating at all times to protect us from the onslaught of microorganisms and molecules with which we live. Examples of genetically insufficient inflammatory response are leukocyte adhesion deficiency-1 and -2 (LAD-1 and -2) (Anderson *et al.*, 1985; Etzioni *et al.*, 1992), caused by a lack of adhesion molecules involved in leukocyte extravasation. These diseases are quite rare, probably because they were fatal at an early age prior to the discovery of sulfa and antibiotic drugs. An acquired defect of immunity is caused by the human immunodeficiency virus. Loss of CD4+ T cells due to the virus results in a wide variety of opportunistic infections of bacteria, fungi, and viruses that are relatively harmless to the population at large.

Overactivity or inappropriate triggering of the immune/inflammatory response causes a large number of common diseases such as allergies, rheumatoid arthritis, inflammatory bowel disease, and autoimmune disorders such as lupus erythematosis and multiple sclerosis. These diseases currently are treated with a variety of anti-inflammatory drugs which alleviate but do not cure them. Some of the most effective drugs in this catagory, such as steroids, have undesirable side-effects and must be rotated with other, less effective medications. Therefore there is still a large unmet need for new medicines which can block the inflammatory process. New points of intervention have been suggested by our increasing knowledge of the molecular mechanisms of leukocyte extravasation, as described below.

Apart from its role in inflammation, extravasation of lymphocytes is part of the process by which lymphocytes are recirculated through healthy tissues and returned to lymphoid tissue. Specialized endothelial cells and adhesion molecules are utilized in this process, though there is some overlap with molecules involved in inflammation. Lymphocyte recirculation is described elsewhere in this volume and in recent reviews (Butcher and Picker, 1996; Springer, 1995); this chapter will focus on emigration due to inflammation.

*Current address: 30 Falmouth Rd., Arlington, MA 02174, USA.

Table 1 Molecules involved in leukocyte recruitment to sites of inflammation.

Mediators of endothelial activation:	
Immediate: Thrombin, histamine	
Delayed: LPS, IL-1, TNF, IL-4, oxidants (induce nuclear factor κB)	
Rolling adhesion:	
Endothelium:	*Leukocyte counter-receptor:*
P-selectin	PSGL-1, sialyl Lewis x
E-selectin	sialyl Lewis x on protein
GlyCAM 1, CD34, MAdCAM	L-selectin
VCAM-1	α4 integrins
Activation of leukocyte integrins:	
Chemokines	
Adhesion molecules: CD3/TCR, CD2, CD7, CD28, CD31, CD44	
Phorbol ester, divalent cations	
Firm adhesion:	
Endothelium:	*Leukocyte counter-receptor:*
ICAM-1,2,3	β2 integrins
VCAM-1	α4 integrins
MAdCAM-1	$\alpha 4\beta 7$ integrin
Migration:	
ICAM-1, β2 integrins, CD31 (PECAM-1), proteases	

A useful multistep model of leukocyte recruitment has been promulgated in the last few years (Butcher, 1991; Schweighoffer and Shaw, 1992). This model incorporates evidence from intravital microscopy, *in vitro* adhesion of cultured endothelial and leukocytic cells, and most recently adhesion studies performed on isolated or recombinant molecules under flow conditions mimicking those of the vasculature. The model proposes a set of consecutive steps that must occur for a leukocyte to leave the bloodstream and enter the underlying tissue. In the case of inflammation-induced migration, required steps include:

(1) Stimulation of endothelium to express adhesion molecules
(2) Initial contact and rolling adhesion of leukocytes on endothelium
(3) Activation of leukocyte integrins
(4) Firm adhesion through integrin-Ig superfamily molecule interaction
(5) Transendothelial migration
(6) Migration into the tissue.

Some of the molecules known to be involved in each of these steps and discussed in this chapter are listed in Table 1. Each of these steps offers a potential avenue for blockade by anti-inflammatory agents, which will be considered in the latter portion of this chapter.

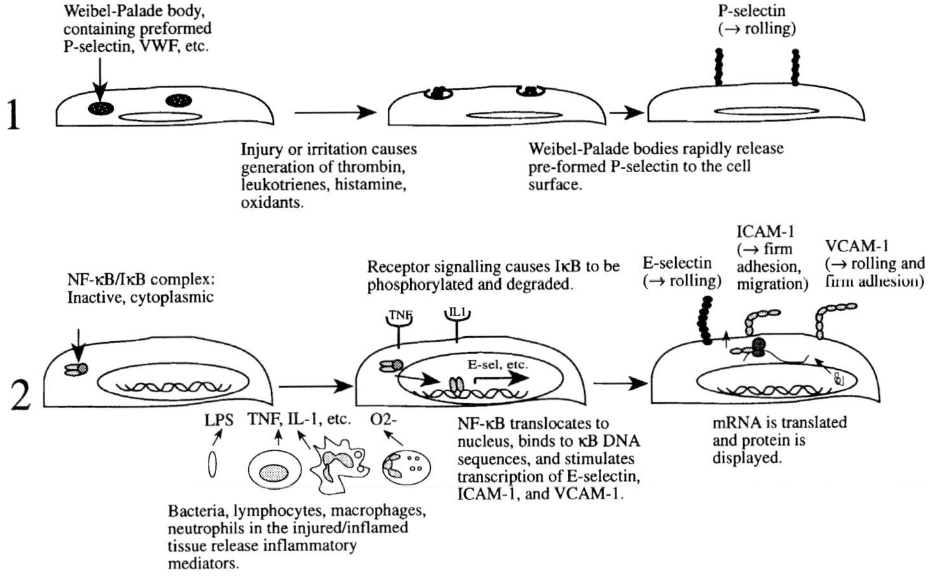

Figure 1 Two pathways of endothelial cell activation. Pathway 1 results in display of P-selectin on endothelial cells within seconds after injury or other stimulation. Pathway 2 requires *de novo* protein synthesis and results in display of E-selectin, ICAM-1, and/or VCAM-1 within 1-4 hours after stimulation. See text for details.

ADHESION OF LEUKOCYTES TO ENDOTHELIUM

Activation of Endothelium

There are at least two pathways by which adhesion molecules are induced on the surface of endothelial cells (Figure 1). The first is a rapid process triggered by mediators such as thrombin or histamine, resulting in translocation of pre-formed P-selectin from granules known as Weibel-Palade bodies to the cell surface in a matter of seconds (Bonfanti *et al.*, 1989; Johnston *et al.*, 1989a; Johnston *et al.*, 1989b). P-selectin mediates rolling of neutrophils. As rolling is the first step in adhesion, and neutrophils are the most numerous, non-specific, and fast-acting leukocytes in the blood stream, it is appropriate that an injury such as a break in the skin, which would activate thrombin, results in immediate display of P-selectin and recruitment of infection-fighting neutrophils.

The second pathway of endothelial activation requires an hour or more, and is stimulated by bacterial products such as LPS, or inflammatory cytokines such as TNF or IL-1. This pathway results in transcriptional activation of the genes for E-selectin, ICAM-1, and VCAM-1, via the nuclear factor κB (NF-κB) (Iademarco *et al.*, 1992; Montgomery *et al.*, 1991; Voraberger *et al.*, 1991; Whelan *et al.*, 1991). Engagement of the endothelial LPS or cytokine receptor causes disengagement of a cytoplasmic inhibitory protein, IκB, from a dimer of Rel or NF-κB family proteins. This exposes nuclear localization signals on the dimer, which migrates into the nucleus and binds to a consensus sequence in the transcriptional enhancer or promoter region of the target genes, where it quickly stimulates transcription (reviewed in Siebenlist *et al.*, 1994). Electrophoretic mobility shift assays of nuclear extracts show active DNA-binding NF-κB within 15 minutes of addition of TNF to the cells' growth medium in human umbilical vein endothelial cells (HUVEC) (Montgomery *et al.*, 1991). Abundant E-selectin mRNA can be detected within an hour after stimulation (Osborn *et al.*, 1989).

Other transcriptional factors are also involved in regulation of E-selectin, ICAM-1, and VCAM-1 expression, probably contributing to the somewhat different time courses of rise and fall of these gene's products, and to the differing non-endothelial cell-type specificities shown by these genes. For example, unlike E-selectin which is quite restricted to endothelial cells, ICAM-1 is expressed constitutively on a wide variety of cell types, while VCAM-1 is expressed on developing skeletal and smooth muscle cells under the control of a separate set of transcriptional activating elements from those used in endothelium (Iademarco *et al.*, 1992).

Rolling Adhesion

Following endothelial activation, the initial adhesive event between a leukocyte and an endothelial cell is usually mediated by interaction of a selectin molecule with its carbohydrate ligand. This interaction is relatively weak and releases quickly, resulting in the rolling of the leukocyte on the endothelial layer. The leukocyte presumably samples the endothelial surface during rolling, and either encounters agents that activate it, leading to firm adhesion and migration (see below), or is not stimulated and returns to the blood stream when it has rolled past the "sticky patch" caused by the local inflammation. The three selectins now known, E-, L-, and P-selectin (CD62E, L, and P) each consist of a lectin domain at the N-terminus, an EGF (epidermal growth factor)-like motif, and a varying number of consensus repeats homologous to complement regulatory protein domains, followed by a single transmembrane-spanning region and a cytoplasmic tail. In inflammation, P- and E-selectin are induced on the endothelial cell surface by the two pathways described above and in Figure 1. Like the lectins from which their name derives, selectins bind to carbohydrate moieties. These are generally presented on the cell surface by certain proteins containing heavily glycosylated mucin-like domains rich in serine and threonine. E-selectin binds to fucose- and sialate-containing tetrasaccharides such as sialyl Lewis x and Lewis a; the affinity of binding is low unless they are displayed on an appropriate scaffold (Nelson *et al.*, 1993). P-selectin binds to sialyl Lewis x-like carbohydrate displayed on P-selectin glycoprotein ligand-1 (PSGL-1). Although P- and E-selectin were original thought to be specific for neutrophils, it is now clear that certain subclasses of lymphocytes will bind and roll on these selectins (Alon *et al.*, 1994; Damle *et al.*, 1992; Moore and Thompson, 1992; Shimizu *et al.*, 1991). L-selectin on leukocytes can bind and roll on carbohydrates displayed on the mucin-like domain of MAdCAM-1 (Berg *et al.*, 1993).

Rolling adhesion is not exclusive to selectins, however; it has recently been shown that lymphocytes can roll on VCAM-1 and MAdCAM-1, endothelial Ig-superfamily (IGSF) molecules which, like ICAM-1, also mediate the later step of firm adhesion (Berlin *et al.*, 1995). Rolling has not been observed on ICAM-1, and perhaps would be undesirable, as ICAM-1 is present at a low to moderate level constitutively on endothelium, which should preserve a non-adhesive phenotype except in cases of inflammation or infection. Rolling on VCAM-1 and the IGSF domains of MAdCAM-1, like firm adhesion, is mediated by $\alpha4$ integrins on the leukocyte surface.

Activation of Leukocyte Integrins

Cells that roll on the endothelium are thought to be brought into contact with stimulatory molecules called chemokines that are secreted by the endothelium, and then held near the endothelial surface, perhaps by binding to carbohydrate matrices (Tanaka *et al.*, 1993). Each chemokine, of which more than 20 are now known, belongs to one of two related structural families, defined according to the spacing of the two amino-proximal cysteine residues: the a or C-X-C and the b or C-C chemokines. T lymphocytes are stimulated by the b or C-C chemokines MCP-1, -2, -3, MIP-1α and β, and RANTES (Carr *et al.*, 1994; Loetscher *et al.*, 1994; Schall *et al.*, 1993; Schall *et al.*, 1990; Taub *et al.*, 1993). Each of these acts on specific subpopulations of lymphocytes: the rolling leukocytes can sample the chemokines in a particular inflammatory display, and be selectively recruited depending on their ability to be stimulated by the particular chemokines present.

One important function of chemokine stimulation is to increase the affinity or avidity of leukocyte integrins for their endothelial counter-receptors. The integrins are a family of heterodimers consisting of an alpha and a beta chain, which bind to cell surface molecules of the IGSF and to extracellular matrix components such as laminin, fibronectin and collagen. Integrins $\alpha L\beta2$ (LFA-1), $\alpha4\beta1$ (VLA-4), and $\alpha4\beta7$ (LPAM) are important in inflammatory lymphocyte extravasation. Both $\beta2$ and $\alpha4$ integrins are constitutively present in a relatively inactive state on most resting leukocytes. When activated, $\alpha L\beta2$ binds to ICAM-1, one of the important trio of molecules up-regulated on the surface of endothelium in response to inflammatory mediators, described previously (Figure 1). Both $\alpha4\beta1$ and $\alpha4\beta7$ bind to VCAM-1 and to fibronectin, which are up-regulated in parallel with ICAM-1. The $\alpha4$ integrin relevant to lymphocyte recruitment during most cases of inflammation is probably $\alpha4\beta1$, although $\alpha4\beta7$, found mainly on a subclass of gut-homing lymphocytes, also can bind to VCAM-1 and fibronectin, and may play a role in inflammation as well as normal lymphocyte recirculation in the gut. MAdCAM-1, a selective ligand for $\alpha4\beta7$ whose IGSF integrin-binding domains resemble VCAM-1 and ICAM-1 in molecular structure, is a component of normal intestinal lymphoid tissue and may be overexpressed or otherwise up-regulated in activity in circumstances such as inflammatory bowel disease (IBD). The omnipresent but inactive leukocyte integrins are rapidly up-regulated upon contact with stimulatory molecules that include chemokines, adhesion molecules such as CD3/TCR, CD2, CD7, CD28, CD31, and CD44 (Chan *et al.*, 1991; Dustin and Springer, 1989; Koopman *et al.*, 1990; Shimizu *et al.*, 1992; Shimizu *et al.*, 1990; Tanaka *et al.*, 1992; van Kooyk *et al.*, 1989), and others (Hermanowski-Vosatka *et al.*, 1992).

It is not yet clear whether this up-regulation of affinity or avidity is accomplished by a change in the 3-dimensional structure or arrangement of the heterodimeric subunits, or to some kind of clustering effect. No crystal- or NMR-generated structure of a complete integrin chain or dimer has been solved to date. However, the structure of an isolated functional domain present in some alpha chains and perhaps (judging by amino acid homologies) in all known beta chains, called the I or A domain, has been solved by X-ray crystallography (Lee *et al.*, 1995). Furthermore, functional epitope mapping and mutagenic studies have been done on leukocyte integrins, allowing some understanding of the regions of the peptide chains likely to be involved in binding counter-receptor (Irie *et al.*, 1995; Kamata *et al.*, 1995; Landis *et al.*, 1993; Schiffer *et al.*, 1995).

"Inside-out" and "Outside-in" Integrin Signalling

Up-regulation of integrin avidity stimulated by phorbol ester (which presumable mimics that caused by binding to chemokines or adhesion molecules) depends upon the presence of the alpha chain cytoplasmic tails. Because this type of regulation depends upon the cytoplasmic portion of the molecule, but shows its effect in the ability of the extracellular portion of the molecule to bind ligand, this type of regulation is known as "inside-out" signalling. Alpha chain tails can be exchanged among (at least some) integrins without loss of phorbol ester up-regulation, but cannot be deleted . *In vitro*, affinity or avidity of integrins on lymphocytic cell lines can also be rapidly influenced by the presence of certain divalent cations: Mn++ causes maximum up-regulation of most integrins tested, while Mg++ and Ca++ have variable affects ranging from moderate up-regulation in some integrin/ligand combinations (eg. $\alpha4\beta1$ binding to VCAM-1) to down-regulation resulting in complete lack of ligand binding activity (eg. $\alpha4\beta1$ binding to MAdCAM-1). Divalent cation activation is sometimes but not always dependent upon the composition of the cytoplasmic tail (Kassner and Hemler, 1993; Kawaguchi and Hemler, 1993). Due to the difficulty of measuring cation concentrations in the microenvironment of the cell surface, it has not been determined whether this mechanism of activation is used *in vivo*.

Beta chain cytoplasmic tails also influence integrin function. When the $\beta1$ tail is replaced with that of $\beta5$ in transfected cells, proliferation in response to binding of ligand is eliminated, and haptotactic migration of the cells is increased (Pasqualini and Hemler, 1994). Effects on cell behavior caused by integrin binding to an extracellular ligand or antibody is termed "outside-in" signalling.

Firm Adhesion and Migration to the Endothelial Cell-cell Junction

When its integrins are activated, the rolling leukocyte binds firmly to the endothelium via $\beta2$ integrin/ICAM and/or $\alpha4$ integrin/VCAM-1. The leukocyte comes to a halt, and eventually crawls or is transported toward the junction between endothelial cells. It is unclear whether the active mechanism of movement is located in the T cell, in the endothelial cell, or both. In the case of neutrophils and monocytes, CD31 (PECAM), an IGSF molecule localized at the junctions between endothelial cells, is required for transmigration through the endothelial monolayer (Muller *et al.*, 1993).

T cells have been shown to lose $\alpha4$ integrin from their surface during transmigration (Romanic *et al.*, 1997). At the same time, proteases are activated, which are thought to aid the T cell in migration across the extracellular matrix underlying the endothelium, and through the tissue (Romanic and Madri, 1994). The adhesion molecules used by leukocytes in their migration through the tissue to the source of the chemotactic mediators are not well defined at present.

Table 2 Potential targets for anti-inflammatory drugs inhibiting lymphocyte recruitment.

- Cytokines (block TNFα, IL-1, etc. activation of endothelial cells)
- NF-κB (block transcriptional induction of endothelial-leukocyte adhesion molecules)
- Chemokines (block activation of lymphocyte integrins involved in emigration)
- Lymphocyte-endothelial adhesion molecules (block binding of lymphocytes to inflamed endothelium)
- Pro-migratory molecules (proteases, chemotactic factors, etc.) (block migration of lymphocytes through endothelium or subendothelial tissue)

THERAPEUTIC PROSPECTS

Because the number of patients affected by inflammatory disorders is large, much effort has been devoted to developing anti-inflammatory therapeutic agents which exhibit less severe side effects and more specific anti-inflammatory effects than those currently available. The multi-step nature of lymphocyte recruitment suggests that intervention might be possible at any of several steps; some of the possibilities are listed in Table 2. Potential therapies targetting these steps are in various stages of research and/or development in academic laboratories, biotechnology companies, and pharmaceutical firms world-wide. Only therapeutics directed toward adhesion molecules will be considered in the following discussion.

Monoclonal Antibodies

Historically, the use of function-blocking monoclonal antibodies (mAbs) specific for cell adhesion molecules has been the first strategy pursued. MAbs, with their great specificity and relative ease of production, are useful as experimental treatments in animal models of disease, and as reagents for experiments on the cell and molecular biology of cell surface molecules. Neutralizing antibody to the $\beta 2$ integrins such as LFA-1 ($\alpha L\beta 2$) and Mac-1 ($\alpha M\beta 2$) has demonstrated remarkable anti-inflammatory properties in numerous animal models of injury or disease (for review see Talbott *et al.*, 1994). Especially striking are its effects in injury due to ischemia followed by reperfusion. Anti-$\beta 2$ integrin mAb prevents emigration of activated neutrophils, which harm tissue under these circumstances, into the damaged tissue by blocking binding of neutrophil integrin $\alpha M\beta 2$ to ICAM-1 on endothelium. Models in which anti-$\beta 2$ mAb has been effective in preventing reperfusion injury include ischemia of the rabbit ear, hypovolemic shock in baboons (whole body ischemia caused by blood loss after injury), frostbite of extremities in the rabbit, and ischemia of the kidney. Treatment may be given concurrent with reperfusion, as might occur when an accident victim is given blood transfusions to replace lost blood, or even several hours after reperfusion, as might occur in the treatment of severe frostbite .

MAbs to several other leukocyte-endothelial adhesion molecules have been tested in various models of inflammation. A summary of these studies and references are given in Table 3. Blockade of ICAM-1, a counter-receptor of $\beta 2$ integrins, not surprisingly is effective in some of the same models of inflammation as $\beta 2$ blockade, described above; it has been extensively tested in models of acute and chronic inflammation as well as ischemia/reperfusion. P-, L-, or E-selectin blockade have been shown to decrease neutrophil recruitment in numerous models of acute inflammation.

Table 3 Animal Models of Anti-Adhesion Therapy.
Models in Which Mononuclear Leukocytes Mediate Inflammation.

MAb Used	Observation	References
1. Organ transplantation		
ICAM-1	Partial prevention of rejection in cardiac allograft	(Bashuda *et al.*, 1996; Isobe and Ihara, 1993; Isobe *et al.*, 1992; Komori *et al.*, 1993; Willenborg *et al.*, 1993)
ICAM-1	Partial prevention of renal allograft rejection	(Cosimi *et al.*, 1990)
ICAM-1	Delayed skin graft rejection	(Gorczynski *et al.*, 1996; Gorczynski and Wojcik, 1994)
ICAM-1	With anti-LFA-1 mAb, prevented rejection of nerve allograft	(Nakao *et al.*, 1995)
ICAM-1	With anti-LFA-1 mAb, prevented rejection of bowel allograft	(Kato *et al.*, 1995; Kato *et al.*, 1996)
ICAM-1	With anti-LFA-1 mAb, reduced cardiac allograft-associated atherosclerosis	(Russell *et al.*, 1995)
ICAM-1	Delayed leukocyte infiltration of islet xenograft	(Zeng *et al.*, 1994)
VCAM-1	Prevention of graft rejection in cardiac allograft	(Orosz *et al.*, 1993; Pelletier *et al.*, 1993)
$\alpha 4$ integrin	Slight prolongation of cardiac allograft survival	(Paul *et al.*, 1993; Paul *et al.*, 1993)
$\alpha 4$ integrin	Decreased leukocyte infiltration and vasculitis in cardiac allograft	(Paul *et al.*, 1993)
$\alpha 4$ integrin	With mAbs to VCAM-1, delayed skin graft rejection	(Gorczynski *et al.*, 1996)
$\alpha 4$ integrin	With mAbs to LFA-1, delayed islet graft rejection	(Yang *et al.*, 1995)
$\alpha 4$ integrin	Partially prevented mucosal pathology of intestinal graft-versus-host disease	(Tanaka *et al.*, 1995)
$\alpha 1$ integrin	Partially prevented mucosal pathology of intestinal graft-versus-host disease	(Tanaka *et al.*, 1995)
$\beta 7$ integrin	Partially prevented mucosal pathology of intestinal graft-versus-host disease	(Tanaka *et al.*, 1995)
2. Autoimmune/chronic inflammatory disorder		
ICAM-1	Partial inhibition of leukocyte recruitment in collagen- or adjuvant- induced arthritis	(Iigo *et al.*, 1991; Kakimoto *et al.*, 1992)
ICAM-1	Partial inhibition of mononuclear leukocyte recruitment in antiglomerular basement membrane form of glomerulonephritis	(Kawasaki *et al.*, 1993)

Table 3 *Continued.*

MAb Used	Observation	References
ICAM-1	Decreased leukocytic infiltration and suppressed the development of experimental autoimmune encephalitis (EAE)	(Archelos *et al.*, 1993)
ICAM-1	Decreased the development of active, but not passively transferred, EAE	(Willenborg *et al.*, 1993)
ICAM-1	When administered with CD11a MoAb, prevented progression of crescentic glomerulonephritis	(Nishikawa *et al.*, 1993)
ICAM-1	Reduced leukocytic infiltration and proteinuria in model of autoimmune nephritis	(Harning *et al.*, 1992)
ICAM-1	Partial reduction in number of mononuclear leukocytes at site of delayed-type hypersensitivity (DTH) reaction	(Scheynius *et al.*, 1993)
ICAM-1	Loss of DTH reactivityand clinical improvement in rheumatoid arthritis patients	(Davis *et al.*, 1995; Kavanaugh *et al.*, 1994)
ICAM-1	With anti-LFA-1 mAb, prevented development of insulitis in viral induced diabetes-like syndrome	(Hayashi *et al.*, 1995)
ICAM-1	With anti-LFA-1 mAb, decreased leukocytic infiltration and suppressed the development of experimental autoimmune encephalitis (EAE)	(Kobayashi *et al.*, 1995)
ICAM-1	Inhibited development of experimental uveitis	(Uchio *et al.*, 1994; Whitcup *et al.*, 1995)
ICAM-1	With anti-LFA-1 mAb, reduced ischemia-reperfusion injury of the myocardium	(Tamiya *et al.*, 1995)
ICAM-1	Inhibited acute immune complex injury of lung	(Mulligan *et al.*, 1995)
ICAM-1	Prevented injury due to ischemia/reperfusion in kidney	(Kelly *et al.*, 1994)
ICAM-1	Inhibited acute IgG complex injury of lung	(Mulligan *et al.*, 1993)
VCAM-1	Decreased leukocyte entry into CNS in EAE model	(Yednock *et al.*, 1992)
VCAM-1	With anti-$\alpha 4$ integrin mAbs, prevented lymphocytic infiltration of islets and overt diabetes in non-obese diabetic (NOD) mice	(Tsukamoto *et al.*, 1995)
$\alpha 4$ integrin	Decreased migration to sites of DTH reaction	(Chisholm *et al.*, 1993; Elices *et al.*, 1993; Issekutz and Issekutz, 1991)
$\alpha 4$ integrin	Decreased lymphocyte accumulation in arthritic joints	(Issekutz and Issekutz, 1991)
$\alpha 4$ integrin	Decreased leukocytic infiltration in model of colitis	Podolsky, 1993 #11

Table 3 *Continued.*

MAb Used	Observation	References
$\alpha 4$ integrin	Decreased accumulation of leukocytes in model of EAE	(Baron *et al.*, 1993)
$\alpha 4$ integrin	Decreased blood-brain barrier permeability changes in model of EAE	(Kent *et al.*, 1995)
$\alpha 4$ integrin	Partially decreased migration of smooth muscle cells and leukocytes into damaged arterial intima	(Kling *et al.*, 1995)
$\alpha 4$ integrin	Inhibited lymphocytic infiltration of islets in NOD mice	(Yang *et al.*, 1993; Yang *et al.*, 1994; Yang *et al.*, 1994)
$\alpha 4$ integrin	With anti-LFA-1 mAbs, decreased monocyte accumulation in arthritic joints	(Issekutz and Issekutz, 1995)
$\alpha 4$ integrin	Decreased mercury-induced glomerular and interstitial nephritis	(Molina *et al.*, 1994)
$\alpha 4$ integrin	Inhibited acute IgG complex injury of lung	(Mulligan *et al.*, 1993)
E-selectin	Inhibited acute IgG complex injury of lung	(Mulligan *et al.*, 1993)
L-selectin	Decreased lymphocyte migration to sites of DTH reaction	(Dawson *et al.*, 1992)
L-selectin	Inhibited lymphocytic infiltration of islets in NOD mice	(Yang *et al.*, 1993; Yang *et al.*, 1994a; Yang *et al.*, 1994b)
L-selectin	Inhibited acute immune complex injury of lung	(Mulligan *et al.*, 1995)
L-selectin	Reduced leukocyte accumulation in thioglycollate-induced inflammation of the peritoneum	(Pizcueta and Luscinskas, 1994)
Models in Which Eosinophils Mediate Inflammation		
ICAM-1	Decreased accumulation of eosinophils in model of antigen-induced asthma	(Wegner *et al.*, 1990)
ICAM-1	No inhibition of eosinophil recruitment in primate model of asthma unless animals were withdrawn from dexamethasone	(Gundel *et al.*, 1992)
$\alpha 4$ integrin	Decreased late phase airway hyperresponsiveness after antigen challenge	(Abraham *et al.*, 1994)
$\alpha 4$ integrin	Decreased eosinophilic infiltration in passive cutaneous anaphylaxis reaction	(Weg *et al.*, 1993)
$\alpha 4$ integrin	Decreased inflammation and lymphocyte and eosinophil infiltration into airway lumen in antigen-induced lung inflammation	(Richards *et al.*, 1996)

Adapted with permission from Carlos and Harlan, 1994.
For review of anti-adhesion therapy directed to CD11/CD18, see Talbott *et al.*, 1994.

Chronic, mononuclear cell-mediated inflammation of the DTH type has also been reduced by administration of L-selectin mAbs. Chronic inflammation, which tends to be mediated not by neutrophils but by mononuclear cells, can be reduced by treatment with mAbs to ICAM-1 (presumably by blocking interaction with integrins $\alpha L\beta 2$ on lymphocytes and/or $\alpha M\beta 2$ on monocytes). MAbs to VCAM-1 or its counter-receptor integrin $\alpha 4\beta 1$ are especially successful in attenuating DTH reactions, experimental allergic encephalomyelitis (EAE, a model for multiple sclerosis) (Yednock *et al.*, 1992), and airway hyperresponsiveness associated with antigen-induced asthma (Abraham *et al.*, 1994). Better results are often seen with $\alpha 4$ mAbs than with VCAM-1 mAbs. It is unclear whether this is due to an additional blockade of $\alpha 4$ integrin interactions with molecules other than VCAM-1 (such as MAdCAM-1 or fibronectin), or whether there is some functional advantage to targeting the leukocyte instead of the endothelial cell.

Although mAbs are excellent reagents and prototypes for "proof of concept" in animal disease models, they may not make ideal drugs, expecially for chronic use. Murine mAbs are always immunogenic, and humanized mAbs (made by transferring the antigen combining sites of murine mAbs to a human Ig framework) often elicit an anti-idiotype response directed against the antigen recognition sequences, which would render the drug useless in a particular patient after the first few doses. Production of any large protein such as an Ig is relatively expensive, and the only practical administration route is injection. Finally, the half-life of an Ig in the serum is usually at least 24 hours: this long half-life would be a disadvantage in case of an adverse reaction by a patient to the mAb. To counter these problems, alternative means of specifically blocking adhesion pathways have been explored.

Ig Chimerae

Ig chimerae consisting of the active portion of an adhesion molecule fused to a human Ig constant region (Fc) have been used with some success as competitive inhibitors of leukocyte-endothelial binding (Jakubowski *et al.*, 1995; Mulligan *et al.*, 1993; Watson *et al.*, 1991). These molecules are thought to inhibit better than the recombinant soluble adhesion molecule moiety alone due to increased avidity provided by their bivalence. The Fc portion is also convenient for purification and detection purposes, and provides a long half-life in the serum, a mixed blessing as noted above. Finally they are not expected to be immunogenic, as only the join between the adhesion molecule and Fc moiety is novel to the recipient's immune system. Ig chimerae, like any large protein, would be quite expensive to produce and administer as drugs.

Antisense Oligonucleotides

Compounds with a shorter half-life would include anti-sense oligomers for preventing the synthesis of those adhesion molecules which are made *de novo* in inflammation, such as ICAM-1, VCAM-1, and E-selectin. All three of these pathways have successfully been inhibited *in vitro* (Bennett *et al.*, 1994), but no *in vivo* studies have yet been reported, although animal studies showing successful anti-sense blocking of vascular smooth muscle proliferatiion has been reported using *myb* antisense oligonucleotide (Simons *et al.*, 1992).

Oligosaccharides

In the case of the selectins, soluble oligosaccharides displaying sialyl Lewis x can block adhesion. These compounds have shown protective effects in two models of inflamatory lung injury (Mulligan *et al.*, 1993; Mulligan *et al.*, 1993), and in models of reperfusion injury (Buerke *et al.*, 1994; Han *et al.*, 1995).

Peptides and Small Molecules

Peptides derived from the binding sites of selectins and of ICAM and VCAM have been shown to block binding of these molecules to their counter-receptors, *in vitro* (Geng *et al.*, 1992; Heavner *et al.*, 1993; Ross *et al.*, 1992; Stanley *et al.*, 1994; Wang *et al.*, 1995). Peptides are relatively inexpensive to manufacture, and have a short half-like in serum. They are usually low in potency, and so may best be used as lead compounds in synthesizing specific, potent small molecules that mimic the peptide's effect. Such small molecules have many advantages: in addition to increased potency, they will be less likely to prompt a host immune response, they may be orally available, and they are usually relatively inexpensive to manufacture, store, and administer compared to a large protein or a peptide. However, the process of developing such small inhibitors, as distinguished from manufacturing them, is even more laborious and expensive than that for proteins, and is beyond the resources of an academic laboratory or of many small companies. Therefore we might see a "first generation" of large protein drugs, whose development costs are less, before the arrival of sophisticated small molecule inhibitors to the marketplace.

CONCLUSION

In the last 10 years many molecules involved in the recruitment of leukocytes to sites of inflammation have been identified and characterized. Understanding leukocyte recruitment at the molecular level has allowed the generation and testing of reagents that influence the recruitment process. Academic laboratories, biotechnology companies and pharmaceutical firms have cooperated in the process of turning these discoveries into useful therapeutics. The next five years should see the arrival of the first generation of drugs based on this technology into general medical practice, closely followed by more refined medicines addressing patients' needs for lower costs and convenient administration, as well as increased efficacy.

REFERENCES

Abraham, W.M., Sielczak, M.W., Ahmed, A., Cortes, A., Lauredo, I.T., Kim, J., Pepinsky, B., Benjamin, C.D., Leone, D.R., Lobb, R.R. and Weller, P.F. (1994). Alpha 4-integrins mediate antigen-induced late bronchial responses and prolonged airway hyperresponsiveness in sheep. *J. Clin. Invest.*, **93**, 776–787.

Alon, R., Rossiter, H., Wang, X., Springer, T.A. and Kupper, T.S. (1994). Distinct cell surface ligands mediate T lymphocyte attachment and rolling on P and E selectin under physiological flow. *J. Cell Biol.*, **127**, 1485–1495.

Anderson, D.C., Schmalsteig, F.C., Finegold, M.J., Hughes, B.J., Rothlein, R., Miller, L.J., Kohl, S., Tosi, M.F., Jacobs, R.L., Waldrop, T.C., *et al.* (1985). The severe and moderate phenotypes of heritable Mac-1, LFA-1 deficiency: their quantitative definition and relation to leukocyte dysfunction and clinical features. *J. Infect. Dis.*, **152**, 668–689.

Archelos, J.J., Jung, S., Maurer, M., Schmied, M., Lassmann, H., Tamatani, T., Miyasaka, M., Toyka, K.V. and Hartung, H.P. (1993). Inhibition of experimental autoimmune encephalomyelitis by an antibody to the intercellular adhesion molecule ICAM-1. *Ann. Neurol.*, **34**, 145–154.

Archelos, J.J., Maurer, M., Jung, S., Miyasaka, M., Tamatani, T., Toyka, K.V. and Hartung, H.P. (1994). Inhibition of experimental autoimmune neuritis by an antibody to the lymphocyte function-associated antigen-1. *Lab. Invest.*, **70**, 667–675.

Baron, J.L., Madri, J.A., Ruddle, N.H., Hashim, G. and Janeway, C.A., Jr. (1993). Surface expression of alpha 4 integrin by CD4 T cells is required for their entry into brain parenchyma. *J. Exp. Med.*, **177**, 57–68.

Bashuda, H., Takazawa, K., Tamatani, T., Miyasaka, M., Yagita, H. and Okumura, K. (1996). Induction of persistent allograft tolerance in the rat by combined treatment with anti-leukocyte function-associated antigen-1 and anti-intercellular adhesion molecule-1 monoclonal antibodies, donor-specific transfusion, and FK506. *Transplantation*, **62**, 117–122.

Bennett, C.F., Condon, T.P., Grimm, S., Chan, H. and Chiang, M.Y. (1994). Inhibition of endothelial cell adhesion molecule expression with antisense oligonucleotides. *J. Immunol.*, **152**, 3530–3540.

Berg, E.L., McEvoy, L.M., Berlin, C., Bargatze, R.F. and Butcher, E.C. (1993). L-selectin-mediated lymphocyte rolling on MAdCAM-1. *Nature*, **366**, 695–698.

Berlin, C., Bargatze, R.F., Campbell, J.J., von Andrian, U.H., Szabo, M.C., Hasslen, S.R., Nelson, R.D., Berg, E.L., Erlandsen, S.L. and Butcher, E.C. (1995). Alpha 4 integrins mediate lymphocyte attachment and rolling under physiologic flow. *Cell*, **80**, 413–422.

Blazar, B.R., Taylor, P.A., Panoskaltsis-Mortari, A., Gray, G.S. and Vallera, D.A. (1995). Coblockade of the LFA1:ICAM and CD28/CTLA4:B7 pathways is a highly effective means of preventing acute lethal graft-versus-host disease induced by fully major histocompatibility complex-disparate donor grafts. *Blood*, **85**, 2607–2618.

Bloemen, P.G., Buckley, T.L., van den Tweel, M.C., Henricks, P.A., Redegeld, F.A., Koster, A.S. and Nijkamp, F.P. (1996). LFA-1, and not Mac-1, is crucial for the development of hyperreactivity in a murine model of nonallergic asthma. *Am. J. Respir. Crit. Care Med.*, **153**, 521–529.

Bonfanti, R., Furie, B.C., Furie, B. and Wagner, D.D. (1989). PADGEM (GMP140) is a component of Weibel-Palade bodies of human endothelial cells. *Blood*, **73**, 413–422.

Buerke, M., Weyrich, A.S., Zheng, Z., Gaeta, F.C., Forrest, M.J. and Lefer, A.M. (1994). Sialyl Lewisx-containing oligosaccharide attenuates myocardial reperfusion injury in cats. *J. Clin. Invest.*, **93**, 1140–1148.

Butcher, E.C. (1991). Leukocyte-endothelial cell recognition: three (or more) steps to specificity and diversity. *Cell*, **67**, 1033–1036.

Butcher, E.C. and Picker, L.J. (1996). Lymphocyte homing and homeostasis. *Science*, **272**, 60–66.

Cannella, B., Cross, A.H. and Raine, C.S. (1993). Anti-adhesion molecule therapy in experimental autoimmune encephalomyelitis. *J. Neuroimmunol.*, **46**, 43–55.

Carlos, T.M. and Harlan, J.M. (1994). Leukocyte-endothelial adhesion molecules. *Blood*, **84**, 2068–2101.

Carr, M.W., Roth, S.J., Luther, E., Rose, S.S. and Springer, T.A. (1994). Monocyte chemoattractant protein 1 acts as a T-lymphocyte chemoattractant. *Proc. Natl. Acad. Sci. USA*, **91**, 3652–3656.

Chan, B.M., Wong, J.G., Rao, A. and Hemler, M.E. (1991). T cell receptor-dependent, antigen-specific stimulation of a murine T cell clone induces a transient, VLA protein-mediated binding to extracellular matrix. *J. Immunol.*, **147**, 398–404.

Chisholm, P.L., Williams, C.A. and Lobb, R.R. (1993). Monoclonal antibodies to the integrin alpha-4 subunit inhibit the murine contact hypersensitivity response. *Eur J. Immunol.*, **23**, 682–688.

Cosimi, A.B., Conti, D., Delmonico, F.L., Preffer, F.I., Wee, S.L., Rothlein, R., Faanes, R. and Colvin, R.B. (1990). *In vivo* effects of monoclonal antibody to ICAM-1 (CD54) in nonhuman primates with renal allografts. *J. Immunol.*, **144**, 4604–4612.

Damle, N.K., Klussman, K., Dietsch, M.T., Mohagheghpour, N. and Aruffo, A. (1992). GMP-140 (P-selectin/CD62) binds to chronically stimulated but not resting CD4+ T lymphocytes and regulates their production of proinflammatory cytokines. *Eur. J. Immunol.*, **22**, 1789–1793.

Davis, L.S., Kavanaugh, A.F., Nichols, L.A. and Lipsky, P.E. (1995). Induction of persistent T cell hyporesponsiveness in vivo by monoclonal antibody to ICAM-1 in patients with rheumatoid arthritis. *J. Immunol.,* **154**, 3525–3537.

Dawson, J., Sedgwick, A.D., Edwards, J.C. and Lees, P. (1992). The monoclonal antibody MEL-14 can block lymphocyte migration into a site of chronic inflammation. *Eur J. Immunol.,* **22**, 1647–1650.

Dustin, M.L. and Springer, T.A. (1989). T-cell receptor cross-linking transiently stimulates adhesiveness through LFA-1. *Nature,* **341**, 619–624.

Elices, M.J., Tamraz, S., Tollefson, V. and Vollger, L.W. (1993). The integrin VLA-4 mediates leukocyte recruitment to skin inflammatory sites in vivo. *Clin. Exp. Rheumatol.,* **11**, S77–80.

Etzioni, A., Frydman, M., Pollack, S., Avidor, I., Phillips, M.L., Paulson, J.C. and Gershoni-Baruch, R. (1992). Brief report: recurrent severe infections caused by a novel leukocyte adhesion deficiency. *N. Engl. J. Med.*, **327**, 1789–1792.

Fischer, A., Landais, P., Friedrich, W., Gerritsen, B., Fasth, A., Porta, F., Vellodi, A., Benkerrou, M., Jais, J.P. and Cavazzana-Calvo, M., *et al.* (1994). Bone marrow transplantation (BMT) in Europe for primary immunodeficiencies other than severe combined immunodeficiency: a report from the European Group for BMT and the European Group for Immunodeficiency. *Blood,* **83**, 1149–1154.

Flavin, T., Ivens, K., Rothlein, R., Faanes, R., Clayberger, C., Billingham, M. and Starnes, V.A. (1991). Monoclonal antibodies against intercellular adhesion molecule 1 prolong cardiac allograft survival in cynomolgus monkeys. *Transplant Proc.*, **23**, 533–534.

Geng, J.G., Heavner, G.A. and McEver, R.P. (1992). Lectin domain peptides from selectins interact with both cell surface ligands and Ca2+ ions. *J. Biol. Chem.*, **267**, 19846–19853.

Gorczynski, R.M., Chung, S., Hoang, Y., Sullivan, B. and Chen, Z. (1996). Altered patterns of migration of cytokine-producing T lymphocytes in skin-grafted naive or immune mice following in vivo administration of anti-VCAM-1 or -ICAM-1. *Immunology,* **87**, 573–580.

Gorczynski, R.M. and Wojcik, D. (1994). A role for nonspecific (cyclosporin A) or specific (monoclonal antibodies to ICAM-1, LFA-1, and IL-10) immunomodulation in the prolongation of skin allografts after antigen-specific pretransplant immunization or transfusion. *J. Immunol.*, **152**, 2011–2019.

Gotoh, M., Fukuzaki, T., Monden, M., Dono, K., Kanai, T., Yagita, H., Okumura, K. and Mori, T. (1994). A potential immunosuppressive effect of anti-lymphocyte function-associated antigen-1 monoclonal antibody on islet transplantation. *Transplantation,* **57**, 123–126.

Gundel, R.H., Wegner, C.D., Torcellini, C.A. and Letts, L.G. (1992). The role of intercellular adhesion molecule-1 in chronic airway inflammation. *Clin. Exp. Allergy,* **22**, 569–575.

Han, K.T., Sharar, S.R., Phillips, M.L., Harlan, J.M. and Winn, R.K. (1995). Sialyl Lewis(x) oligosaccharide reduces ischemia-reperfusion injury in the rabbit ear. *J. Immunol.*, **155**, 4011–4015.

Harning, R., Pelletier, J., Van, G., Takei, F., Merluzzi, V.J. (1992). Monoclonal antibody to MALA-2 (ICAM-1) reduces acute autoimmune nephritis in kdkd mice. *Clin. Immunol. Immunopathol.*, **64**, 129–134.

Hayashi, T., Yamamoto, S. and Onodera, T. (1995). Prevention of reovirus type 2–induced diabetes-like syndrome in DBA/1 suckling mice by treatment with antibodies against intercellular adhesion molecule-1 and lymphocyte function-associated antigen-1. *Int. J. Exp. Pathol.*, **76**, 403–409.

He, Y., Mellon, J., Apte, R. and Niederkorn, J.Y. (1994). Effect of LFA-1 and ICAM-1 antibody treatment on murine corneal allograft survival. *Invest. Ophthalmol. Vis. Sci.*, **35**, 3218–3225.

Heavner, G.A., Falcone, M., Kruszynski, M., Epps, L., Mervic, M., Riexinger, D. and McEver, R.P. (1993). Peptides from multiple regions of the lectin domain of P-selectin inhibiting neutrophil adhesion. *Int. J. Pept. Protein Res.*, **42**, 484–489.

Hermanowski-Vosatka, A., Van Strijp, J.A., Swiggard, W.J. and Wright, S.D. (1992). Integrin modulating factor-1: a lipid that alters the function of leukocyte integrins. *Cell*, **68**, 341–352.

Hershkoviz, R., Greenspoon, N., Mekori, Y.A., Hadari, R., Alon, R., Kapustina, G. and Lider, O. (1994). Inhibition of CD4+ T lymphocyte binding to fibronectin and immune-cell accumulation in inflammatory sites by non-peptidic mimetics of Arg-Gly-Asp. *Clin. Exp. Immunol.*, **95**, 270–276.

Iademarco, M.F., McQuillan, J.J., Rosen, G.D. and Dean, D.C. (1992). Characterization of the promoter for vascular cell adhesion molecule-1 (VCAM-1). *J. Biol. Chem.*, **267**, 16323–16329.

Iigo, Y., Takashi, T., Tamatani, T., Miyasaka, M., Higashida, T., Yagita, H., Okumura, K. and Tsukada, W. (1991). ICAM-1-dependent pathway is critically involved in the pathogenesis of adjuvant arthritis in rats. *J. Immunol.*, **147**, 4167–4171.

Irie, A., Kamata, T., Puzon-McLaughlin, W. and Takada, Y. (1995). Critical amino acid residues for ligand binding are clustered in a predicted beta-turn of the third N-terminal repeat in the integrin alpha 4 and alpha 5 subunits. *EMBO J.*, **14**, 5550–5556.

Isobe, M. and Ihara, A. (1993). Tolerance induction against cardiac allograft by anti-ICAM-1 and anti-LFA-1 treatment: T cells respond to in vitro allostimulation. *Transplant Proc.* **25**, 1079–1080.

Isobe, M., Yagita, H., Okumura, K. and Ihara, A. (1992). Specific acceptance of cardiac allograft after treatment with antibodies to ICAM-1 and LFA-1. *Science*, **255**, 1125–1127.

Issekutz, A.C. and Issekutz, T.B. (1995). Monocyte migration to arthritis in the rat utilizes both CD11/CD18 and very late activation antigen 4 integrin mechanisms. *J. Exp. Med.*, **181**, 1197–1203.

Issekutz, T.B. and Issekutz, A.C. (1991). T lymphocyte migration to arthritic joints and dermal inflammation in the rat: differing migration patterns and the involvement of VLA-4. *Clin. Immunol. Immunopathol.*, **61**, 436–447.

Jakubowski, A., Ehrenfels, B.N., Pepinsky, R.B. and Burkly, L.C. (1995). Vascular cell adhesion molecule-Ig fusion protein selectively targets activated alpha 4–integrin receptors in vivo. Inhibition of autoimmune diabetes in an adoptive transfer model in nonobese diabetic mice. *J. Immunol.*, **155**, 938–946.

Johnston, G.I., Cook, R.G. and McEver, R.P. (1989a). Cloning of GMP-140, a granule membrane protein of platelets and endothelium: sequence similarity to proteins involved in cell adhesion and inflammation. *Cell*, **56**, 1033–1044.

Johnston, G.I., Kurosky, A. and McEver, R.P. (1989b). Structural and biosynthetic studies of the granule membrane protein, GMP- 140, from human platelets and endothelial cells. *J. Biol. Chem.*, **264**, 1816–1823.

Kakimoto, K., Nakamura, T., Ishii, K., Takashi, T., Iigou, H., Yagita, H., Okumura, K. and Onoue, K. (1992). The effect of anti-adhesion molecule antibody on the development of collagen-induced arthritis. *Cell. Immunol.*, **142**, 326–337.

Kamata, T., Puzon, W. and Takada, Y. (1995). Identification of putative ligand-binding sites of the integrin alpha 4 beta 1 (VLA-4, cd49d/CD29). *Biochem. J.*, **305**, 945–951.

Kassner, P.D. and Hemler, M.E. (1993). Interchangeable alpha chain cytoplasmic domains play a positive role in control of cell adhesion mediated by VLA-4, a beta 1 integrin. *J. Exp. Med.*, **178**, 649–660.

Kato, Y., Yamataka, A., Yagita, H., Bashuda, H., Okumura, K. and Miyano, T. (1995). Prevention of fetal bowel allograft rejection by combined treatment with anti-ICAM-1 and anti-LFA-1 antibodies. *J. Pediatr. Surg.*, **30**, 1093–1097.

Kato, Y., Yamataka, A., Yagita, H., Okumura, K., Fujiwara, T. and Miyano, T. (1996). Specific acceptance of fetal bowel allograft in mice after combined treatment with anti-intercellular adhesion molecule-1 and leukocyte function-associated antigen-1 antibodies. *Ann. Surg*, **223**, 94–100.

Kavanaugh, A.F., Davis, L.S., Nichols, L.A., Norris, S.H., Rothlein, R., Scharschmidt, L.A. and Lipsky, P.E. (1994). Treatment of refractory rheumatoid arthritis with a monoclonal antibody to intercellular adhesion molecule 1. *Arthritis Rheum.*, **37**, 992–999.

Kawaguchi, S. and Hemler, M.E. (1993). Role of the alpha subunit cytoplasmic domain in regulation of adhesive activity mediated by the integrin VLA-2. *J. Biol. Chem.*, **268**, 16279–16285.

Kawasaki, K., Yaoita, E., Yamamoto, T., Tamatani, T., Miyasaka, M. and Kihara, I. (1993). Antibodies against intercellular adhesion molecule-1 and lymphocyte function-associated antigen-1 prevent glomerular injury in rat experimental crescentic glomerulonephritis. *J. Immunol.*, **150**, 1074–1083.

Kelly, K.J., Williams, W.W., Jr., Colvin, R.B. and Bonventre, J.V. (1994). Antibody to intercellular adhesion molecule 1 protects the kidney against ischemic injury. *Proc. Natl. Acad. Sci. U S A*, **91**, 812–816.

Kent, S.J., Karlik, S.J., Rice, G.P., Horner, H.C. (1995). A monoclonal antibody to α4-integrin reverses the MR-detectable signs of experimental allergic encephalomyelitis in the guinea pig. *J. Magn. Reson. Imaging*, **5**, 535–540.

Kling, D., Fingerle, J., Harlan, J.M., Lobb, R.R. and Lang, F. (1995). Mononuclear leukocytes invade rabbit arterial intima during thickening formation via CD18- and VLA-4-dependent mechanisms and stimulate smooth muscle migration. *Circ. Res*, 77, 1121–1128.

Kobayashi, Y., Kawai, K., Honda, H., Tomida, S., Niimi, N., Tamatani, T., Miyasaka, M. and Yoshikai, Y. (1995). Antibodies against leukocyte function-associated antigen-1 and against intercellular adhesion molecule-1 together suppress the progression of experimental allergic encephalomyelitis. *Cell.Immunol.*, **164**, 295–305.

Komori, A., Nagata, M., Ochiai, T., Nakajima, K., Hori, S., Asano, T., Isono, K., Tamatani, T. and Miyasaka, M. (1993). Role of ICAM-1 and LFA-1 in cardiac allograft rejection of the rat. *Transplant Proc.*, **25**, 831–832.

Koopman, G., van Kooyk, Y., de Graaff, M., Meyer, C.J., Figdor, C.G. and Pals, S.T. (1990). Triggering of the CD44 antigen on T lymphocytes promotes T cell adhesion through the LFA-1 pathway. *J. Immunol.*, **145**, 3589–3593.

Landis, R.C., Bennett, R.I. and Hogg, N. (1993). A novel LFA-1 activation epitope maps to the I domain. *J. Cell Biol.*, **120**, 1519–1527.

Lee, J.O., Rieu, P., Arnaout, M.A. and Liddington, R. (1995). Crystal structure of the A domain from the alpha subunit of integrin CR3 (CD11b/CD18). *Cell*, **80**, 631–638.

Loetscher, P., Seitz, M., Clark-Lewis, I., Baggiolini, M. and Moser, B. (1994). Monocyte chemotactic proteins MCP-1, MCP-2, and MCP-3 are major attractants for human CD4+ and CD8+ T lymphocytes. *FASEB J.*, **8**, 1055–1060.

Molina, A., Sanchez-Madrid, F., Bricio, T., Martin, A., Barat, A., Alvarez, V. and Mampaso, F. (1994). Prevention of mercuric chloride-induced nephritis in the brown Norway rat by treatment with antibodies against the alpha 4 integrin. *J. Immunol.*, **153**, 2313–2320.

Montgomery, K.F., Osborn, L., Hession, C., Tizard, R., Goff, D., Vassallo, C., Tarr, P.I., Bomsztyk, K., Lobb, R., Harlan, J.M., *et al.* (1991). Activation of endothelial-leukocyte adhesion molecule 1 (ELAM-1) gene transcription. *Proc. Natl. Acad. Sci. U S A*, **88**, 6523–6527.

Moore, K.L. and Thompson, L.F. (1992). P-selectin (CD62) binds to subpopulations of human memory T lymphocytes and natural killer cells. *Biochem. Biophys. Res. Commun.*, **186**, 173–181.

Morikawa, M., Tamatani, T., Miyasaka, M., Uede, T. (1994). Cardiac allografts in rat recipients with simultaneous use of anti-ICAM-1 and anti-LFA-1 monoclonal antibodies leads to accelerated graft loss. *Immunopharmacology*, **28**, 171–182.

Muller, W.A., Weigl, S.A., Deng, X. and Phillips, D.M. (1993). PECAM-1 is required for transendothelial migration of leukocytes. *J. Exp. Med.* , **178**, 449–460.

Mulligan, M.S., Lowe, J.B., Larsen, R.D., Paulson, J., Zheng, Z.L., DeFrees, S., Maemura, K., Fukuda, M. and Ward, P.A. (1993). Protective effects of sialylated oligosaccharides in immune complex-induced acute lung injury. *J. Exp. Med.* , **178**, 623–631.

Mulligan, M.S., Paulson, J.C., De Frees, S., Zheng, Z.L., Lowe, J.B. and Ward, P.A. (1993). Protective effects of oligosaccharides in P-selectin-dependent lung injury. *Nature*, **364**, 149–151.

Mulligan, M.S., Smith, C.W., Anderson, D.C., Todd, R.F.d., Miyasaka, M., Tamatani, T., Issekutz, T.B. and Ward, P.A. (1993). Role of leukocyte adhesion molecules in complement-induced lung injury. *J. Immunol.*, **150**, 2401–2406.

Mulligan, M.S., Vaporciyan, A.A., Warner, R.L., Jones, M.L., Foreman, K.E., Miyasaka, M., Todd, R.F.R. and Ward, P.A. (1995). Compartmentalized roles for leukocytic adhesion molecules in lung inflammatory injury. *J. Immunol.*, **154**, 1350–1363.

Mulligan, M.S., Watson, S.R., Fennie, C. and Ward, P.A. (1993). Protective effects of selectin chimeras in neutrophil-mediated lung injury. *J. Immunol.*, **151**, 6410–6417.

Mulligan, M.S., Wilson, G.P., Todd, R.F., Smith, C.W., Anderson, D.C., Varani, J., Issekutz, T.B., M iyasaka, M., Tamatani, T., Myasaka, M., *et al.* (1993). Role of beta 1, beta 2 integrins and ICAM-1 in lung injury after deposition of IgG and IgA immune complexes [published erratum appears in J Immunol 1993 Jun 1;150(11):5209]. *J. Immunol.*, **150**, 2407–2417.

Nakao, Y., Mackinnon, S.E., Strasberg, S.R., Hertl, M.C., Isobe, M., Susskind, B.M., Mohanakumar, T. and Hunter, D.A. (1995). Immunosuppressive effect of monoclonal antibodies to ICAM-1 and LFA-1 on peripheral nerve allograft in mice. *Microsurgery*, **16**, 612–620.

Nelson, R.M., Dolich, S., Aruffo, A., Cecconi, O. and Bevilacqua, M.P. (1993). Higher-affinity oligosaccharide ligands for E-selectin. *J. Clin. Invest..*, **91**, 1157–1166.

Nishikawa, K., Guo, Y.J., Miyasaka, M., Tamatani, T., Collins, A.B., Sy, M.S., McCluskey, R.T. and Andres, G. (1993). Antibodies to intercellular adhesion molecule 1/lymphocyte function- associated antigen 1 prevent crescent formation in rat autoimmune glomerulonephritis. *J. Exp. Med.* , 177, 667–677.

Orosz, C.G., Ohye, R.G., Pelletier, R.P., Van Buskirk, A.M., Huang, E., Morgan, C., Kincade, P.W. and Ferguson, R.M. (1993). Treatment with anti-vascular cell adhesion molecule 1 monoclonal antibody induces long-term murine cardiac allograft acceptance. *Transplantation*, **56**, 453–460.

Osborn, L., Hession, C., Tizard, R., Vassallo, C., Luhowskyj, S., Chi-Rosso, G. and Lobb, R. (1989). Direct expression cloning of vascular cell adhesion molecule 1, a cytokine-induced endothelial protein that binds to lymphocytes. *Cell*, **59**, 1203–1211.

Pasqualini, R., Hemler, M.E. (1994). Contrasting roles for integrin beta 1 and beta 5 cytoplasmic domains in subcellular localization, cell proliferation, and cell migration. *J. Cell Biol.*, **125**, 447–460.

Paul, L.C., Davidoff, A., Benediktsson, H. and Issekutz, T.B. (1993). The efficacy of LFA-1 and VLA-4 antibody treatment in rat vascularized cardiac allograft rejection. *Transplantation*, 55, 1196–1199.

Paul, L.C., Davidoff, A., Paul, D.W., Benediktsson, H. and Issekutz, T.B. (1993). Monoclonal antibodies against LFA-1 and VLA-4 inhibit graft vasculitis in rat cardiac allografts. *Transplant Proc.*, 25, 813–814.

Pelletier, R., Ohye, R., Kincade, P., Ferguson, R. and Orosz, C. (1993). Monoclonal antibody to anti-VCAM-1 interferes with murine cardiac allograft rejection. *Transplant Proc.*, 25, 839–841.

Pizcueta, P. and Luscinskas, F.W. (1994). Monoclonal antibody blockade of L-selectin inhibits mononuclear leukocyte recruitment to inflammatory sites in vivo. *Am. J. Pathol.*, **145**, 461–469.

Podolsky, D.K., Lobb, R., King, N., Benjamin, C.D., Pepinsky, B., Sehgal, P. and deBeaumont, M. (1993). Attenuation of colitis in the cotton-top tamarin by anti-alpha 4 integrin monoclonal antibody. *J. Clin. Invest.*, **92**, 372–380.

Pretolani, M., Ruffie, C., Lapa e Silva, J.R., Joseph, D., Lobb, R.R. and Vargaftig, B.B. (1994). Antibody to very late activation antigen 4 prevents antigen-induced bronchial hyperreactivity and cellular infiltration in the guinea pig airways. *J. Exp. Med.* , **180**, 795–805.

Richards, I.M., Kolbasa, K.P., Hatfield, C.A., Winterrowd, G.E., Vonderfecht, S.L., Fidler, S.F., Griffin, R.L., Brashler, J.R., Krzesicki, R.F., Sly, L.M., Ready, K.A., Staite, N.D. and Chin, J.E. (1996). Role of very late activation antigen-4 in the antigen-induced accumulation of eosinophils and lymphocytes in the lungs and airway lumen of sensitized brown Norway rats. *Am J. Respir. Cell. Mol.Biol.*, **15**, 172–183.

Romanic, A.M., Graesser, D., Visintin, I., L., B.J., Janeway, C.A. and Madri, J.A. (1997). T cell adhesion to endothelial cell adhesion molecules and extracellular matrix is modulated upon transendothelial cell migration. *Lab. Invest..*, in press.

Romanic, A.M. and Madri, J.A. (1994). The induction of 72-kD gelatinase in T cells upon adhesion to endothelial cells is VCAM-1 dependent. *J. Cell Biol.*, **125**, 1165–1178.

Ross, L., Hassman, F. and Molony, L. (1992). Inhibition of Molt-4-endothelial adherence by synthetic peptides from the sequence of ICAM-1. *J. Biol. Chem.*, **267**, 8537–8543.

Russell, P.S., Chase, C.M. and Colvin, R.B. (1995). Coronary atherosclerosis in transplanted mouse hearts. IV Effects of treatment with monoclonal antibodies to intercellular adhesion molecule-1 and leukocyte function-associated antigen-1. *Transplantation*, **60**, 724–729.

Schall, T.J., Bacon, K., Camp, R.D., Kaspari, J.W. and Goeddel, D.V. (1993). Human macrophage inflammatory protein alpha (MIP-1 alpha) and MIP-1 beta chemokines attract distinct populations of lymphocytes. *J. Exp. Med.* , **177**, 1821–1826.

Schall, T.J., Bacon, K., Toy, K.J. and Goeddel, D.V. (1990). Selective attraction of monocytes and T lymphocytes of the memory phenotype by cytokine RANTES. *Nature*, **347**, 669–671.

Scheynius, A., Camp, R.L. and Pure, E. (1993). Reduced contact sensitivity reactions in mice treated with monoclonal antibodies to leukocyte function-associated molecule-1 and intercellular adhesion molecule-1. *J. Immunol.*, **150**, 655–663.

Schiffer, S.G., Hemler, M.E., Lobb, R.R., Tizard, R. and Osborn, L. (1995). Molecular mapping of functional antibody binding sites of alpha 4 integrin. *J. Biol. Chem.*, **270**, 14270–14273.

Schweighoffer, T. and Shaw, S. (1992). Adhesion cascades: diversity through combinatorial strategies. *Curr. Opin. Cell Biol..*, **4**, 824–829.

Shimizu, Y., Shaw, S., Graber, N., Gopal, T.V., Horgan, K.J., Van Seventer, G.A. and Newman, W. (1991). Activation-independent binding of human memory T cells to adhesion molecule ELAM-1. *Nature*, **349**, 799–802.

Shimizu, Y., van Seventer, G.A., Ennis, E., Newman, W., Horgan, K.J. and Shaw, S. (1992). Crosslinking of the T cell-specific accessory molecules CD7 and CD28 modulates T cell adhesion. *J. Exp. Med.*, **175**, 577–582.

Shimizu, Y., Van Seventer, G.A., Horgan, K.J. and Shaw, S. (1990). Regulated expression and binding of three VLA (beta 1) integrin receptors on T cells. *Nature*, **345**, 250–253.

Siebenlist, U., Franzoso, G. and Brown, K. (1994). Structure, regulation and function of NF-kappa B. *Annu. Rev. Cell Biol.*, **10**, 405–455.

Simons, M., Edelman, E.R., DeKeyser, J.L., Langer, R. and Rosenberg, R.D. (1992). Antisense c-myb oligonucleotides inhibit intimal arterial smooth muscle cell accumulation in vivo. *Nature*, **359**, 67–70.

Springer, T.A. (1995). Traffic signals on endothelium for lymphocyte recirculation and leukocyte emigration. *Annu. Rev. Physiol.*, **57**, 827–872.

St. Clair, E.W. and Haynes, B.F. (1993). The future of rheumatoid arthritis treatment. *Bull. Rheum. Dis.*, **42**, 1–4.

Stanley, P., Bates, P.A., Harvey, J., Bennett, R.I. and Hogg, N. (1994). Integrin LFA-1 alpha subunit contains an ICAM-1 binding site in domains V and VI. *EMBO J.*, **13**, 1790–1798.

Stepkowski, S.M., Tu, Y., Condon, T.P. and Bennett, C.F. (1995). Induction of transplantation tolerance by treatment with ICAM-1 antisense oligonucleotides and anti-LFA-1 monoclonal antibodies. *Transplant Proc.*, **27**, 113.

Talbott, G.A., Sharar, S.R., Harlan, J.M. and Winn, R.K. (1994). Leukocyte-endothelial interactions and organ injury: the role of adhesion molecules. *New Horiz.*, **2**, 545–554.

Talento, A., Nguyen, M., Blake, T., Sirotina, A., Fioravanti, C., Burkholder, D., Gibson, R., Sigal, N.H., Springer, M.S., Koo, G.C. (1993). A single administration of LFA-1 antibody confers prolonged allograft survival. *Transplantation*, **55**, 418–422.

Tamiya, Y., Yamamoto, N. and Uede, T. (1995). Protective effect of monoclonal antibodies against LFA-1 and ICAM-1 on myocardial reperfusion injury following global ischemia in rat hearts. *Immunopharmacology*, **29**, 53–63.

Tanaka, T., Ohtsuka, Y., Yagita, H., Shiratori, Y., Omata, M. and Okumura, K. (1995). Involvement of alpha 1 and alpha 4 integrins in gut mucosal injury of graft-versus-host disease. *Int .Immunol.*, **7**, 1183–1189.

Tanaka, Y., Adams, D.H., Hubscher, S., Hirano, H., Siebenlist, U. and Shaw, S. (1993). T-cell adhesion induced by proteoglycan-immobilized cytokine MIP-1 beta. *Nature*, **361**, 79–82.

Tanaka, Y., Albelda, S.M., Horgan, K.J., van Seventer, G.A., Shimizu, Y., Newman, W., Hallam, J., Newman, P.J., Buck, C.A. and Shaw, S. (1992). CD31 expressed on distinctive T cell subsets is a preferential amplifier of beta 1 integrin-mediated adhesion. *J. Exp. Med.*, **176**, 245–253.

Taub, D.D., Conlon, K., Lloyd, A.R., Oppenheim, J.J. and Kelvin, D.J. (1993). Preferential migration of activated CD4+ and CD8+ T cells in response to MIP-1 alpha and MIP-1 beta. *Science*, **260**, 355–358.

Tsukamoto, K., Yokono, K., Amano, K., Nagata, M., Yagi, N., Tominaga, Y., Moriyama, H., Miki, M., O kamoto, N., Yoneda, R., *et al.* (1995). Administration of monoclonal antibodies against vascular cell adhesion molecule-1/very late antigen-4 abrogates predisposing autoimmune diabetes in NOD mice. *Cell. Immunol.*, **165**, 193–201.

Uchio, E., Kijima, M., Tanaka, S. and Ohno, S. (1994). Suppression of experimental uveitis with monoclonal antibodies to ICAM-1 and LFA-1. *Invest. Ophthalmol. Vis. Sci.*, **35**, 2626–2631.

van Kooyk, Y., van de Wiel-van Kemenade, P., Weder, P., Kuijpers, T.W. and Figdor, C.G. (1989). Enhancement of LFA-1-mediated cell adhesion by triggering through CD2 or CD3 on T lymphocytes. *Nature*, **342**, 811–813.

Voraberger, G., Schafer, R. and Stratowa, C. (1991). Cloning of the human gene for intercellular adhesion molecule 1 and analysis of its 5'-regulatory region. Induction by cytokines and phorbol ester. *J. Immunol.*, **147**, 277–286.

Wang, J.H., Pepinsky, R.B., Stehle, T., Liu, J.H., Karpusas, M., Browning, B. and Osborn, L. (1995). The crystal structure of an N-terminal two-domain fragment of vascular cell adhesion molecule 1 (VCAM-1): a cyclic peptide based on the domain 1 C-D loop can inhibit VCAM-1–alpha 4 integrin interaction. *Proc. Natl. Acad. Sci. U S A*, **92**, 5714–5718.

Watson, S.R., Fennie, C. and Lasky, L.A. (1991). Neutrophil influx into an inflammatory site inhibited by a soluble homing receptor-IgG chimaera. *Nature,* **349**, 164–167.

Weg, V.B., Williams, T.J., Lobb, R.R. and Nourshargh, S. (1993). A monoclonal antibody recognizing very late activation antigen-4 inhibits eosinophil accumulation *in vivo. J. Exp. Med.* , 177, 561–566.

Wegner, C.D., Gundel, R.H., Reilly, P., Haynes, N., Letts, L.G. and Rothlein, R. (1990). Intercellular adhesion molecule-1 (ICAM-1) in the pathogenesis of asthma. *Science,* **247**, 456–459.

Whelan, J., Ghersa, P., Hooft van Huijsduijnen, R., Gray, J., Chandra, G., Talabot, F. and DeLamarter, J.F. (1991). An NF kappa B-like factor is essential but not sufficient for cytokine induction of endothelial leukocyte adhesion molecule 1 (ELAM-1) gene transcription. *Nucleic Acids Res.,* **19**, 2645–2653.

Whitcup, S.M., DeBarge, L.R., Rosen, H., Nussenblatt, R.B. and Chan, C.C. (1993). Monoclonal antibody against CD11b/CD18 inhibits endotoxin-induced uveitis. *Invest. Ophthalmol. Vis. Sci.,* **34**, 673–681.

Whitcup, S.M., Hikita, N., Shirao, M., Miyasaka, M., Tamatani, T., Mochizuki, M., Nussenblatt, R.B. and Chan, C.C. (1995). Monoclonal antibodies against CD54 (ICAM-1) and CD11a (LFA-1) prevent and inhibit endotoxin-induced uveitis. *Exp. Eye Res.,* **60**, 597–601.

Willenborg, D.O., Simmons, R.D., Tamatani, T. and Miyasaka, M. (1993). ICAM-1-dependent pathway is not critically involved in the inflammatory process of autoimmune encephalomyelitis or in cytokine-induced inflammation of the central nervous system. *J. Neuroimmunol.,* **45**, 147–154.

Yang, H., Issekutz, T.B. and Wright, J.R., Jr. (1995). Prolongation of rat islet allograft survival by treatment with monoclonal antibodies against VLA-4 and LFA-1. *Transplantation,* **60**, 71–76.

Yang, X.D., Karin, N., Tisch, R., Steinman, L. and McDevitt, H.O. (1993). Inhibition of insulitis and prevention of diabetes in nonobese diabetic mice by blocking L-selectin and very late antigen 4 adhesion receptors. *Proc. Natl. Acad. Sci. U S A,* **90**, 10494–10498.

Yang, X.D., Michie, S.A., Tisch, R., Karin, N., Steinman, L. and McDevitt, H.O. (1994a). Cell adhesion molecules: a selective therapeutic target for alleviation of IDDM. *J. Autoimmun.,* 7, 859–864.

Yang, X.D., Michie, S.A., Tisch, R., Karin, N., Steinman, L. and McDevitt, H.O. (1994b). A predominant role of integrin alpha 4 in the spontaneous development of autoimmune diabetes in nonobese diabetic mice. *Proc. Natl. Acad. Sci. U S A,* **91**, 12604–12608.

Yednock, T.A., Cannon, C., Fritz, L.C., Sanchez-Madrid, F., Steinman, L. and Karin, N. (1992). Prevention of experimental autoimmune encephalomyelitis by antibodies against alpha 4 beta 1 integrin. *Nature,* **356**, 63–66.

Zeng, Y., Gage, A., Montag, A., Rothlein, R., Thistlethwaite, J.R. and Bluestone, J.A. (1994). Inhibition of transplant rejection by pretreatment of xenogeneic pancreatic islet cells with anti-ICAM-1 antibodies. *Transplantation,* **58**, 681–689.

INDEX

RECOMMENDED FURTHER READING

GENERAL OVERVIEWS

Ford, W.L. (1980). The lymphocyte — its transformation from a frustrating enigma to a model of cellular function. In M.M. Wintrobe, (ed.), *Blood, Pure and Eloquent: a Story of the Discovery of People and Ideas*, McGraw Hill, New York, pp. 457–508.

Parrott, D.M.V. and Wilkinson, P.C. (1981). Lymphocyte locomotion and migration. *Prog. Allergy*, **28**, 193–284.

Butcher, E.C. (1986). The regulation of lymphocyte traffic. *Curr. Top. Microbiol. Immunol.*, **128**, 85–122.

Husband, A.J., Editor (1988). Migration and homing of lymphoid cells (Boca Raton: CRC Press), Volume I, 194 pages, Volume II, 174 pages.

Springer, T.A. (1994). Traffic signals for lymphocyte recirculation and leukocyte emigration: the multistep paradigm. *Cell*, **76**, 301–14.

Girard, J.P. and Springer, T.A. (1995). High endothelial venules (HEV): specialized endothelium for lymphocyte migration. *Immunology Today*, **16**, 449–457.

Springer, T.A. (1995) Traffic signals on endothelium for lymphocyte recirculation and leukocyte emigration. *Annu. Rev. Physiol.*, **57**, 827.

Imhof, B.A. and Dunon, D. (1995). Leukocyte migration and adhesion. *Advances in Immunology*, **58**, 345–416.

Butcher, E.C. and Picker, L.J. (1996). Lymphocyte homing and homeostasis. *Science*, **272**, 60–6.

Bradley, L.M. and Watson, S.R. (1996). Lymphocyte migration into tissue: the paradigm derived from CD4 subsets. *Curr. Opinion Immunol.*, **8**, 312–320.

Ebnet, K., Kaldijan, E.P., Anderson, A.O. and Shaw, S. (1996). Orchestrated information transfer underlying leukocyte endothelial interactions. *Annu. Rev. Immunol.*, **14**, 155–177.

MOLECULES

Hemler, M.E. (1990). VLA proteins in the integrin family. Structures, functions and their role on leukocytes. *Annu. Rev. Immunol.*, **8**, 365–400.

Hynes, R.O. (1992). Integrins: versatility, modulation, and signaling in cell adhesion. *Cell*, **69**, 11.

Bevilaqua M.P. (1993). Endothelial-leukocyte adhesion molecules. *Annu. Rev. Immunol.*, **11**, 767–804.

Carlos, T.M. and Harlan, J.M. (1994). Leukocyte-endothelial adhesion molecules. *Blood*, **84**, 2068–2101.

Rosen, S.D. and Bertozzi, C.R. (1994). The selectins and their ligands. *Curr. Opin. Cell Biol.*, **6**, 663–73.

Hogg, N. and Berlin, C. (1995). Structure and function of adhesion receptors in leukocyte trafficking. *Immunol. Today*, **16**, 327–30.

Tedder, T.F., Steeber, D.A., Chen, A. and Engel, P. (1995). The selectins: vascular adhesion molecules. *Faseb J.*, **9**, 866–73.

Dedhar, S., Hannigan, G.E. (1996). Integrin cytoplasmic interactions and bidirectional transmembrane signaling. *Curr. Op. Cell Biol.*, **8**, 657–667.

Immune Receptor Supplement. *Immunol. Today*, October 1996.

Hamann, A. and Jonas, P. (1997). The selectins in lymphocyte homing. In D. Vestweber (ed.), The Selectins: Initiators of Leucocyte Endothelial Adhesion, Harwood Acad. Publ., Amsterdam, pp. 105–115.

TECHNIQUES

Butcher, E.C. and Ford, W.L. (1986). Following cellular traffic: methods of labelling lymphocytes and other cells to trace their migration *in vivo*. In D.M. Weir (ed.), *Handbook of experimental immunology*, Blackwell Scientific Publications, Oxford, pp. 57.1–57.23.

Hamann, A. and Jonas, P. (1996). Lymphocyte migration *in vivo*: the mouse model. In I. Lefkovits (ed.), *Immunology Methods Manual*, Academic Press, London, pp. 1333–1341.

Ager, A. (1996). Transendothelial migration of lymphocytes *in vitro*. In I. Lefkovits (ed.), *Immunology Methods Manual*, Academic Press, London.

1. Specific trafficking: Which cells, which function?

Smith, M.E., Martin, A.F. and Ford, W.L. (1980). Migration of lymphoblasts in the rat. *Monogr. Allergy*, **16**, 203–232.

Smith, M.E. and Ford, W.L. (1983). The recirculating lymphocyte pool of the rat: a systematic description of the migratory behaviour of recirculating lymphocytes. *Immunology*, **49**, 83–94.

Young, A.J., Hay, J.B. and Mackay, C. R. (1993). Lymphocyte recirculation and life span *in vivo*. *Curr. Top. Microbiol. Immunol.*, **184**, 161–73.

Hamann, A. and Rebstock, S. (1993). Migration of activated lymphocytes. In D. Dunon, C.R. Mackay and B. Imhof (eds.), *Adhesion in leucocyte homing and differentiation*, Springer Verlag, Berlin-Heidelberg-New York, pp. 109–124.

Mackay, C.R. (1993). Homing of naive, memory and effector lymphocytes. *Curr. Opin. Immunol.*, **5**, 423–7.

Bradley, L.M. and S.R. Watson (1996). Lymphocyte migration into tissue: the paradigm derived from CD4 subsets. *Curr. Opinion Immunol.*, **8**, 312–320.

Westermann, J., and Pabst, R. (1996). How organ-specific is the migration of "naive" and "memory" T cells? *Immunol. Today*, **17**, 278–282.

Tietz, W. and Hamann, A. (1997, in press). The migratory behavior of murine CD4+ cells of memory phenotype. *Eur. J. Immunol.*

Austrup, F., Vestweber, D., Borges, E., Löhning, M., Bräuer, R., Herz, U., Renz, H., Hallmann, R., Scheffold, A., Radbruch, A. and Hamann, A. (1997). P-and E-selectin mediate recruitment of T helper 1 but not T helper 2 cells into inflamed tissues. *Nature*, **385**, 81–83.

2. Lymphocyte traffic to lymphoid and non-lymphoid organs in different species is regulated by several mechanisms

Pabst, R. and Binns, R.M. (1989). Heterogeneity of lymphocyte homing physiology: several mechanisms operate in the control of migration to lymphoid and non-lymphoid organs *in vivo*. *Immunological Reviews*, **108**, 83–109.

Westermann, J. and Pabst, R. (1990). Lymphocytes in the blood. A diagnostic window on the lymphoid system? *Immunology Today*, **11**, 406–410.

Hamann, A. and Rebstock, S. (1993). Migration of activated lymphocytes. *Current Topics in Microbiology and Immunology*, **184**, 109–124.

Pabst, R. and Tschernig, T. (1995). Lymphocytes in the lung: an often neglected cell. Numbers, characterization and compartmentalization. *Anatomy and Embryology*, **192**, 293–299.

Griebel, P.J. and Hein, W.R. (1996). Expanding the role of Peyer's patches in B cell ontogeny. *Immunology Today*, **17**, 30–39.

Westermann, J. and Pabst, R. (1996). How organ specific is the migration of "naive" and "memory" T cells? *Immunology Today*, 17, 278–282.

3. The multistep paradigm and the control of lymphocyte trafficking: Homing to mucosal tissues as a model

Butcher E.C. (1991). Leukocyte-endothelial cell recognition: Three (or more) steps for specificity and diversity. *Cell* 67: 1033–1036.
Springer, T.A. (1994). Traffic signals for lymphocyte recirculation and leukocyte emigration: the multistep paradigm. *Cell*, **76**, 301–14.
Bargatze, R.F., Jutila, M.A. and Butcher, E.C. (1995). Distinct roles of L-selectin and integrins $\alpha_4\beta_7$ and LFA-1 in lymphocyte homing to Peyer's patch-HEV *in situ*: the multistep model confirmed and refined. *Immunity* 3, 99–108.
Butcher, E.C. and Picker, L.J. (1996). Lymphocyte homing and homeostasis. *Science*, 272, 60–6.

4. Adhesion molecules and the regulation of the (trans-) migration of lymphocytes

Molecules: see first section.

Transmigration:

Smith, C.W. (1993). Transendothelial migration. *Curr. Top. Microbiol. Immunol.*, **184**, 201–212.
Oppenheimer-Marks N. and Lipsky, P.E. (1994). Transendothelial migration of T cells in chronic inflammation. *The Immunologist*, **2**, 58–64.
Muller, W.M. (1996). Transendothelial migration of leucocytes. In: Peltz, G., Ed. (1996). Leukocyte recruitment in inflammatory disease. Molecular Biology Intelligence Unit: Springer/R.G. Landes Company.
Ager, A. (1997). Regulation of lymphocyte migration into lymph nodes by high endothelial venules. *Biochem. Soc. Trans.*, **25**, 421–428.

5. Chemokines as regulators of leucocyte adhesion and migration

Baggiolini, M., Dewald, B. and Moser, B. (1994). Interleukin-8 and related chemotactic cytokines – CXC and CC Chemokines. *Adv. Immunol.*, **55**, 97.
Schall, T.J. and Bacon, K.B. (1994). Chemokines, leukocyte trafficking, and inflammation. *Curr. Opin. Immunol.*, **6**, 865–73.
Murphy, P.M. (1994). The molecular biology of leukocyte chemoattractant receptors. *Annu. Rev. Immunol.*, **12**, 593.
Taub, D.D. and Oppenheim, J.J. (1994). Chemokines, inflammation and the immune system. *Ther. Immunol.*, **1**, 229.
Taub, D.D., Lloyd, A.R., Wang, J.M. and Oppenheim, J.J. (1995) The α and β chemokine superfamily. In *Current Protocols in Immunology*, Unit 6.12, eds. Coligan, J.E., Kruisbeek, A., Margulies, D., Shevach, E. and Strober, W., Greene Publishing Associates and Wiley Interscience, New York.
Bacon, K.B. and Schall, T.J. (1996). Chemokines as mediators of allergic inflammation. *Int. Arch. Allergy. Immunol.*, **109**, 97–109.
Mantovani, A., Locati, M., Allavena, P. and Sozzani, S. (1996). The chemokine superfamily: crosstalk with the Il-1 system. *Immunobiology*, **195**, 522.

Hedrick, J.A. and Zlotnik, A. (1996). Chemokines and lymphocyte biology. *Curr. Opinion Immunol.*, **8**, 343–347.

Bates, P. (1996). Chemokine receptors and HIV: and attractive pair? *Cell*, **86**, 1.

Unutmaz, D. and Litmann, D.R. (1996). Expression pattern of HIV coreceptors on T cells: implications for viral transmission and lymphocyte homing. *Proc. Natl. Acad. Sci. USA*, **94**, 1615–1618.

Mackay, C. (1996). Chemokine receptors and T cell chemotaxis. *J. Exp. Med.*, **184**, 799–802.

Power, C.A. and Wells, T.N.C. (1996). Cloning and characterization of human chemokine receptors. *Trends Pharmacol. Sci.*, **17**, 209–213.

Strieter, R.M., Standiford, T.J., Huffnagle, G.B., Colletti, L.M., Lukacs, N.W. and Kunkel, S.L. (1996). The good, the bad, and the ugly. The role of chemokines in models of human disease. *J. Immunol.*, **156**, 3583.

Vaddi, K., Keller, M. and Newton, R. eds. (1997). The chemokine facts book. Academic Press, London. 200 pages.

6. The ontogeny of lymphocytes and homing to the thymus

Van Ewijk, W. (1991). T-cell differentiation is influenced by thymic microenvironments. *Annu. Rev. Immunol.*, **9**, 591–615.

Rodewald, H.R. (1995). Pathways from hemopoietic stem cells to thymocytes. *Curr. Opin. Immunol.*, **7**, 176–187.

Kisielow, P. and von Boehmer, H. (1995). Development and selection of T cells: facts and puzzles. *Adv. Immunol.*, **58**, 87–209.

Zuniga-Pflücker, J.C. and Lenardo, M.J. (1996). Regulation of thymocyte development from immature progenitors. *Curr. Opin. Immunol.*, **8**, 215–224.

Imhof, B.A. and Dunon, D. (1995). Leukocyte migration and adhesion. *Adv. Immunol.*, **58**, 345–416.

7. Roles of $\alpha_4\beta_7$ and $\alpha_E\beta_7$ integrins in the interactions of lymphocytes with the intestinal mucosa

Guy-Grand, D. and Vassalli, P. (1988). Origin and traffic of gut mucosal lymphocytes and mast cells. In: *Migration and homing of lymphoid cells*. A. Husband, Ed., CRC Press, 99–111.

Brantzaeg, P., Halstensen, T.S., Kett, K., Krajci, P., Kvale, D., Rognum, T.O., Scott, H. and Sollid, L.M. (1989). Immunobiology and immunopathology of human gut mucosa: humoral immunity and intraepithelial lymphocytes. *Gastroenterology*, **97**, 1562–1584.

Zeitz, M., Schiefferdecker, H.L., Ullrich, R., Jahn, H-U., James, S.P., Riecken, E-O. (1991). Phenotype and function of lamina propria T lymphocytes. *Immunol. Res.*, **10**, 199–206.

Cerf-Bensussan, N., Cerf, M. and Guy-Grand, D. (1993). Gut intraepithelial lymphocytes and gastrointestinal diseases. *Cur. Op. Gastroenterol.*, **9**, 953–962.

Guy-Grand, D. and Vassalli, P. (1993). Gut intraepithelial T lymphocytes. *Curr. Opin. Immunol.*, **5**, 247–252.

Handbook of Mucosal Immunology Eds. Ogra, P.L., Mestecky, J., Lamm, M.E., Strober, W., McGhee, J.R., Bienenstock, J. Academic Press Inc., 1994.

Phillips-Quagliata, J.M. and Lamm, M.E. (1994). Lymphocyte homing to mucosal sites. In: *Handbook of Mucosal Immunology*, pp. 225–239.

Hamann A., Andrew D.P., Jablonski-Westrich D., Holzmann B., Butcher E.C. (1994). Role of α4-integrins in lymphocyte homing to mucosal tissues *in vivo*. *J. Immunol.*, **152**, 3282–3293.

Bargatze, R.F., Jutila, M.A. and Butcher, E.C. (1995). Distinct roles of L-selectin and integrins $\alpha_4\beta_7$ and LFA-1 in lymphocyte homing to Peyer's patch-HEV *in situ*: the multistep model confirmed and refined. *Immunity*, **3**, 99–108.

Arroyo, A.G., Yang, J.T., Rayburn, H. and Hynes, R.O. (1996). Differential requirements for α_4 integrins during fetal and adult hematopoiesis. *Cell*, **85**, 997–1008.
Wagner, N., Löhler, J., Kunkel, E.J., Ley, K., Leung, E., Krissanssen, G., Rajewsky, K. and Müller, W. (1996). Critical role for β_7 integrins in formation of the gut-associated lymphoid tissue. *Nature*, **382**, 366–370.
Hesterberg, P.E., Winsor-Hines, D., Briskin, M.J., Soler-Ferran, D., Merrill, C., Mackay, C.R., Newman, W. and Ringler D.J. (1996). Rapid resolution of chronic colitis in the cotton-top tamarin with an antibody to a gut-homing integrin $\alpha_4\beta_7$. *Gastroenterology*, **111**, 1373–1380.
Karecla, P.I., Green, S.J., Simon, S.J., Coadwell, J. and Kilshaw, P.J. (1996). Identification of a binding site for integrin $\alpha_E\beta_7$ in the N-terminal domain of E-cadherin. *J. Biol. Chem.*, **271**, 30909–30915.

8. T cell trafficking to skin: a paradigm for tissue-selective lymphocyte homing

Picker L.J. and Butcher E.C. (1992). Physiological and Molecular Mechanisms of Lymphocyte Homing. *Annu. Rev. Immunol.*, **10**, 561–591.
Picker L.J. (1994). Control of Lymphocyte Homing. *Curr. Opin. Immunol.*, **6**, 394–406.
Groves, R.W. and Kupper, T.S. (1996). Leucocyte recruitment in Cutaneous Inflammation. In: Peltz, G., Ed. (1996). Leukocyte recruitment in inflammatory disease. Molecular Biology Intelligence Unit: Springer/R.G. Landes Company.

9. Lymphocyte trafficking through the central nervous system

Weller, R.O., Shinya, K. and Zhang, E.-T. (1992). Pathways of fluid drainage from the brain-morphological aspects and immunological significance in rat and man. *Brain Pathol.*, **2**, 277–284.
Kuchroo, V.K., Martin, C.A., Greer, J.M., Ju, S.-T., Sobel, R.A. and Dorf, M.E. (1993). Cytokines and adhesion molecules contribute to the ability of myelin proteolipid protein-specific T cell clones to mediate experimental allergic encephalomyelitis. *J. Immunol.*, **151**, 4371–4382.
Engelhardt, B. and Risau, W. (1995). Development of the blood-brain barrier. "New concepts of a Blood-Brain Barrier". Eds. J. Greenwood, D. Begley, M. Segal and S. Lightman. Plenum Press, pp. 11–31.
Kent, S.J., Karlik, S.J., Cannon, C., Hines, D.K., Yednock, T.A., Fritz, L.C. and Horner, H.C. (1995). A monoclonal antibody to α_4 integrin suppresses and reverses active experimental allergic encephalomyelitis. *Journal of Neuroimmunology*, **58**, 1–10.
Keszthelyi, E., Karlik, S., Hyduk, S., Rice, G.P.A., Gordon, G., Yednock. T. and Horner, H. (1996). Evidence for a prolonged role of α_4 integrin throughout active experimental allergic encephalomyelitis. *Neurol.*, 47: 1053–1059.
Weller, R.O., Engelhardt, B. and Phillips, M.J. (1996). Lymphocyte targeting of the central nervous system (CNS): A review of afferent and efferent CNS-immune pathways. *Brain Pathol.*, **6**, 275–288.

10. Molecules involved in trafficking of NK cells and dendritic cells: Implications for tumor immunotherapy

Allavena, P. and Mantovani, A. (1994). Therapeutic strategies targeted to tumor-associated leukocytes in human tumors, in: Tumor-associated leukocytes: Pathophysiology and therapeutic applications, ed. A. Mantovani (R.G. Landes Publisher, Austin, USA) p. 90.

Mantovani, A., Locati, M., Allavena, P. and Sozzani, S. (1996). The chemokine superfamily: crosstalk with the Il-1 system. *Immunobiology*, **195**, 522.

Austyn, J.M. (1996). New insights into the mobilization and phagocytic activity of dendritic cells. *J. Exp. Med.*, **183**, 1287.

Girolomoni, G. and Ricciardi-Castagnoli, P. (1997). Dendritic cells hold promise for immunotherapy. *Immunol. Today*, **18**, 102.

Hedrick, J.A., Saylor, V. Figueroa, D., Mizoue, L., Xu, Y., Menon, S., Abrams, J., Handel, T. and Zlotnik, A. (1997). Lymphotactin is produced by NK cells and attracts both NK cells and T cells *in vivo*. *J. Immunol.*, **158**, 1533.

11. Endothelium and inflammation: Molecules and therapeutic prospects

Carlos, T.M. and Harlan, J.M. (1994). Leukocyte-endothelial adhesion molecules. *Blood*, **84**, 2068–2101.

McMurray, R.W. (1996). Adhesion molecules in autoimmune disease. *Semin. Arthritis Rheum.*, **25**, 215–233.

Peltz, G., Editor (1996). Leukocyte recruitment in inflammatory disease. Molecular Biology Intelligence Unit: Springer/R.G. Landes Company, 317 pages.

MOLECULES — QUICK REFERENCE CHART

Lymphocyte adhesion molecules

Molecule	Other names	Superfamily	Structure	Ligands	Implicated in
L-selectin	CD62L, LECAM-1, LAM-1,	Selectin		Carbohydrate (PNAd on CD34, sgp200, GlyCAM-1 and others)	Lymphocyte migration into lymph nodes
$\alpha_4\beta_7$	LPAM-1	β_7 Integrin		MadCAM-1, VCAM-1, Fibronectin	Lymphocyte homing to mucosal tissues
$\alpha_4\beta_1$	CD49d/CD29, VLA-4, LPAM-2	β_1 Integrin		VCAM-1, Fibronectin (CS-1)	Lymphocyte/ECM interaction, migration into inflamed tissues
$\alpha_E\beta_7$	HML-1, $M290_{IEL}$, CD103	β_7 Integrin		E-Cadherin	Gut IEL/intestinal epithelium interaction
$\alpha 5\beta 1$	CD49e/CD29, VLA-5, FNR	β_1 Integrin		Fibronectin, L1 cell adhesion molecule	Lymphocyte/ECM interaction
$\alpha_6\beta_1$	CD49f/CD29, VLA-6	β_1 Integrin		Laminin, others ?	Lymphocyte/ECM interaction, pro T-cell homig to thymus
LFA-1	CD11a/CD18, $\alpha_L\beta_2$	β_2 Integrin		ICAM-1, ICAM-2, ICAM-3	General role in leukocyte extravasation
Mac-1	CD11b/CD18, CR3, $\alpha_M\beta_2$	β_2 Integrin		ICAM-1, Fibrinogen, C3bi, Glucanes	Migration to inflamed tissues
CD44	Pgp-1, H-CAM, ECMRIII	Proteoglycan link proteins	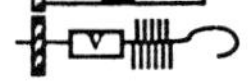	Hyaluronate, Fibronectin, Collagen	Lymphocyte/ECM interactions, others ?
PSGL-1	-	Sialomucin	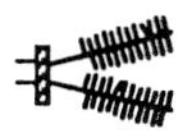	P-selectin, E-selectin	skin - and inflammation - specific migration of effector and memory cells
ESL-1	-	unassigned		E-selectin	

Endothelial adhesion molecules

Molecule	Other names	Superfamily	Structure	Ligands	Implicated in
P-selectin	CD62P, PADGEM, GMP-140	Selectin		PSGL-1, HSA	Lymphocyte/myeloid cell recruitment into inflamed tissues
E-selectin	CD62E, ELAM-1, INCAM-110	Selectin		Carbohydrate (CLA): on ESL-1, PSGL-1	Lymphocyte/myeloid cell recruitment into inflamed tissues
ICAM-1	CD54	Immunoglobulin	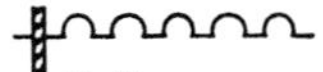	LFA-1, MAC-1, CD43	General role in leukocyte extravasation
ICAM-2	CD102	Immunoglobulin		LFA-1, $\alpha_D\beta_2$	General role in leukocyte extravasation
VCAM-1	CD106	Immunoglobulin		$\alpha_4\beta_1$, $\alpha_4\beta_7$	Lymphocyte recruitment into inflamed tissues
MadCAM-1	-	Immunoglobulin		$\alpha_4\beta_1$, $\alpha_4\beta_7$, L-selectin	Lymphocyte recruitment into mucosal sites
PNAd					
CD34	Sgp90	Sialomucin		L-selectin	Lympkocyte recruitment into lymph nodes
Sgp200	-	?	?	?	Lympkocyte recruitment into lymph nodes
GlyCAM-1	Sgp50	Sialomucin	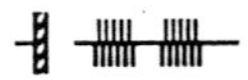	L-selectin	Regulation of lymphocyte migration into lymph nodes
VAP-1	-	?		unknown	Lymphocyte/endothelium interactions
CD31	PECAM-1	Immunoglobulin		CD31, $\alpha_V\beta_3$	Transendothelial migration

complement binding domain
EGF domain
lectin-like domain
chicken FGF receptor-like region
cysteine-rich domain
metal-binding domain
cleavage site
I-domain
variation insertion site
Ig repeat
carbohydrate residues
mucin-like region

Thanks to Sirpa Jalkanen. Table prepared by Jochen Lehmann.